Advances in Intelligent Systems and Computing

Volume 1362

The series "Advances in Intelligent Systems and Computing" contains publications on theory, applications, and design methods of Intelligent Systems and Intelligent Computing. Virtually all disciplines such as engineering, natural sciences, computer and information science, ICT, economics, business, e-commerce, environment, healthcare, life science are covered. The list of topics spans all the areas of modern intelligent systems and computing such as: computational intelligence, soft computing including neural networks, fuzzy systems, evolutionary computing and the fusion of these paradigms, social intelligence, ambient intelligence, computational neuroscience, artificial life, virtual worlds and society, cognitive science and systems, Perception and Vision, DNA and immune based systems, self-organizing and adaptive systems, e-Learning and teaching, human-centered and human-centric computing, recommender systems, intelligent control, robotics and mechatronics including human-machine teaming, knowledge-based paradigms, learning paradigms, machine ethics, intelligent data analysis, knowledge management, intelligent agents, intelligent decision making and support, intelligent network security, trust management, interactive entertainment, Web intelligence and multimedia.

The publications within "Advances in Intelligent Systems and Computing" are primarily proceedings of important conferences, symposia and congresses. They cover significant recent developments in the field, both of a foundational and applicable character. An important characteristic feature of the series is the short publication time and world-wide distribution. This permits a rapid and broad dissemination of research results.

Indexed by DBLP, EI Compendex, INSPEC, WTI Frankfurt eG, zbMATH, Japanese Science and Technology Agency (JST), SCImago.

All books published in the series are submitted for consideration in Web of Science.

More information about this series at http://www.springer.com/series/11156

Szczepan Paszkiel
Editor

Control, Computer Engineering and Neuroscience

Proceedings of IC Brain Computer Interface 2021

 Springer

Editor
Szczepan Paszkiel
Faculty of Electrical Engineering,
Automatic Control and Informatics
Opole University of Technology
Opole, Poland

ISSN 2194-5357   ISSN 2194-5365 (electronic)
Advances in Intelligent Systems and Computing
ISBN 978-3-030-72253-1   ISBN 978-3-030-72254-8 (eBook)
https://doi.org/10.1007/978-3-030-72254-8

This Springer imprint is published by the registered company Springer Nature Switzerland AG
The registered company address is: Gewerbestrasse 11, 6330 Cham, Switzerland

Preface

With the next issue of the volume "Advances in Intelligent Systems and Computing," I am pleasure to present the proceedings of the 4th International Scientific Conference on Brain–Computer Interfaces IC BCI 2021, Opole, Poland. The event was held at Opole University of Technology in Poland 21 September 2021. Since 2014, the conference has taken place every two years at the Faculty of Electrical Engineering, Automatic Control and Informatics, Opole University of Technology. During the conference, the speakers focussed on the issues regarding new trends in the modern brain–computer interfaces and control engineering, including neurobiology, neurosurgery, cognitive science, bioethics, biophysics, biochemistry, modelling, neuroinformatics, BCI technology, biomedical engineering, control and robotics, computer engineering, neurorehabilitation and biofeedback.

The previous three IC BCI conferences brought into focus several important solutions with regard to both scientific and engineering problems. I can also emphasize that the three last events attracted over 1500 followers representing the biggest national academic centres and industrial companies. In 2018, the collection of the papers was published in the form of a scientific monograph entitled "Biomedical Engineering and Neuroscience, Proceedings of the 3rd International Scientific Conference on Brain–Computer Interfaces, BCI 2018, March 13–14, Opole, Poland," Wojciech P. Hunek and Szczepan Paszkiel Eds., Springer Publishing, 2018. In 2016, the collection of the papers was published in the form of a scientific monograph entitled "Contemporary problems in biomedical engineering and neurosciences," Szczepan Paszkiel and Jan Sadecki, Eds., Opole University of Technology Press, 2016. In all previous editions, the awards were sponsored by D-Link Corp.

I can also note that IC BCI 2021 takes place under the honorary auspices of the Minister of Education and Science Republic of Poland. The second time Janusz Kacprzyk from Systems Research Institute of Polish Academy of Sciences was appointed to be the chairman of the scientific committee, while Szczepan Paszkiel, PhD, is the chairman of the organizing committee. Three last BCI meetings gathered a group of outstanding professionals such as Wlodzislaw Duch, Marcin

Czerwiński, Włodzislaw Duch, Grzegorz Francuz, Tomasz Halski, Michał Kuczyński, Dariusz Łątka, Mirosław Łątka, Dariusz Man, Rev. Piotr Morciniec, Tadeusz Skubis and Jan Szczegielniak. This year the conference was overseen by a local organizing committee having the technical skills, and the team consists of the following staff: Paweł Dobrakowski, PhD; Wiktoria Frącz; Michał Gajewski, Eng.; Adam Łysiak, MSc. Eng.; Rafał Mzyk, Eng.; Andrzej Olczak, MSc. Eng.

Szczepan Paszkiel

Organization

The 4th International Scientific Conference on Brain–Computer Interfaces IC BCI 2021 is organized by the Faculty of Electrical Engineering, Automatic Control and Informatics, Opole University of Technology.

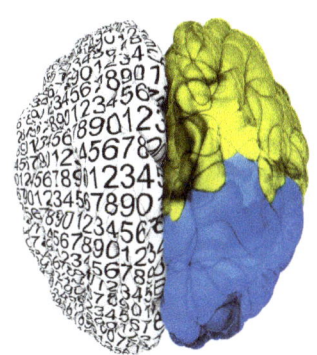

Scientific Committee

Chairman of Scientific Committee

Janusz Kacprzyk	Systems Research Institute Polish Academy of Sciences, Warsaw, Poland

Members of Scientific Committee

Dawid Bączkowicz	Opole University of Technology, Poland
Ryszard Beniak	Opole University of Technology, Poland
Tomasz Boczar	Opole University of Technology, Poland
Sebastian Borucki	Opole University of Technology, Poland
Andrzej Cichoń	Opole University of Technology, Poland

Marcin Czerwiński   Institute of Immunology and Experimental
                             Therapy Polish Academy of Sciences, Poland
Paweł Dobrakowski   Humanitas University, Institute of Psychology,
                             Sosnowiec, Poland
Włodzisław Duch   Nicolaus Copernicus University in Toruń, Poland
Paweł Frącz   Opole University of Technology, Poland
Grzegorz Francuz   University of Opole, Poland
Tomasz Halski   University of Opole, Poland
Wojciech Hunek   Opole University of Technology, Poland
Jerzy Klamka   Silesian University of Technology, Poland
Jozef Korbicz   University of Zielona Gora, Poland
Krzysztof Latawiec   Opole University of Technology, Poland
Mikołaj Leszczuk   AGH University of Science and Technology,
                             Poland
Dariusz Łątka   University of Opole, Department of
                             Neurosurgery, Medical Centre, Opole, Poland
Marian Łukaniszyn   Opole University of Technology, Poland
Feng-Huei Lin   National Health Research Institutes, Taiwan
Krystyna Macek-Kamińska   Opole University of Technology, Poland
Dariusz Man   University of Opole, Poland
Karol Miller   The University of Western Australia;
                             Cardiff University, UK
Piotr Morciniec   University of Opole, Poland
Mieczysław Pokorski   Mossakowski Medical Research Centre Polish
                             Academy of Sciences, Warsaw, Poland
Remigiusz Rak   Warsaw University of Technology, Poland
Leszek Rutkowski   Czestochowa University of Technology, Poland
Ryszard Rojek   Opole University of Technology, Poland
Daniel Sánchez-Morillo   University of Cadiz, Spain
Jan Sadecki   Opole University of Technology, Poland
Jerzy Skubis   Opole University of Technology, Poland
Tadeusz Skubis   Silesian University of Technology, Poland
Robert Sochacki   University of Opole, Poland
Rafał Stanisławski   Opole University of Technology, Poland
Mirosław Szmajda   Opole University of Technology, Poland
Jan Szczegielniak   Hospital of the Ministry of the Interior
                             and Administration in Glucholazy, Poland
Jerzy Świątek   Wroclaw University of Technology, Poland
Wiesław Tarczyński   Opole University of Technology, Poland
Ryszard Tadeusiewicz   AGH University of Science and Technology,
                             Cracow, Poland
Krzysztof Tomczewski   Opole University of Technology, Poland
Władysław Torbicz   Nalecz Institute of Biocybernetics
                             and Biomedical Engineering Polish Academy
                             of Sciences, Warsaw, Poland

Eleuterio Toro	University of Trento, Italy
Yudong Zhang	Nanjing Normal University, China; Columbia University, USA
Anna Walaszek-Babiszewska	University of Zielona Gora, Poland
Dariusz Zmarzły	Opole University of Technology, Poland

Organizing Committee

Chairman of Organizing Committee

Szczepan Paszkiel

Members of Organizing Committee

Paweł Dobrakowski
Wiktoria Frącz
Michał Gajewski
Adam Łysiak, MSc. Eng.
Rafał Mzyk, Eng.
Andrzej Olczak, MSc. Eng.

Contents

About the Editor

Szczepan Paszkiel PhD, works as an assistant professor at the Faculty of Electrical, Control and Computer Engineering at Opole University of Technology. He obtained PhD in Technical Sciences (automatics and robotics) in 2011. His research interests focus on modern methods of control systems, brain–computer technology and modelling of neuronal cell fractions. Szczepan Paszkiel, PhD, is the author of more than one hundred publications including those from ISI Master Journal List and a lecturer on several dozen scientific conferences, festivals and panel discussions. He is the winner of many prizes and competitions including those organized by Ministry of Science and Higher Education, Republic of Poland, and the recipient of grants for young scientists. He is also the co-editor of the monograph entitled: "Biomedical Engineering and Neuroscience, Proceedings of the 3rd International Scientific Conference on Brain–Computer Interfaces, BCI 2018, March 13-14, Opole, Poland," Springer Publishing, 2018; "Contemporary problems of biomedical engineering and neurosciences," Opole University of Technology Publishing, 2016; the author of the first handbook in Poland entitled "Brain–computer Interfaces. Neuroinformatics," 2014; a member of Metrology Commission, Polish Academy of Sciences, Katowice Branch; an expert of the National Centre for Research and Development in Poland, the Centre of OPI Processing Information in the National Research

Centre (PIB Poland), an expert of European Commission and Wroclaw Research Centre EIT+; a member of expertise panels and pre-panels in the National Centre for Research and Development including being a chairman of expert teams and a head expert.

Robotics in Neurosurgery – Past, Presence and Future

Olbrycht Tomasz[1,2], Kołodziej Waldemar[1,2], Łątka Kajetan[1,2], Chowaniec Jacek[1,2], Sobolewski Tomasz[1,2], and Łątka Dariusz[1,2(✉)]

[1] Department of Neurosurgery, University Hospital of in Opole, Opole, Poland
dariusz.latka@usk.opole.pl
[2] Institute of Medicine, University of Opole, Opole, Poland

Abstract. In multiple ways, neurosurgery is the perfect field for the implementation of robotic assisted procedures. Neurosurgical operations require precise and fine manipulation of deeply located critical neural structures that are accessed through a small corridor. The concept of robots has evolved from "human-like" machines to programmable, multifunctional specialized devices. To this day, the majority of robotic-assisted neurosurgical operations involve a shared-control system. They have involved a robotic arm that moves an instrument to a specific location based on Cartesian coordinates and is then locked in place. The operating neurosurgeon proceeds with the instrument along the path defined by the robot. One of the most important goals is to promote active cooperation between engineers and neurosurgeons. However, unfamiliarity with robot technology and the high costs of maintenance and purchasing the few available robotic systems can discourage their use. Nevertheless, improvements in the quality of healthcare should eventually surpass the inherent costs of robotic surgery systems. While we witness the dawn of artificial intelligence and brain-machine interfaces, neurosurgery and treatment of previously incurable diseases has already surpassed the ideas of science fiction writers. As with every tool created by man, we must ensure as scientists and medical doctors that robotics is used as means to benefits individuals and all of mankind.

Keywords: Robotized surgery · AI-guided surgery · Neuronavigation · Minimally invasive spine surgery · Image-guided surgery

1 Introduction

In order to understand the current and future trends in neurosurgery, we must briefly introduce the history of the topic. Trepanning (also known as trepanation, trephination, trephining, or making a burr hole) is a surgical intervention in which a hole is drilled or scraped into the human skull. Trephination of the human skull is the oldest documented surgical procedure. Trephined skulls have been found in the Old World and the New World (particularly Peru), from the Neolithic age to the very dawn of history [1]. Most reported series show that 2.5 to 19% of skulls from the Neolithic period have been trephined with single or multiple skull openings of various sizes [2]. The exact reason

© The Author(s), under exclusive license to Springer Nature Switzerland AG 2021
S. Paszkiel (Ed.): ICBCI 2021, AISC 1362, pp. 1–8, 2021.
https://doi.org/10.1007/978-3-030-72254-8_1

why these prehistoric "neurosurgeons" performed this procedure is not known for certain. In "A History of Medicine," Plinio Prioreschi proposed a theory that this procedure was an attempt to resurrect the dead. More frequently than other injuries, they might have observed that small injuries to the head resulted in "dying" (i.e., loss of consciousness with a concussion or a contusion resulting in coma) and "undying" (i.e., spontaneous recovery). Thus, according to Prioreschi, they must have come to believe that "something in the head had to do with undying" [1].

The Renaissance in Europe saw dramatic improvement in the understanding of anatomy and the development of surgical techniques [3]. In 1950, a sketchbook of Leonardo da Vinci with design notes for a robot was rediscovered. The robot was designed as a knight and was capable of standing, sitting, and independently maneuvering its arms. The entire robotic system was operated by a series of pulleys and cables. Since the discovery of the sketchbook, the robot has been built faithfully based on Leonardo's design and was found to be fully functional [4]. Therefore, it is not surprising that the most common and widely known robotic platform for surgery today is named after the legendary Renaissance inventor – the Da Vinci System created by Intuitive Surgical.

One of the first major breakthroughs in neurosurgery was the introduction of the microscope. The era of microneurosurgery began in 1957. Theodore Kurze performed the first microscope-assisted neurological surgery, which involved the removal of a neurilemmoma of the cranial nerve VII [7]. Further advances in microneurosurgery were made by M.G. Yasargil in Zurich and R.M.P Dinghy in the United States [8]. Robotics may represent the next technological leap. In support of this, one of the most modern neurosurgical operating microscopes, the Zeiss Kinevo, is in fact a robotic device equipped with an arm that remembers its position in the operating room space. This feature greatly facilitates the spatial orientation of the surgeon in operations requiring attacks from different trajectories. Nevertheless, it may be just a first step and a preview of future applications.

In multiple ways, neurosurgery is the perfect field for the implementation of robotic assisted procedures. Neurosurgical operations require precise and fine manipulation of deeply located critical neural structures that are accessed through a small corridor. Neurosurgical procedures are frequently long and tedious, leading to significant surgeon fatigue. Robotic enhancement may improve microsurgical dexterity by minimizing physiological tremor and scaling down hand motion [9]. A surgeon's physiological tremor is around 40 μm, which can be reduced to 4 μm by the use of a robotic interface [11]. Robots are indefatigable and able to perform repetitious tasks with precision and reproducible outcomes [10]. In the field of neurosurgery, it was not until the mid-1980s that surgeons performed a precise biopsy and utilized the concept of robotics for the first time [6].

The concept of robots has evolved from "human-like" machines to programmable, multifunctional specialized devices [5]. The term "robot" itself comes from a Slavic root, *robot*, which has meanings associated with labor. It was first used by a Czech playwright, Karl Capek, in a science fiction play called Rossum's Universal Robots, which was written in 1920 and premiered on January 25, 1921. Based on information from the Robot Institute of America, we propose the following definition of the term "surgical robot": "A reprogrammable, multifunctional manipulator designed to move material, parts, tools, or specialized devices through various programmed motions for the performance of a variety of tasks" [12].

Nathoo et al. proposed a classification of surgical robots into three categories: supervisory controlled systems, tele-surgical systems, and shared control systems. In a supervisory controlled system, the surgeon plans the operation offline and precisely establishes the motions that the robot must follow to perform the operation. The robot proceeds with the surgery autonomously with the surgeon closely supervising. Tele-surgical systems enable surgeons to control instruments held by the robot via a robotic manipulator using hand controls or a joystick. This control system has also been called a master-slave control system. The surgeon performs the surgical manipulations via an on-line input device (typically a joystick as the master) and a manipulator (a "slave") that copies the motions of the input device to perform the surgery. Lastly, shared-control systems enable a robot and surgeon to control the surgical instrument together. They are synergistic systems in which the robot provides the surgeon with steady-hand manipulation of an instrument [13, 14].

To this day, the majority of robotic-assisted neurosurgical operations involve a shared-control system. They have involved a robotic arm that moves an instrument to a specific location based on Cartesian coordinates and is then locked in place. The operating neurosurgeon proceeds with the instrument along the path defined by the robot. This has been successfully implemented in stereotactic procedures, spinal pedicle screw placement, and endoscopy [15]. Each type of surgical control system will be described with greater detail with examples of clinical application.

2   The Tele-surgical Robot (Master-Slave)

In this type of robotic solution surgeon remotely controls a tele-surgical robot. A very promising example is the NeuroArm project (University of Calgary), which began in September 2001 [18]. The NeuroArm was developed by a group at the University of Calgary led by Garnette Sutherland in cooperation with MDA Robotics (Brampton, Canada). MDA previously developed robotic systems for the US Space Shuttle Program, including teleoperated robotic technology [9]. The company's experience benefited the NeuroArm project through knowledge related to prior aerospace achievements [18].

The NeuroArm system comprises two robotic arms that are capable of manipulating both specially designed and established microsurgical tools. They are connected via a main system controller to a workstation with a sensory immersive human-machine interface. The surgeon is positioned at a workstation and uses the human-machine interface to interact with the surgical site [18]. It is an MRI-compatible robotic arm that mimics the movements of a surgeon's hands. It uses piezoelectric motors and has eight degrees of freedom (DOF) [17]. The human-machine interface combines real-time HD and 3-D images of the surgical site with MRIs [18]. Force feedback is provided by two modified phantom hand controllers that are each linked to two MR safe force sensors (ATI, NC, USA), which are both in contact with the tool. The interaction force between the surgical tool and the environment (e.g., brain tissue) is measured by 2 force sensors mounted on each arm and applied to the surgeon's hand by a haptic device [20]. This robot is the first to provide tactile feedback and is controlled by the neurosurgeon. It has been reported to be involved in more than 1,000 neurosurgical procedures [17].

A classical type of master-slave robotic system is the Da Vinci robot. The development of this tele-manipulator was initiated by SRI International as a research project funded by the United States Department of Defense. The goal of the program was to create a platform by which surgeons could operate on injured soldiers from a secure and remote location [22]. The company Intuitive Surgical was established to modify the manipulator into a format that is compatible for use in minimal-access surgery and acquired the rights to SRI patents in 1995 [22, 23]. After obtaining the prototype, they commercialized the robotic system. To this day, the system has been upgraded and improved multiple times. In 2000, the Da Vinci System became the first robotic surgical system to be implemented in an American operating room after the US FDA approved it for laparoscopic procedures [24].

The Da Vinci robot consists of two major units. The first is the surgeon's console, which consists of a display system, user interface, and electronic controllers. The other unit has four slave manipulators: three for telemanipulation of the surgical tools and one for the endoscopic camera [22]. From a functional viewpoint, the system offers two features: visualization of the surgical field with an endoscope connected to the 3D display, and transformation of the surgeon's hand movement to that of the surgical instruments [22], thereby allowing the robot to serve as an extension of the surgeon's arm. The 3D video is provided for the surgeon by a dual set of cameras.

In neurosurgery, the Da Vinci robot has been used in many procedures, including transoral odontoidectomies, thoracolumbar neurofibroma and paraspinal schwannoma resection, and anterior interbody lumbar fusion (ALIF). The advanced capabilities of newer-generation Da Vinci systems make the ALIF procedure possible and efficient. Case series have shown successful dissection, exposure, and interbody placement without any vessel or ureteral complications [25, 26].

3   The Shared Control Robot System

In a shared control robot system, the robot and the surgeon control the instruments together in an operation. Such systems were developed to augment the manipulative skills and dexterity of neurosurgeon with the precise actions of the robot. Despite the shared control, the surgeon remains in charge of the decision-making related to the procedure, and the robot provides steady-hand manipulation of the instruments [27]. A good example is the Steady Hand System designed at John Hopkins University. The instrument is held by the surgeon and the robot, and it allows fine dissection and eliminates tremor and muscle fatigue [17].

Another example is the Robotic Stereotactic Assistance system, or ROSA (Medtech, Montpellier, France). It is a computer-controlled robotic arm with an integrated platform that combines image-guided neurosurgical planning software with robotic navigation to assist neurosurgeons with minimally invasive procedures [21]. ROSA has four main components: a robot stand, a touch-screen, a retractable telescopic support arm, and a robotic arm. Preoperative images are registered by laser scanning, which is very similar to how standard neuronavigation systems perform. After registration, ROSA aligns to the appropriate planned trajectory.

The system allows for planning of multiple trajectories using computerized tomography and magnetic resonance imaging [21]. It allows the surgeon to navigate and guide the instruments held by the robot arm in a dedicated mode, which is aided by haptic feedback. Positioning of the robotic arm requires minimal exertion [17]. The main field in which the ROSA robot has been used is stereotactic and functional neurosurgery. A very useful feature in the software is the highlighting of important neural and vascular structures so that they can be avoided. Some of the procedures in which the ROSA robot may applied are the implantation of DBS electrodes, biopsy of complex tumors, placement of microcathethers for targeted chemotherapy, and insertion of pedicle screws in the spine.

In 2004, the SpineAssist (Mazor Robotics, Israel) became the first robot to be approved by the FDA for use in spinal surgery in the United States and remains one of the most widely used [28]. It uses a shared-control system that allows for superior navigation in comparison to traditional intraoperative computer assisted navigation. Traditional CAN requires the surgeon to follow pre-planned trajectories manually, which often requires significant hand-eye coordination [28]. The SpineAssist system positions its arm along a fixed trajectory, which allows for optimal positioning of the screws. The surgeon performs the drilling manually.

4 Supervisory Controlled Systems

In supervisory controlled systems, the robot helps the surgeon to carry out precise tasks. The robot is preprogrammed by the surgeon and then monitored while performing the specified steps autonomously. These systems are used mainly to support stereotactic guided brain biopsy needles and for planning pedicle screw trajectories [17]. Stereotactic brain biopsy represents one of the earliest applications of surgical robotics. On April 11, 1985, a team at the Memorial Medical Center used a modified PUMA industrial robot (Advance Research & Robotics, CT, USA) to perform a robot-assisted stereotactic brain biopsy on a 52-year old man [16]. Multiple other robotic systems have since been developed.

The Renaissance Robotic System (Mazor Robotics, Israel) was developed for pedicle screw insertion in the spine. It allows tool guiding and drill assistance and has more recently been approved for use in the brain. The Excelsius GPS (Globus Medical, Audubon, Pennsylvania) was FDA approved in 2017. It has great potential in spinal surgery and features real-time intraoperative imaging, automatic compensation for patient movement, and direct screw insertion through a rigid external arm, which obviates the need for K-wires or clamps [25]. The surgeon is provided with instant feedback with the robot's monitor if there is drill malalignment or the reference frame moves. However, due to only recent clearance by the FDA, little research has been done on the accuracy of the system. In the following years, there should be more data available. This device is intended to work with artificial intelligence in the future, thus prompting the surgeon with appropriate solutions based on a collected database of cases from around the world and stored in the cloud. The brain surgery software is also yet to come (Fig. 1).

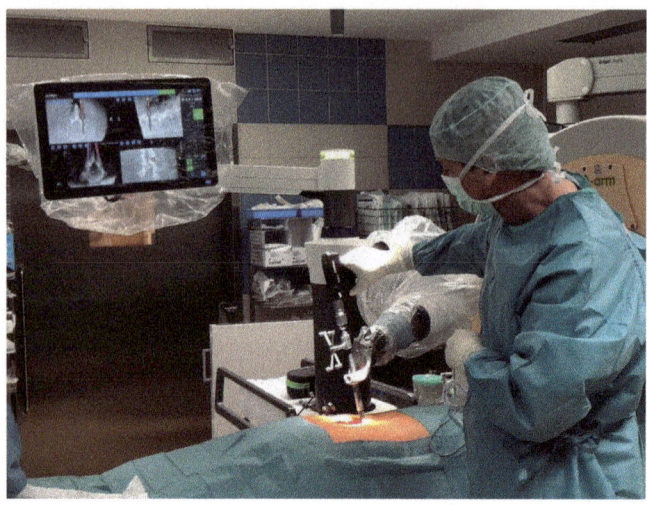

Fig. 1. Robot-assisted spinal surgery (Excelsius GPS robot – transpedicular screws application - performed by prof. Peter Douglas Klassen, Bonifatius Hospital, Lingen, Germany. (own archive).

5   Conclusions

"There certainly will be job disruption. Because what's going to happen is robots will be able to do everything better than us... I mean all of us." – Elon Musk. The most significant tool in neurosurgery in the 20th century has been undoubtedly the microscope. It allowed for a major upgrade in the quality of neurosurgical procedures and clinical outcomes of patients. It also brought tremendous progress by creating a whole new type of surgical procedure – microsurgery. In our opinion, robotics will be the hallmark of neurosurgery in the 21st century. At the present time, surgical robots are mainly used as a stand that frees a surgeon's hands and reduces muscle fatigue. Surgical robots have already contributed significantly to improving surgical practice by increasing precision and allowing the integration of state-of-the-art technology. The neutralization of the physical limitations is a major benefit of robotics. Hand tremor, tiredness, and correction of errors in the assessment of the depth of the surgical field will enhance the performance of neurosurgeons, reduce complications, improve outcomes, and provide shorter recovery and hospitalization periods.

One of the most important goals is to promote active cooperation between engineers and neurosurgeons. However, unfamiliarity with robot technology and the high costs of maintenance and purchasing the few available robotic systems can discourage their use. Nevertheless, improvements in the quality of healthcare should eventually surpass the inherent costs of robotic surgery systems. While we witness the dawn of artificial intelligence and brain-machine interfaces, neurosurgery and treatment of previously incurable diseases has already surpassed the ideas of science fiction writers. As with every tool created by man, we must ensure as scientists and medical doctors that robotics is used as means to benefits individuals and all of mankind.

References

1. Faria, M.A.: Violence, mental illness, and the brain—A brief history of psychosurgery: Part 1—From trephination to lobotomy. Surg. Neurol. Int. **4**, 49 (2013)
2. Prioreschi, P.: A History of Medicine. Primitive and Ancient Medicine. Horatius Press, Omaha, Nebraska, vol. 1, pp. 21–30 (1995)
3. Ormond, et al.: The history of neurosurgery and its relation to the development and refinement of the frontotemporal craniotomy. Neurosurg. Focus **36**(4), E12 (2014)
4. Rosheim, M.E.: Leonardo's Lost Robots. Springer, vol. 69 (2006)
5. Doulgeris, J.J., Gonzalez-Blohm, S.A., Filis, A.K., Shea, T.M., Aghayev, K., Vrionis, F.D.: Robotics in Neurosurgery: Evolution, Current Challenges, and Compromises
6. Lanfranco, A.R., Castellanos, A.E., Desai, J.P., et al.: Robotic surgery: a current perspective. Ann. Surg. **239**(1), 14–21 (2004)
7. Kriss, T.C., Kriss, V.M.: History of the operating microscope: from magnifying glass to microneurosurgery. Neurosurgery **42**, 899–907 (1998)
8. Van Zuylen, J.: The microscopes of Antoni van Leeuwenhoek. J. Microsc. **121**, 309–328 (1981)
9. Louw, D.F., Fielding, T., McBeth, P.B., et al.: Surgical robotics: a review and neurosurgical prototype development. Neurosurgery **54**, 525–537 (2004)
10. Int. J. Med. Robot. Comput. Assist. Surg. **2**, 105–106 (2006). (www.interscience.wiley.com), https://doi.org/10.1002/rcs.93
11. Apuzzo, M.L.J.: In the realm of ideas: the advent of advanced surgery of the human cerebrum and neurosurgical education. Acta Neurochir. Suppl. (Wien) **69**, 145–150 (1997)
12. http://www.frc.ri.cmu.edu/robotics-faq/1.html#1.1. Accessed 2 Apr 2005
13. Nathoo, et al.: In touch with robotics: neurosurgery for the future. Neurosurgery **56**, 421–433 (2005)
14. Cavusoglu, M.C., Williams, W., Tendick, F., Sastry, S.S.: Robotics for telesurgery: Second generation Berkeley/UCSF laparoscopic telesurgical workstation and looking towards the future applications—Special issue on medical robotics. Ind Rob **30**, 22–29 (2003)
15. Holly, L.T.: Int. J. Med. Robot. Comput. Assist. Surg. **2**, 105–106 (2006)
16. Kwoh, Y.S., Hou, J., Jonckheere, E.A., Hayati, S.: A robot with improved absolute positioning accuracy for CT guided stereotactic brain surgery. IEEE Trans. Biomed. Eng. **35**, 153–160 (1988)
17. Veejay, B., et al.: Robotics in Neurosurgery. Ann. R. Coll. Surg. Engl. **100**(6), 23–26 (2018)
18. Sutherland, et al.: The Evolution of NeuroArm. Neurosurgery **72**, S1 (2013)
19. Attenello, F., Lee, B., Yu, C., Liu, C.Y., Apuzzo, M.L.: Supplementing the neurosurgical virtuoso: evolution of automation from mythology to operating room adjunct. World Neurosurg. (2014). https://doi.org/10.1016/j.wneu.2014.03.011
20. Hokayem, P.F., Spong, M.W.: Bilateral teleoperation: an historical survey. Automatica **42**(12), 2035–2057 (2006)
21. Nelson, J.H.: Robotic stereotactic assistance (ROSA) for pediatric epilepsy: a single-center experience of 23 consecutive cases. Children **7**, 94 (2020). https://doi.org/10.3390/children7080094
22. Fresci, C., et al.: Int. J. Med. Robot. Comput. Assist. Surg. **9**, 396–406 (2013). https://doi.org/10.1002/rcs.1468
23. Ballantyne, G.H., Moll, F.: The da Vinci telerobotic surgical system: the virtual operative field and telepresence surgery. Surg. Clin. North Am. **83**(1293–1304), vii (2003)
24. Bonsor, K., Strickland, J.: How Robotic Surgery Will Work (2000). http://science.howstuffworks.com/roboticsurgery.html

25. D'Souza, M., et al.: Robotic-assisted spine surgery: history, efficacy, cost, and future trends. Robot. Surg. Res. Rev. **6**, 9–23 (2019)
26. Lee, J.Y.K., Bhowmick, D.A., Eun, D.D., Welch, W.C.: Minimally invasive, robot-assisted, anterior lumbar interbody fusion: a technical note. J. Neurol. Surg. Part A Cent. Eur. Neurosurg. **74**, 258–261 (2013). https://doi.org/10.1055/s-0032-1330121
27. Kaushal., et al.: Robotic-Assisted Systems for Spinal Surgery. http://dx.doi.org/10.5772/int echopen.88730
28. Dreval, O.N., Rynkov, I.P., Kasparova, K.A., Bruskin, A., Aleksandrovskii, V., Bernstein, V.Z.I.L.: Results of using spine assist mazor in surgical treatment of spine disorders. Zhurnal Vopr Neirokhirugii Im N N Burdenko (2014)

MATLAB® Utility for Small Invasive Procedure to Confirm Objectively the New Disease – Chronic Pain

Elzbieta Skorupska[1,2(✉)], Tomasz Dybek[2], Marta Jokiel[1,3], Michał Rychlik[4],
Paweł Dobrakowski[5], Jarosław Szyszka[6], and Dariusz Zmarzły[7]

[1] Department of Physiotherapy, Poznan University of Medical Sciences,
Fredry 10, 61-701 Poznan, Poland
skorupska@ump.edu.pl
[2] Faculty of Physical Education and Physiotherapy,
Opole University of Technology, Opole, Poland
[3] Department of Orthopedics, Traumatology and Hand Surgery,
Poznan University of Medical Sciences, Poznan, Poland
[4] Department of Virtual Engineering, Poznan University of Technology, Poznan, Poland
[5] Humanitas University in Sosnowiec, Psychology Institute, Sosnowiec, Poland
[6] Orthopedic Surgery Department, Opole Rehabilitation Center, Korfantów, Poland
[7] Faculty of Electrical Engineering Automatic Control and Informatics,
Opole University of Technology, Opole, Poland

Abstract. A new disease, named a chronic pain, which affects more than 20% of adult population, has been recognized. The chronic pain patients need to cope with the effects of that treatment resistant being to an ailment such as emotional overloading, disability, sleep and mood disturbances, etc. A novel idea for the effective chronic pain control is focused on identifying one of the three possible pain mechanisms: a nociceptive, neuropathic, or a nociplastic pain related to the central sensitization (CS). This differentiation means that the treatment should follow the pain type. However, the problem of determining the dominant pain component has not yet been solved. The current trend in the research related to pain diagnostics is a desire to define an objective tool for the CS confirmation. It is hypothesized that the autonomic nervous system, dysregulation measurement (response to noxious stimuli) can be one of the parameters for the objective CS confirmation. The new validated and awarded method to confirm the myofascial pain syndrome (MPS) dependent on the central sensitisation process has been presented. The protocol of the method is considered a new type of the active dynamic thermography. The involvement of the CS related to muscles is confirmed objectively being based on the visualization of the perceived pain area by presenting the vasomotor response to stimulation. Unfortunately, the main disadvantages of that method is time consuming of the manual thermal data analysis. The use of the MATLAB® software for the thermal data analysis seems interesting for consideration.

Keywords: Active dynamic thermography · Autonomic nervous system · Myofascial pain · Trigger points · Infrared thermography camera

© The Author(s), under exclusive license to Springer Nature Switzerland AG 2021
S. Paszkiel (Ed.): ICBCI 2021, AISC 1362, pp. 9–18, 2021.
https://doi.org/10.1007/978-3-030-72254-8_2

1   Medical Background

In the 11th revision of the International Statistical Classification of Diseases and Related Health Problems (ICD-11), the chronic pain has been categorized for the first time as a separate disease [1]. A systematic classification of clinical conditions associated with chronic pain created by an interdisciplinary group of the International Association for the Study of Pain (IASP) has been exhibited. The fact of including the chronic pain as a distinct disease in the ICD-11 is supported by the tripart division of a pain pathomechanism into a nociceptive, neuropathic, and nociplastic respectively with the necessity of applying different treatment strategies for each subtype. Generally a nociceptive pain is considered physiological whereas the other two pathological.

Furthermore, the neuropathic and nociplastic pain is characterized by the abnormal functioning of the somatosensory nervous system. According to the established neurological diagnostic criteria, the neuropathic pain is due to a demonstrable lesion or disease while a nociplastic pain is associated with no such signs. The nociceptive and neuropathic pain is commonly known, but the nociplastic pain is currently poorly recognized by most clinicians.

Clinically, patients with the nociplastic pain patomechanism are manifested with an amplified pain perception considering its intensity, duration, and distribution. The hypersensitivity of the somatosensory system is associated with a series of symptoms such as referred pain, a widespread pain, a tactile allodynia, the heightened response to the nonnoxious stimuli, a maladaptive psychosocial factors, and the low vagal nerve activity [2]. The autonomic nervous system (ANS) imbalance as well as the hypothalamic-pituitary-adrenal (HPA) axis dysregulation are indicated as leading mechanisms for the CS development in the musculoskeletal system [3, 4]. The ANS impact is considered to occur via the reflexes that affect the muscle circulation and contractility, the sensory motor control, and inflammatory processes. Changes in the ANS regulation, mainly through the sympathetic branch, provoke a nociceptor activation indirectly by vasoconstriction-vasodilatation imbalance or directly by sympathetic-nociceptor activation resulting in a widespread pain, the hyperalgesia, as well as allodynia.

The autonomic nervous system involvement is also taken into consideration with regards to the myofascial pain syndrome (MPS) development and maintenance, which has been recently classified as related to a central sensitization process [5]. The MPS is characterized by: (i) trigger points (TrPs), i.e. limited sites of the severe muscle tenderness or hypersensitivity, (ii) a determined area of the referred pain, and (iii) the characteristic motor, sensory, and autonomic dysfunctions and symptoms. A trigger point (TrP) is an area with the multiple "TrP loci" composed of a sensory component: the sensitized nociceptor, i.e. a free nerve ending (LTR locus), and a motor component: a dysfunctional end-plate in the vicinity of the sensitized nociceptor (SEA locus). From the clinical point of view, the trigger points are divided into active, causing MPS symptoms, and latent referring to no pain until mechanical stimulation is applied.

The most accepted theoretical concept of the TrPs activation is the Simons' integrated hypothesis which assumes 5 or 6 steps of the positive-feedback cycle [6]. The hypothesis suggests that the abnormal depolarization of the post-junctional membrane of the motor end-plates cause the sarcomere shortening, the lower concentration of adenosine triphosphate, and the localized hypoxia leading to a bradykinin release and severe sensitization of the intramuscular nociceptors.

A series of studies suggesting the link between the ANS and the nociplastic pain among low back leg pain (LBLP) patients has been published [7–11]. The studies documented the intensive vasodilatation in the perceived sciatic pain area provoked by a small-invasive method applied to trigger points within gluteus minimus. The observed impaired blood flow in the pain region is said to be related to the sympathetic hyperactivity and diminished parasympathetic tone. Thus, the ANS dysregulation characteristic of the central sensitization process can be presumed for trigger points. It is not known, if TrPs in other muscles would provoke a similar reaction, which – if positive – would be consistent with CS studies.

In pain characteristic of the MPS (central sensitization) involvement, the knowledge of the referred pain is crucial because its location is neither compatible with innervation nor with the commonly known dermatomes. The referred pain pattern has a key diagnostic meaning and it is one of the additional criteria confirming MPS because there is a strong correlation between a given trigger point location and the referred pain pattern defined for this trigger point and occurring in the same area with all patients. Moreover – in the majority of cases – the referred pain does not occur in the area of the trigger point presence. That is why the therapy is usually applied in the area different from the location corresponding to the daily complaint. Moreover, the MPS symptoms are reversible which is compatible with Hart et al. [12] hypothesis concerning the "bottom up" CS subtype existence.

2 Thermography in Pain Medicine

One of the technologies applied in modern medicine is imaging performed with a thermographic camera. The biological background for the infrared thermography (IRT) measurement is a complex reaction between the blood-flow rate and the local structures of the subcutaneous tissues under the regulation of the sympathetic part of the ANS. The skin temperature in healthy individuals is symmetric. Thus, thermal symmetry assessment is considered a valuable method for assessing the physiological normality/abnormality. Thermal asymmetries greater than 0.5–0.7 °C are usually associated with a dysfunction of the musculoskeletal system [13].

Currently, two types of thermography are used in medicine: the Static Thermography (ST) and the *Active Dynamic Thermography* (ADT). The ST is a qualitative method of non-objective visual analysis of a single image. This method has raised many doubts and is considered to be of a limited use. The most valuable data are obtained by the ADT. It is observed that the quality of the obtained data demanded the precisely defined protocols, an advanced thermogram analysis software, and a high-quality camera. The ADT has been applied in different branches of medicine, e.g. oncology, rheumatology, sports medicine, etc. In the pain medicine, the ADT utility has been indicated for neural diseases, musculoskeletal diseases, inflammatory diseases, and vascular diseases. An infrared thermovision (IRT) camera can objectively support the diagnosis of patients' pain, especially when the autonomic nervous system (ANS) is actively involved [14].

The high IRT reliability has been also confirmed for the muscle examination [15–17]. The ADT protocols are based on the image sequence analysis conducted for the subjects undergoing a short-term stimulation, e.g. cooling, warming, or the exercise under the IRT control followed by a recovery phase [18]. This shows that a noxious stimulus provokes a more intensive ANS response observed in the pathological area [19].

3   A Small Invasive Procedure to Confirm Myofascial Pain Syndrome – A New Type of Active Dynamic Thermography

The small invasive procedure to confirm the MPS is a new type of the ADT where the needling of sensitized nociceptors (trigger points within the muscles) under infrared camera control was first timed used as a noxious stimulant. The method is aimed to visualise the patients perceived pain area hypothetically linked with the stimulated muscles. The vasomotor response is characteristic of pain patients with myofascial pain co-existence, exclusively. The MPS theory is based on the studies where the referred pain pattern for each muscle was defined according to patients' reports of a trigger points stimulation. The other ADT method allowed to observe the ANS response despite the pain patomechanism. Further on, a response was observed to a temperature stimulant (hot, cold) or mechanical stimulant (exercise) provoking global or local thermoregulatory reactions [20].

Initially, the method was named Thermovision Technique of Dry Needling (TTDN) and then renamed to a small invasive procedure [10]. The full protocol of the method was published in the validation paper of the sciatica patients [20]. The diagnostic value was indicated for two parameters of the procedure, namely a significant increase in the average temperature and the autonomic referred pain (AuRP). The term AuRP described the new isotherm above the initial maximum temperature resulting from a small invasive procedure. The highest value of the average temperature and the AURP size are characteristic for the observation phase of the procedure. The final thermogram 'visualise' the area of the felt pain area coincidental with a referred pain pattern of the stimulated gluteus minimus muscle (Fig. 1a). The gray picture only visualizes the AuRP (Fig. 1b).

Summing up, the small invasive procedure allows registering autonomic dysregulation in the pain region in real time [10]. It seems that the autonomic response depends on the stimulant intensity but not on the level of the felt pain. Recent studies focused on the ANS response to a noxious stimulation in pain patients have revealed that the ANS response is more related to the stimulus intensity than to felt pain intensity [19, 21]. That assumption is consistent with the central sensitization pathomechanism where autonomic imbalance is one of the processes leading to central sensitization. Additionally, it seems interesting to combine thermal data with other parameters allowing to confirm the ANS activity e.g. heart rate variability, both under electroencephalography control. Every mentioned parameters can be simultaneously analysed by a MATLAB®, thus allowing to strengthen the method results [22–25].

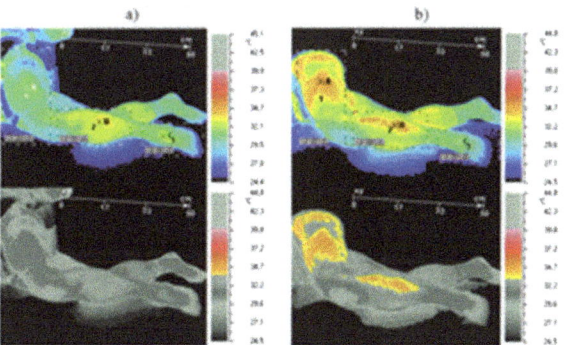

Fig. 1. Results of the small invasive procedure applied to the gluteus minimus trigger point in sciatic patients. Gray picture: the isotherm reflecting autonomic referred pain (AuRP). Legend: (a) state before a procedure, (b) a post-procedure phase (vasodilatation in the felt pain area).

3.1 A Short Description of the Minimally Invasive Procedure

The whole procedure consisted of: A. The examination part: (1) a pre-stimulation phase, (2) a stimulation phase, (3) a post-stimulation phase, and B. The thermogram analysis.

The Examination (Procedure Part A); A Pre-stimulation Phase: A side-to-side comparison of the patient at rest. The average temperature decrease of more than 0.5 °C in the painful area relative to the opposite asymptomatic side is confirmed, the neuropathic pain is considered possible. A Stimulation Phase: The noxious stimulation of the two most sensitive points within the examined muscle lasted 5 min, each with a one-minute break in between (dry needling of every point within the examined muscles). The area of a referred pain defined for the examined muscles is registered by an infrared thermovision (IRT) camera. During the whole noxious stimulation, the patients reported the localization of the pain sensation provoked by needling. A Post-stimulation Phase: The next step of the procedure is the IRT observation of the patient at rest for 6 min (it the region of the referred pain distribution). At the end of the whole procedure, the patient is asked if the pain evoked by needling was similar to his/her daily pain?

An example of the video of the procedure performed on the low back leg pain patients is available in one of the papers [10].

The Thermogram Analysis; A Description of The Manual, Numerical Thermogram Analysis (Procedure Part B); Another innovation of the small invasive procedure is a new thermogram analysis where the technician manually quantifies the thermogram and calculates the size in cm^2 of the vasodilatation and the average temperature increases in that particular isothermal area. The calculation of the isothermal area in the Thermography studio software was used. Moreover, each image is calibrated using a length standard. In the software for the analysis of thermograms, the length standard (visible in the image) is assigned the actual length value. The 25 cm long aluminum standard in research was used. The thermogram calibration image is made individually for each set of data (each patient) before starting the stimulation phase. The length standard is placed in the scope of the camera, immediately next to the examined patient's body part (e.g. lower limb), and then a thermogram is recorded.

To determine the isothermal areas, masks are applied firstly on the thermogram to define the boundaries of the analyzed body parts (in this case; the thighs, calves and feet). Then a median filter with a 3 × 3 matrix is used. This filter determines the averaging of the values around the 'middle' measuring pixel, removing the so-called noise "salt and pepper" and generating the isothermal areas (regions of constant temperature). In the next step, for each of the analyzed parts of the body, the values of the size of the isothermal areas (in cm^2) are read for individual temperatures of each of the isotherms.

Average temperature of isothermal-area

Avr °C		Thermogram	005	239	359	
26,5°C	Thigh	Area 1	0,0000	0,0000	0,0000	
	Calf	Area 2	0,0000	0,0000	0,0000	26,5°C
	Foot	Area 3	31,3501	32,2872	34,8519	
27,3°C	Thigh	Area 1	137,3230	0,0000	82,6274	
	Calf	Area 2	19,7605	0,0000	0,0000	27,3°C
	Foot	Area 3	23,8461	24,7632	26,4308	
28,2°C	Thigh	Area 1	370,3640	119,8140	165,2550	
	Calf	Area 2	26,7643	24,5131	36,5195	28,2°C
	Foot	Area 3	19,0935	22,1785	37,5200	
28,9°C	Thigh	Area 1	267,8090	256,3030	208,1110	
	Calf	Area 2	90,5483	44,0235	130,6530	28,9°C
	Foot	Area 3	31,4334	43,1897	53,6119	
29,8°C	Thigh	Area 1	84,7118	200,8570	134,4880	
	Calf	Area 2	191,7690	97,0518	261,3890	29,8°C
	Foot	Area 3	26,3474	37,8535	19,0935	
30,2°C	Thigh	Area 1	67,0358	106,2230	101,2210	
	Calf	Area 2	68,3698	226,7880	28,7653	30,2°C
	Foot	Area 3	25,9305	15,1748	9,4217	
31,1°C	Thigh	Area 1	27,7648	137,2400	153,2480	
	Calf	Area 2	2,0011	37,6868	8,0043	31,1°C
	Foot	Area 3	17,7595	7,0871	1,2507	
32,1°C	Thigh	Area 1	0,0000	130,9870	155,5000	
	Calf	Area 2	0,0000	1,9177	2,2512	32,1°C
	Foot	Area 3	1,0839	0,0000	0,0000	
33,2°C	Thigh	Area 1	0,0000	43,9401	24,6798	
	Calf	Area 2	0,0000	0,0000	0,0000	33,2°C
	Foot	Area 3	0,0000	0,0000	0,0000	
33,9°C	Thigh	Area 1	0,0000	17,4260	5,1694	
	Calf	Area 2	0,0000	0,0000	0,0000	33,9°C
	Foot	Area 3	0,0000	0,0000	0,0000	
34,5°C	Thigh	Area 1	0,0000	0,0000	0,0000	
	Calf	Area 2	0,0000	0,0000	0,0000	34,5°C
	Foot	Area 3	0,0000	0,0000	0,0000	
34,8°C	Thigh	Area 1	0,0000	0,0000	0,0000	
	Calf	Area 2	0,0000	0,0000	0,0000	34,8°C
	Foot	Area 3	0,0000	0,0000	0,0000	
35,5°C	Thigh	Area 1	0,0000	0,0000	0,0000	
	Calf	Area 2	0,0000	0,0000	0,0000	35,5°C
	Foot	Area 3	0,0000	0,0000	0,0000	
36,2°C	Thigh	Area 1	0,0000	0,0000	0,0000	
	Calf	Area 2	0,0000	0,0000	0,0000	36,2°C
	Foot	Area 3	0,0000	0,0000	0,0000	
36,9°C	Thigh	Area 1	0,0000	0,0000	0,0000	
	Calf	Area 2	0,0000	0,0000	0,0000	36,9°C
	Foot	Area 3	0,0000	0,0000	0,0000	
		Thermogram	005	239	359	
Total area temp. 27,0-40,0	Thigh	Area 1	1180,4600	1236,4100	1220,0700	Total area temp. 27,0-40,0
	Calf	Area 2	501,1010	541,7060	532,8680	
	Foot	Area 3	197,9390	211,9460	196,2710	

Fig. 2. An example of the numerical thermogram analysis of the asymptomatic case.

The whole procedure allowed to obtain 340 thermograms recorded at a 3 s interval for 17 min. One thermogram of every examined part (pre-stimulation, stimulation and a post-stimulation phase) is analyzed numerically (the size of the each iso-thermal area). Then, a comparison of the obtained data is performed.

Currently published data were performed for a lower limb, thus the isothermal area analysis presented below was calculated for that part of the body. Firstly the whole measurement area (thigh, calf, and foot; range of temperature from 23.7 to 40.0 °C) in order to measure changes in the iso-thermal area size were identified for every 0.7 °C. Then, according to the three-sigma rule, the data were grouped for three temperature ranges respectively (low range: 28.4–29.2 °C; middle range: 29.9–32.7 °C, and high range: above 33.4 °C). The calculations were presented for particular subareas (namely thigh, calf, and foot) (Fig. 2). The data obtained for the three ranges of temperature areas in cm^2 were calculated in percentage values.

Some TrPs-positive cases developed vasodilatation which occurred in the middle temperature range according to the three-sigma rule. Then, based on that observation, an idea of the new type of the thermogram analysis and the size calculation of the observed phenomenon were assumed. The occurrence of temperature increase above the initial maximal temperature (Tmax) was most characteristic of trigger points. Thus, for vasodilatation, the size of the iso-area around the Tmax and for vasoconstriction around the minimal temperature (Tmin) was assumed to be calculated in every case. Widespread vessels reactions of the isothermal area above the Tmax at rest or the Tmin at rest were named the autonomic referred pain (AuRP) (Fig. 3).

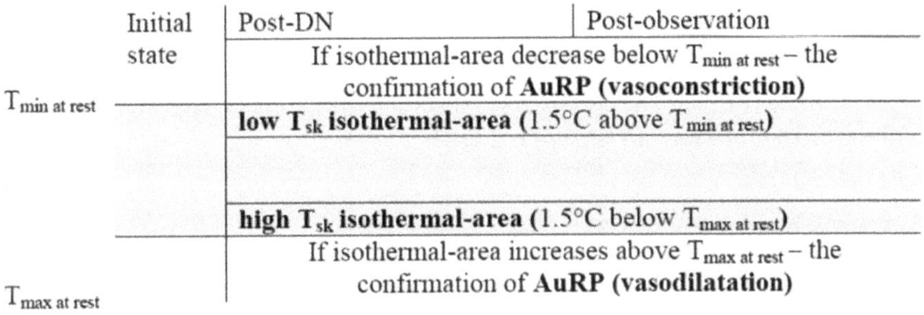

	Initial state	Post-DN	Post-observation
$T_{min \, at \, rest}$		If isothermal-area decrease below $T_{min \, at \, rest}$ – the confirmation of **AuRP (vasoconstriction)**	
		low T_{sk} isothermal-area (1.5°C above $T_{min \, at \, rest}$)	
		high T_{sk} isothermal-area (1.5°C below $T_{max \, at \, rest}$)	
$T_{max \, at \, rest}$		If isothermal-area increases above $T_{max \, at \, rest}$ – the confirmation of **AuRP (vasodilatation)**	

Fig. 3. An isothermal-area analysis related to the TTDN.

4 The Advantages of the Thermograms Analysis Using MATLAB®

The software available for a fast thermal data analysis is thermoHuman, med-hot, etc. However, none of the available thermovision software can be applied to analyze the small invasive procedure. Thus, one of the possibilities for accelerating the thermal data analysis that allows automatic data analysis (average temperature changes and the AuRP size changes) is a MATLAB®. The manual data analysis was based on the four thermograms out of 340 recorded during the whole procedure. The ROI was observed with respect to: (a) a new parameter analysis, i.e. numerical calculation in cm^2 of the whole thermogram in the isothermal area (every 0.7 °C), (b) the presence of the specific isotherm (AuRP) observed within the ROI as a result of trigger points noxious stimulation which indicated the disease. The automatic data analysis by a MATLAB® should be focused on the same two parameters according to a validation study based on the sciatica patients. The automatic data analysis would allow to obtain more precise data and reduce the time necessary for analyzing numerical results, which until now has been done manually. Additionally, it gives the possibility to poses the new data from the procedure. The video file revealed that registered the vasomotor response changes dynamically with a sharp increase and the involution followed by the next increase during the stimulation phase. Next, the exponential growth of the average temperature and the AuRP size with culmination point (then gradual temperature and AuRP size decrease) was characteristic exclusively for a post-stimulation phase of the symptomatic patients. The average time of that cumulative vasomotor response during post-observation phase has not been defined. The precise definition of the thermal recovery after stimulation is indicated as an important aspect of the ADT protocol. The small invasive procedure assumed a six-minute observation in the post-stimulation phase to get the maximum temperature peak. After that time, the recovery phase was not observed. Therefore, the thermal recovery time was not defined. A detailed analysis of the post-stimulation phase would be possible using a MATLAB®.

Additionally, an important step in the improvement of data acquisition quality is the ROI determination. This is one of the most controversial points using the IRT application in humans. Many IRT studies have developed their own criteria for creating and selecting the ROI. The automatic data analysis of the small invasive procedure should use anatomical references as markers, e.g. lower leg segmentation: thigh, calf, and foot. This strategy can be useful in ensuring the reproducibility of the ROI delimitation and it can facilitate further data analysis.

The examples of the MATLAB® analysis of the symptomatic case recorded during a small invasive procedure are shown in Fig. 4.

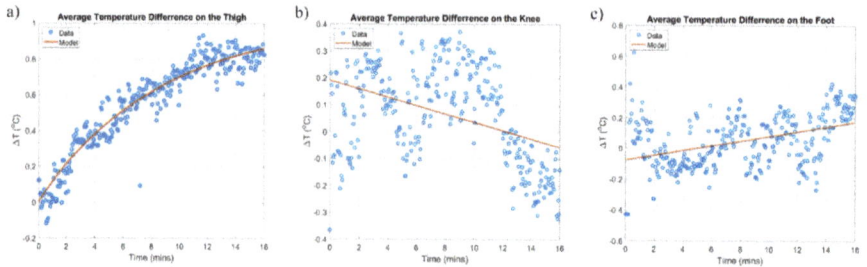

Fig. 4. The average temperature increase in the symptomatic lower leg (a) thigh, (b) calf, (c) foot.

The case presented in Fig. 4 (a, b, c) show the patient suffering from the sciatic pain with the MPS co-existence (the patients daily complains extended down the thigh). The small invasive procedure confirmed the MPS co-existence based on the intensive average temperature increase on a thigh only (Fig. 4a). That temperature increase was coincidental with a patient' reported pain symptoms during procedure and his/ her daily complains.

5 Summary

The pain syndrome coexisting in sciatica patients confirmed by a small invasive procedure has been recently recognized as a gluteal syndrome (in 2019, it was included in the ICD-11 and classified as a chronic pain subtype). Unfortunately, the gluteal syndrome is still recognized on the basis of a non-objective palpatory criteria. Additionally, the method can be applied to other diseases where pain symptoms are related to a coexisting myofascial pain syndrome e.g. tendinopathies, osteoarthritis etc. This fact points out the necessity and importance for further development of the above described small invasive procedure, especially towards a more sophisticated thermogram analysis e.g. a MATLAB®.

References

1. https://www.iasp-pain.org/PublicationsNews/Content.aspx?ItemNumber=1673&navItemNumber=677. Accessed 13 Jan 2021
2. Woolf, C.J.: Central sensitization: Implications for the diagnosis and treatment of pain. Pain **152**, S2–S15 (2011). https://doi.org/10.1016/j.pain.2010.09.030
3. Roatta, S., Farina, D.: Sympathetic actions on the skeletal muscle. Exerc. Sport Sci. Rev. **38**(1), 31–35 (2010)
4. Sjøgaard, G., Lundberg, U., Kadefors, R.: The role of muscle activity and mental load in the development of pain and degenerative processes at the muscle cell level during computer work. Eur. J. Appl. Physiol. **83**(2), 99–105 (2000)
5. Fernández-de-las-Peñas, C., Dommerholt, J.: Myofascial trigger points: peripheral or central phenomenon? Curr. Rheumatol Rep. **16**(1), 395 (2014). https://doi.org/10.1007/s11926-013-0395-2
6. Simons, G.S.: Review of enigmatic MTrPs as a common cause of enigmatic musculoskeletal pain and dysfunction. J. Electromyogr. Kinesiol. **14**(1), 95–107 (2004)
7. Skorupska, E., Rychlik, M., Samborski, W.: Intensive vasodilatation in the sciatic pain area after dry needling. BMC Complement Altern. Med. **15**, 72 (2015)
8. Skorupska, E., Rychlik, M., Pawelec, W., Samborski, W.: Dry needling related short-term vasodilation in chronic sciatica under infrared thermovision. Evid Based Complement Alternat. Med. 214374 (2015)
9. Skorupska, E., Rychlik, M., Pawelec, W., Bednarek, A., Samborski, W.: Trigger point-related sympathetic nerve activity in chronic sciatic leg pain: a case study. Acupunct. Med. **32**(5), 418–422 (2014)
10. Skorupska, E., Jokiel, M., Rychlik, M., Łochowski, R., Kotwicka, M.: Female overrepresentation in low back-related leg pain: a retrospective study of the autonomic response to a minimally invasive procedure. J. Pain Res. **18**(13), 3427–3435 (2020)
11. Skorupska, E., Rychlik, M., Pawelec, W., Bednarek, A., Samborski, W.: Intensive short-term vasodilation effect in the pain area of sciatica patients–case study. BMC Res. Notes. **7**, 620 (2014)
12. Harte, S.E., Harris, R.E., Clauw, D.J.: The neurobiology of central sensitization. J. Appl. Biobehav. Res. **23**, e12137 (2018)

13. Vardasca, R., Ring, F., Plassmann, P., Jones, C.: Thermal symmetry of the upper and lower extremities in healthy subjects. Thermol. Int. **22**, 53–60 (2012)
14. Eddie Ng, Y.K.: Mahnaz Etehadtavakol. Application of Infrared to Biomedical Sciences. Springer, Singapore (2017)
15. Costa, A.C., Dibai Filho, A.V., Packer, A.C., Rodrigues-Bigaton, D.: Intra and inter-rater reliability of infrared image analysis of masticatory and upper trapezius muscles in women with and without temporomandibular disorder. Braz. J. Phys. Ther. **17**(1), 24–31 (2013)
16. James, C.A., Richardson, A.J., Watt, P.W., Maxwell, N.S.: Reliability and validity of skin temperature measurement by telemetry thermistors and a thermal camera during exercise in the heat. J. Therm. Biol. **45**, 141–149 (2014)
17. Sancibrian, R., Gutierrez-Diez, M.C., Redondo-Figuero, C., Llata, J.R., Manuel-Palazuelos, J.C.: Using infrared imaging for assessment of muscular activity in the forearm of surgeons in the performance of laparoscopic tasks. Proc. Inst. Mech. Eng. H. **233**(10), 999–1009 (2019)
18. González, F.J., González, R., López, J.C.: Thermal contrast of active dynamic thermography versus static thermography. Biomed. Spectrosc. Imaging **8**(1–2), 41–45 (2019)
19. Nickel, M.M., May, E.S., Tiemann, L., Postorino, M., Ta Dinh, S., Ploner, M.: Autonomic responses to tonic pain are more closely related to stimulus intensity than to pain intensity. Pain **158**(11), 2129–2136 (2017)
20. Skorupska, E., Rychlik, M., Samborski, W.: Validation and test-retest reliability of new thermographic technique called thermovision technique of dry needling for gluteus minimus trigger points in sciatica subjects and TrPs-negative healthy volunteers. Biomed. Res. Int. **2015**, 546497 (2015)
21. Szyguła, R., Dybek, T., Klimek, A., Tubek, S.: Impact of 10 sessions of whole body cryostimulation on cutaneous microcirculation measured by laser Doppler flowmetry. J. Hum. Kinet. **30**, 75–83 (2011)
22. Maszczyk, A., Dobrakowski, P., Żak, M., et al.: Differences in motivation during the bench press movement with progressive loads using EEG analysis. Biol. Sport. **36**(4), 351–356 (2019)
23. Dobrakowski, P., Łebecka, G.: Individualized neurofeedback training may help achieve long-term improvement of working memory in children with ADHD. Clin. EEG Neurosci. **51**(2), 94–101 (2020)
24. Mali, B., Zulj, S., Magjarevic, R., Miklavcic, D., Tomaz, J.T.: MATLAB®-based tool for ECG and HRV analysis. Biomed. Sign. Proces. Contr. **10**, 108–116 (2014)
25. Herdman, A.T.: SimMEEG software for simulating event-related MEG and EEG data with underlying functional connectivity. J. Neurosci. Methods **350**, 109017 (2020)

Impact of Selected Factors on the "Early Error" Phenomenon Evaluated in the Mirror Test

Agnieszka Machowska-Majchrzak[1], Anna Starostka-Tatar[2,3], Katarzyna Pyrkosz[4], Aleksandra Napieralska[4], Katarzyna Kurczyna[1], Paweł Dobrakowski[3(✉)] (iD), Andrzej Kras[4], Beata Łabuz-Roszak[5,6] (iD), and Krystyna Pierzchała[4]

[1] Department of Neurology, City Hospitals of Chorzów, Chorzow, Poland
[2] MS Therapy Centre, Katowice, Poland
[3] Humanitas University, Institute of Psychology, Sosnowiec, Poland
[4] Department of Neurology Zabrze, Medical University of Silesia, Katowice, Poland
[5] Faculty of Health Sciences in Bytom, Department of Basic Medical Sciences, Medical University of Silesia in Katowice, Katowice, Poland
[6] Department of Neurology, St. Jadwiga's Provincial Specialist Hospital, Opole, Poland

Abstract. Aim. Evaluation of the occurrence of the early error (overestimation of perception of mirror reflections) and its size in the selected groups of individuals, as well as the analysis of the relationship between the size of the error and spatial orientation, evaluation of one's own figure, time spent in front of the mirror and age. One hundred and ninety seven individuals were enrolled in the study, i.e. 46 school-going children, 30 hairdressers, 30 fine arts students, 30 medical students, 31 architecture students, and 30 individuals who were not related to any of the above activities aged 45–70 (Q group). An interview questionnaire was used with the early error evaluation test, a Standard Figural Stimuli Scale, and the WAIS-R block design subtest. The early error was made most frequently in the Q group. A significantly lower mean error was observed in the group of hairdressers, architecture students and individuals who declared themselves as having a scientific-oriented mind. The size of the error also sex-dependent – the error occurrence was significantly lower in males. A significant relationship was not found between the studied factors (age, error in the Standard Figural Stimuli Scale, subjective evaluation of one's geometric imagination, time spent in front of the mirror, acceptance of one's own body, and results of the WAIS-R block design subtest) and error size. The early error is made more frequently by middle-aged individuals and by females, regardless of their age. Its size was different for individual professional groups.

Keywords: Naive physics · Visual perception · Mirrors

1 Introduction

Mirrors, in any form, have accompanied mankind for ages. The prototype of the mirror was a smooth surface of water in which our ancestors could see their reflections.

The history of mirrors has always had an intrinsic metaphysical aspect. It was believed that a mirror reflected not only a person's countenance, but also their soul. This motif was

S. Paszkiel (Ed.): ICBCI 2021, AISC 1362, pp. 19–26, 2021.
https://doi.org/10.1007/978-3-030-72254-8_3

noticeable in the broad context of magic, and was a subject of philosophical discussions. Its qualities were used by painters and writers. Many times it was a "door" to another, inaccessible world. Furthermore, if it had not been for mirrors and their application in magnifying glasses, microscopes and telescopes, we would be ignorant of our surroundings in terms of biology, and we would not know what is far beyond human reach – in the universe.

Mirrors are present everywhere and are encountered on a daily basis. Therefore, it would seem that each individual knows exactly how they work. However, our predictions regarding the reflections in the mirror are incorrect. Many individuals believe that they will see their reflections before they can actually do it. This is the early error whose important aspect is the overestimation of what we are able to see. This error is made not only with respect to our own reflections. The same situation occurs when individuals are to predict the moment at which the reflection of another object will occur in a mirror. In fact, the viewer's line of vision must be perpendicular to the surface of the mirror in order for the reflection to be visible. If we want to see ourselves in the mirror, we need to approach at least its nearer edge. Light needs to bounce off an object, reach the mirror, bounce off at the same angle as the angle of incidence and reach the viewer's eye. This conclusion can be reached on the basis of the fact that the angle of incidence and the angle of reflection are equal, but even our daily experience should suffice. We often walk past mirrors and in each case we see our reflection only when we stand in front of the mirror.

Research in the area of mirrors and perspective has absorbed Marco Bertamini for many years. His work has inspired us to conduct our own study. In his studies, Bertamini pays special attention to an erroneous belief about the size of objects reflected in mirrors, and to the prediction of the size of projections on the mirror surface, depending on the distance [1–4]. He also describes the Venus effect [5], i.e. the motif of the goddess seeing her reflection in a mirror, employed by Renaissance painters; the effect is unfeasible in terms of physics, but it is sometimes used by artists for the benefit of viewers.

Bertamini introduced the term "early error". He defined it as an overestimation of what we can see in a mirror. He observed that individuals tend to overestimate only what they can see from the right and the left side of the mirror, whereas they do not make the same error when assuming what they can see above and below the edge of the mirror, even though individuals pass in front of the mirror from its left side to its right side more often than from top to bottom [2, 3, 6]. He concluded that this error emerges only in young adults, and does not occur in primary school children [7].

2   Aim of the Study

The aim of the study was to evaluate the occurrence of the early error and its size in selected groups of individuals with different professional background and in children. Also, the aim was to analyse dependencies between the size of the early error and the spatial orientation, to evaluate one's own shilouette, the time spent in front of the mirror, and age.

3 Material and Method

One hundred and ninety seven individuals were enrolled in the study i.e. 46 children aged 12–15 (26 girls and 20 boys), (mean age: 12.7 ± 0.7 years), 30 hairdressers (i.e. 27 females - F and 3 males - M), aged 19–60 (mean age: 32.9 ± 9.7 years), 30 fine arts students (25 F and 5 M) aged 20–35 (mean age: 22.5 ± 2.8 years), 30 medical students (16 F and 14 M) aged 21–26 (mean age: 22.4 ± 0.9 years), 31 architecture students (14 F and 17 M) aged 19–27 (mean age: 22.6 ± 2.0 years) and 30 individuals (15 F and 15 M) who were not associated with the above activities (the Q group) aged 45–70 (mean age: 54.6 ± 9.4 years). The basic characteristics of the studied group are presented in Table 1.

Table 1. The characteristics of the studied group.

Feature	N
Females	123
Males	74
Right-handed	180
Left-handed	17
Humanistic-oriented mind	90
Scientific-oriented mind	99
No declaration	8
Lack of eye refraction disorders	107
Eye refraction disorders	90
Lack of head trauma in anamnesis	181
Previous head trauma (mild)	16

An original interview questionnaire was used with the early error evaluation test, a Standard Figural Stimuli Scale [8] and The Revised Wechsler Adult Intelligence Scale (WAIS-R) performance scale subtest – block design [13].

The questions referred to the following data: age, sex, level and field of education, body mass and height (for the purpose of calculation of the body mass index – BMI), handedness and the previous medical history, with particular attention paid to head trauma, eye diseases, currently used medications. Respondents were also asked to declare whether they were scientific-oriented or a humanistic-oriented, to specify the approximate amount of time spent in front of the mirror daily, to make a subjective evaluation (scale 1–10) of their own geometric imagination, and about their acceptance of their own bodies (scale 1–10). The data obtained from the survey are presented in Table 2.

Individuals who took medications which could influence psychomotor abilities or cause vision problems were excluded from the study.

The Standard Figural Stimuli Scale consists of 9 drawings of figures matched to the sex, corresponding to specific BMI values [8]. The respondent's task was to choose the figure that, in their opinion, best reflected their current body shape.

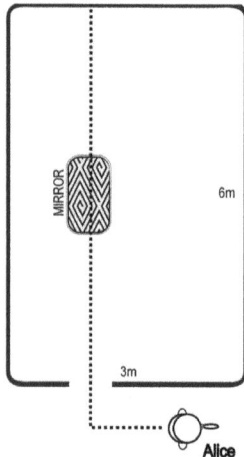

Fig. 1. Alice – entered a room of a certain size, with a mirror.

The WAIS-R performance scale subtest – block design – is aimed to evaluate the visual-motor coordination, spatial abilities, analytical and synthesising skills, as well as the ability to reorganise one's own actions. The study consisted in arranging the patterns shown in the pictures using colored blocks.

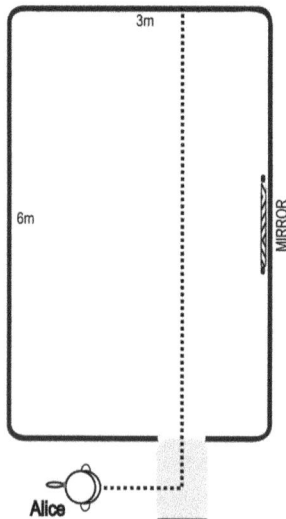

Fig. 2. The mirror is on a side wall.

The test for the early error evaluation, according to Bertamini, consisted of two drawings (Fig. 1 and Fig. 2). Both figures presented the following scene: the main character – Alice – entered a room of a certain size, with a mirror; in Fig. 1, the mirror is on a side wall, in Fig. 2 – on the ceiling. In both cases the respondent was asked to mark the point where Alice would start to see herself in the mirror. Alice was a cardboard cutout that could be moved freely. The drawings were of identical size and appropriate scale. Each time we accurately marked the point Alice was placed by a given participant. By the end of the study, we calculated the distance (in millimeters) between the point marked by a given participant and the actual point where Alice would start to see her reflection in the mirror. The obtained results were statistically analysed, using the STATISTICA software (Student's t-test for independent variables, Pearson's Correlation Coefficient, and Spearman's Rank-Order Correlation). Values of $p \leq 0.05$ were considered statistically significant

4 Results

Errors, both with respect to situation A and situation B, were most often made by respondents from the Q group (Table 2). Out of all the respondents, only 14 individuals (7.1%) gave correct answers regarding situation A, while in the case of situation B – 25 respondents (12.7%), the majority of whom were architects (7 individuals). The biggest error in situation A was 6.7 cm, while in situation B –5 cm. In A of Bertamini test, the lowest mean error is observed in the groups of hairdressers and architecture students (0.56 cm and 0.63 cm respectively).

In situation B the lowest mean error is observed in the groups architecture students and medical students (0.40 cm and 0.59 cm respectively).

Table 2. The occurrence frequency of the early error in particular groups (Error of 1 mm or less meant the character was placed correctly. In case of doubt we asked the participant for clarification).

Group	Error in situation A (%)	Average size of error in situation A (cm) $\pm$ SD	Error in situation B (%)	Average size of error in situation B (cm) $\pm$ SD
The Q group	96.67	1.53 ± 1.29	93.3	1.11 ± 1.07
Children from 12 to 15 years old	93.21	1.14 ± 1.18	78.27	1.15 ± 1.39
Medical students	86.96	0.95 ± 1.05	56.67	0.59 ± 0.73
Architecture students	73.33	0.63 ± 1.05	73.33	0.40 ± 0.65
Hairdressers	70.00	0.56 ± 0.59	73.33	1.10 ± 1.37
Fine arts students	66.67	0.90 ± 1.02	70.0	0.88 ± 1.16

Table 3 presents a comparison of the mean size of the error in each subgroup. Eye refraction disorders had no impact on the size of the error. The error made in situation B was slightly lower in left-handed individuals, as compared with the right-handed individuals, while the group of left-handed respondents was smaller in number. The sex and declaration of a scientific-oriented mind did not influence the size of the error in situation A, however they were significant factors in situation B.

Table 3. The occurrence frequency of the early error in particular groups.

Group	Average size of error in situation A (cm) ± SD	p	Average size of error in situation B (cm) ± SD	p
Female	1,08 ± 1,19	0,86	1,12 ± 1,14	0,05
Male	1,15 ± 1,02		0,74 ± 0,79	
No refraction disorders	1,12 ± 1,15	0,21	1,06 ± 1,17	0,22
Refraction disorders	0,94 ± 0,85		1,12 ± 0,95	
Right-handed	1,10 ± 1,12	0,96	1,12 ± 1,16	0,61
Left-handed	0,98 ± 0,95		0,89 ± 0,73	
Humanistic-oriented	1,12 ± 1,15	0,99	1,33 ± 1,16	0,01
Scientific-oriented	1,18 ± 0,95		0,78 ± 0,75	

The relationship between the selected characteristics and the error was studied for situations A and B. A significant relationship was not found in either situation between the studied factors (age, error in the Standard Figural Stimuli Scale, subjective evaluation of one's geometric imagination, the time spent in front of the mirror, acceptance of one's own body, and the result of the WAIS-R block design subtest) and the size of the error.

5 Discussion

In our study, we assessed the presence of the early error in groups differentiated by age, education, abilities and self-perception. We tried to determine those factors which contribute to the occurrence and the size of the error. It was observed that in situations A and B the early error was less common in students of architecture and fine arts. This is probably due to their familiarity with perspective. Similar results were observed in the group of hairdressers who use mirrors in their everyday work. However, only a third of the group managed to avoid the early error. Children and medical students showed surprising results. In both groups the error was bigger and more common in situation A. This is rather intriguing as mirrors are typically placed on walls rather than ceilings. Bertamini *et al.* compared the occurrence of the early error among students of psychology and physics and they concluded that there was no difference between the students in terms of the size of the error [7].

Our own study showed that a significantly lower early error was noted in the case of males and individuals who declared having a scientific-oriented mind. Spatial skills define one's ability to imagine objects, their shapes, location and proportions. Studies concerning sex differences showed that male spatial imagination and sense of direction are better developed, therefore, they can cope better with such tasks [14]. Obviously, there are females with similar spatial skills, but male dominance in this area is statistically significant. The right hemisphere of the brain is known to be responsible for spatial imagination. It controls abstract thinking processes. An individual with a damaged right hemisphere becomes disoriented and loses their sense of direction. Studies showed that the division of functions between the hemispheres is less distinct in females as compared to males. Both hemispheres participate in verbal and visual actions. On the other hand, the male brain is more specialised – the left hemisphere is almost exclusively responsible for language functions, and the right hemisphere is responsible for spatial functions. This probably stems from a different structure of the corpus callosum – in females, the number of fibres connecting the right and the left hemisphere is greater. Therefore, information exchange between both hemispheres is more extensive.

Bertamini and Wynne concluded that the early error emerges only in young adults, and it does not occur in primary school children [7]. The authors studied a group of children aged 5–11, and found no evidence for the early error in this age group. The erroneous belief regarding the reflection in the mirror is most likely to occur in later school years when individuals develop their system of beliefs. Children aged 12–15 were enrolled in our study. They presented with the occurrence of the early error. Possibly, observation of a larger group of individuals within this age range and the extension of the study to children aged 5–11 would help to identify the age in which this error occurs for the first time.

A more frequent error in the Q group (in the significantly older age) could result from their stronger beliefs, incompatible with the laws of physics. The size of the early error did not depend on age, evaluation of one's shape in the Standard Figural Stimuli Scale, geometric imagination, the time spent in front of the mirror, acceptance of one's own body, or the result of the WAIS-R subtest.

It would appear that as for the most basic forces or phenomena known from our daily experience, our intuitive knowledge of the physical world is generally correct. However, the vast majority of physical theorems are definitely of non-intuitive nature. The similar situation is observed in numerous scientific disciplines.

Research in naive physics, concerning beliefs regarding the physical world, with particular attention paid to systematic errors made by humans, showed that many individuals make serious errors, e.g. when describing the behaviour of bullets, falling objects and levels of fluids [9–12].

The errors made stem from a number of factors, such as overestimation of our knowledge regarding objects we are familiar with, the difficulty of transforming a virtual picture into a real one and the fact that what is seen on the surface of a mirror is only a reflection and it is not perceived in the same way as distant objects. These errors grow bigger when the subject of mirror perspective evaluation is not the observer's field of vision but a camera lens [15].

6   Conclusion

The study constitutes a good complement for Bertamini's papers. We proved that the early error is already very common in teenagers aged 12–15. We have verified the assumption that using mirrors and perspective on a daily basis would reduce the incidence of the early error. This proved true for one third of the participants. We have also observed that males with scientific background scored better when evaluating the reflection in mirrors placed on the ceiling, but not on the wall. What calls for further research is the lack of correlation between the amount and scope of the early error and the results obtained in the block design test by Wechsler. This suggests that mirror reflections are processed differently by the brain than other types of visual-spatial orientation.

References

1. Bertamini, M., Parks, T.E.: On what people know about images on mirrors. Cognition **98**, 85–104 (2005)
2. Bertamini, M., Spooner, A., Hecht, H.: Naive optics: Predicting and perceiving reflections in mirrors. J. Exp. Psychol. Hum. Percept. Perform. **29**, 982–1002 (2003)
3. Croucher, C.J., Bertamini, M., Hecht, H.: Naive optics: understanding the geometry of mirror reflections. J. Exp. Psychol. Hum. Percept. Perform. **28**, 546–562 (2002)
4. Hecht, H., Bertamini, M., Gamer, M.: Naive optics: acting upon mirror reflections. J. Exp. Psychol. Hum. Percept. Perform. **31**, 1023–1038 (2005)
5. Bertamini, M., Latto, R., Spooner, A.: The Venus effect: people's understanding of mirror reflections in paintings. Perception **32**, 593–599 (2003)
6. Bertamini, M.: Mirrors and the mind. Psychologist **23**, 112–114 (2010)
7. Bertamini, M., Wynne, L.A.: The tendency to overestimate what is visible in a planar mirror amongst adults and children. Eur. J. Cogn. Psychol. **22**, 516–528 (2010)
8. Stunkard, A., Sørensen, T., Schulsinger, F.: Use of the Danish Adoption Register for the study of obesity and thinness. In: Kety, S., Roland, L., Sidman, R., Matthysse, S. (eds.) The Genetics of Neurological and Psychiatric Disorders. Raven Press, New York (1983)
9. McCloskey, M.: Intuitive physics. Sci. Am. **248**, 122–130 (1983)
10. Kaiser, M.K., Proffitt, D.R., Whelan, S.M., Hecht, H.: Influence of animation on dynamical judgments. J. Exp. Psychol. Hum. Percept. Perform. **18**, 384–393 (1992)
11. Robert, M., Harel, F.: The gender difference in orienting liquid surfaces and plumb lines: Its robustness, its correlates, and the associated knowledge of simple physics. Can. J. Exp. Psychol. **50**, 280–314 (1996)
12. Hecht, H., Bertamini, M.: Understanding projectile acceleration. J. Exp. Psychol. Hum. Percept. Perform. **26**(2), 730–746 (2000)
13. Brzeziński, J., Gaul, M., Hornowska, E., Jaworowska, A., Machowski, A. i Zakrzewska, M.: Skala Inteligencji D. Wechslera dla dorosłych. Wersja zrewidowana – Renormalizacja, WAIS-R (PL), Pracownia Testów Psychologicznych PTP, Warszawa (2007)
14. Koscik, T., O'Leary, D., Moser, D.J., Andreasen, N.C., Nopoulos, P.: Sex differences in parietal lobe morphology: relationship to mental rotation performance. Brain Cogn. **69**(3), 451–459 (2009). https://doi.org/10.1016/j.bandc.2008.09.004
15. Bertamini, M.: Understanding what is visible in a mirror or through a window before and after updating the position of an object. Front. Hum. Neurosci. **8**, 476 (2014). https://doi.org/10.3389/fnhum.2014.00476

The Computer System Enabling Human-Computer Communication

Anna Sochocka[1](✉), Jan Grodecki[2], and Rafał Starypan[3]

[1] Faculty of Physics, Astronomy and Applied Computer Science, Department of Game Technology, Jagiellonian University, Łojasiewicza 11, 30-059 Krakow, Poland
anna.sochocka@uj.edu.pl
[2] Jan Grodecki is Faculty of Physics, Astronomy and Applied Computer Science, Jagiellonian University, Department of Hot Matter Physics, Łojasiewicza 11, 30-059 Krakow, Poland
[3] Opole, Poland

Abstract. The purpose of this publication is to present the subject of measurement of the brain wave signals and possibilities of communication through such waves. On the basis of the knowledge acquired until present, a system has been created, the purpose of which is to facilitate communication of the people affected with paralysis with the outside world and signalise their needs and emotions to their caregivers and families. The publication describes the modern devices as well as IT technologies used in this system. The system was created bearing in mind mainly Polish patients hence the commands are in the Polish language. The English version is in the process of preparation.

Keywords: EEG · BCI · Cognitive signals

1 Introduction

The human brain is the most complex of the human organs. It controls the senses, body movement, steers the thinking processes and the human behaviour. The brain enables us even to overcome certain instinctive behaviours fixed in the DNA through the ages (such as, for example, a fear of heights or a fear of darkness). The most important reason for evolution of the human being is the adaptability of his brain, in particular, its ability to break the patterns fixed in the genetic code [1].

Through thousands of years, people affected by paralysis have lost their ability to communicate with the world, preserving at the same time mental performance. It has frequently been the case that their lack of reaction was treated as the loss of thinking functions. A paralysed person became prisoner in his own body, deprived of a possibility of effective communication with the surrounding. An invention which has allowed to demonstrate that in certain cases the body paralysis does not disturb the brain activity, is the electroencephalograph. Since electroencephalography is a non-invasive method, it has begun to be applied in all types of patients ill with various diseases, as well as to examine all possible conditions in which a human being may find himself. In time, the developing knowledge and understanding of what the graphs of the EEG show, have

S. Paszkiel (Ed.): ICBCI 2021, AISC 1362, pp. 27–36, 2021.
https://doi.org/10.1007/978-3-030-72254-8_4

allowed not only to read the brain wave signals but also to interpret them. The persons who were paralysed, have received the "second" chance to establish contacts with the world and their closest families [1].

The purpose of this publication is to present the subject of measurement of the brain wave signals and possibilities of communication through such waves. On the basis of the knowledge acquired until present, a system has been created, the purpose of which is to facilitate communication of the people affected with paralysis with the outside world. The system consists of the two applications and the server which fulfils an indirect role in their operation. The first application works on the Windows Platform an is controlled by means of the EEG device. The second application works on a smartphone equipped with the Android system [2] and is used for reception of the information from the first application. This application is designed for a person taking care of the ill person who can send messages to the caregiver's phone. The messages can be defined in any way, depending on the ill person's needs and preferences. Also, the means of the messages activation through the EEG remains flexible and can be adapted to many different cases of physical disability [1].

2 Ways of Imaging Brain Activity

The brain is the central system which controls the functioning and behaviour of a human being. The human encephalon consists of the cerebellum, spinal cord and the brain, part of which is the cerebral cortex. The brain is divided into the two cerebral hemispheres. Although this division is symmetric and even, each of the hemispheres is responsible for various functions. A vast majority of knowledge concerning the functioning of the brain has been derived through the recording of its reactions to external stimuli, or, through correlation between these reactions and the changes which take place in the organism. There are a few methods of recording the brain activity. They can be classified according to the means of interaction with the brain, as non-invasive and invasive methods, the latter of which are always connected with an operation and the risk of complications.

The non-invasive method which is most often applied is the electroencephalography (EEG) – the technique which measures voltage in various areas of the skull surface. The first measurement of the brain activity by means of a galvanometer was carried out by the German neurologist Hans Berger. In 1924 he managed to make a recording of brain waves as potential difference appearing on the human skull [3]. In order to carry out the EEG measurement, it is necessary to have a proper voltage meter, able to read the potential difference between $10 \mu V$–$100 \mu V$. The simplest electroencephalograph system measures the voltage between one point on the skull and the assumed second, neutral point. Devices applied in practice allow for the recording of voltage not from one, but from several, or even from more than one hundred points.

There are two types of the EEG apparatus: the bipolar and the reference ones. Both systems have their advantages and disadvantages. An ideal solution would be complementary functioning of both systems. For over half a century since the first EEG recording (carried out in 1929 by Hans Berger), electroencephalography has been applied mostly to diagnose neurological diseases, to carry out clinical trials, or, to get to know the functioning of the brain. As late as in the 70-ties of the past century, the EEG began to be perceived as a device which can be helpful in reading people's thoughts. Development of the computer technology that took place at the end of the 70-ties, provided the tools which were necessary in order to analyse the complex EEG signals and read specific intentions out of them. The progressing research of the analysis of the EEG graphs has enabled to isolate signals connected with communication and control from stochastic signals. Soon, correlation between the EEG graph and behaviour of the organism was observed. Hans Berger noticed that when a person closed his/her eyes, electrodes placed in the back of the skull receive a cyclic signal [3]. Those waves were named the alpha waves. The continued research of these waves has shown a close connection between these waves and a state of relaxation. There are more such dependencies and they occur in various frequencies and various intensities. They are commonly known as waves, although, in medicine there is a distinction between wave and rhythm. The EEG graph is a superposition of many waves of various frequencies. Among the waves which are read, there are the alpha, beta, delta, mu and theta ones, as well as the normal sleep activity waves. The specific waves or reactions are characterised not only by such parameters as the maximum potential or specific spectral characteristics of the signal, but also by the place of occurrence of the signal in the brain [4]. Among the devices there are a few universal systems of electrodes placement (the most popular is the 10–20 system). Brain examination by means of the EEG is the most popular method used in order to examine the activity of this most complex human organ. This is also the safest and the cheapest method, besides, the EEG devices are compact and mobile [1].

3   Brain-Computer Interface (BCI)

BCI (Brain-Computer Interface) less commonly known as BMI (Brain-Machine Interface) or MMI (Mind-Machine Interface) is a system which translates the brain activity patterns into specific commands or messages. The BCI term was formalised by Vidal in 1973 [5]. In a set with the computer, the brain is to fulfil the role of controller, similar to a keyboard or a computer mouse. Initially, the system was addressed only to people with motor impairment or permanent paralysis, or people affected by numerous nervous system diseases [6]. Today it is a broadly developing branch of knowledge which has gone out of the research laboratories and has more and more possibilities for commercial application. By design, BCI does not have any strictly defined signal receiver from the brain, however, due to the attractiveness of the EEG devices (high time resolution of the received signal, compactness of the device, low price) the BCI systems are usually coupled with these devices. Designing of the BCI system is a task which requires a broad knowledge of many scientific disciplines, among others, neurology, psychology, engineering of signals processing and programming.

Regardless of the BCI system, its manufacturer or use, the work with BCI is divided into two stages [1]:

- The training stage during which the system allows to adjust the impulses emitted by brain activity of a specific user to pre-defined functions of the system.
- The operation stage during which the system maps the brain activity patterns to the functions allocated to them by a programmer.

The second process takes place in the loop and can be split into 3 layers:

- Reading and filtration of the signal from electrodes (hardware).
- Classification of the signal (algorithm of the BCI library).
- Conversion of the recognised signal into a specific action (np. Command in an application, or, sending a signal to an external device controlled by the application- robot, prosthesis).
- Effectiveness of the BCI system depends on all the factors which affect quality of the signal received from the brain. Currents stimulated by brain activity are of the order of a few or several μV. That is why ensuring purity of the signal which reaches the computer as well as its proper filtration are essential to correct interpretation of this signal [7]. Also, selection of the appropriate algorithm of the signal classifier is of equal importance. It has to take into account a wide range of discrepancies in the input signal and its rather poor quality. The last stage must take into account the sum of errors of the device at an input and their classification [1].

4   Emotiv Epoc NeuroHeadset – An Example of Brain-Computer Interface (BCI)

The main disadvantage of BCI systems based on the EEG is high price of the EEG devices. For a long time, price has been the factor which has restricted the area of development of BCI in scientific laboratories. Some companies decided to create products at an attractive price, thus increasing the spectrum of interest in application of the EEG devices.

According to the authors of the paper "Towards inexpensive BCI control for wheelchair navigation in the enabled environment a hardware survey", the most useful low budget device available on the market is Emotiv Epoc [8].

Emotiv Epoc [9] is a device manufactured by the Emotiv company, founded in 2011 and specialising in construction of EEG headsets for the needs of BCI. The company does not only build the EEG headsets, but it also equips them in development software and SDK library for C/C++ language. Emotiv Epoc enables to collect 14–bit samples at the frequency of 128 Hz. The transmitted signal band is between 0.2 Hz–43 Hz [10], limited by high and low-pass filters. Emotiv is additionally equipped in filters which block mains feedback for the frequencies of 50 Hz i 60 Hz. The device has 16 electrodes placed in the skull symmetrically in relation to the hemispheres. Placing the device on the patient's head, we install 8 electrodes in frontal and central part of the skull and one pair on parietal lobe and one on temporal lobe, and two pairs above occipital lobe. In fact, only one pair on occipital lobe serves as the EEG signal. The remaining pair of P3 and P4 electrodes [11] are the so called CMS/DRL electrodes (Common Mode Sense / Driven Right Leg). It is an internal system used in order to stabilise the signal. The CMS electrode is the reference one and the DRL is the electrode grounding the system. Apart from the test software, the Emotiv Epoc set contains documented library of programming functions called the SDK (*Software Development Kit*). The API (Application Programming Interface) functions for the system were placed in dynamic libraries designed for the Windows system (edk.dll, edk_utils.dll) [9]. Besides, Emotiv has prepared a "wrapper" for these libraries, enabling to write an application in the C# language. There are ongoing works on the library for the Java language. The communication port between the EEG headset and the software is EmoEngine which reads the information received by EEG (signals from gyroscopes and accelerometers), and then, by means of its own algorithm, translates them into pre-defined values and places in the EmoState structure where the signals are classified into one of the three groups:

- Expressive (demonstrate facial expressions, e.g. eye movements, a blink, a smile),
- Cognitive (express conscious intention to carry out physical action),
- Emotional (e.g. agitation, joy).

5 Application

Undoubtedly, one of the most important means of use of the BCI systems is an application supporting rehabilitation of patients. Bearing that in mind, the scientists worked out a system of support for persons affected by paralysis, allowing them to communicate with their caregivers. Persons affected with complete paralysis often require 24-h care, which is very onerous to caregivers and happens to be bothersome also to the patient, as he or she may take such care as constant surveillance. The created system may solve the above problems by ensuring the ill person a little independence and offering a moment of rest to the caregiver.

5.1 Communication Scheme Between Elements of the System

The application is created in a form of a communicator. It allows you to send a message from one device to another. The FireBase server [12] only fulfils a role of a communication buffer (Fig. 1).

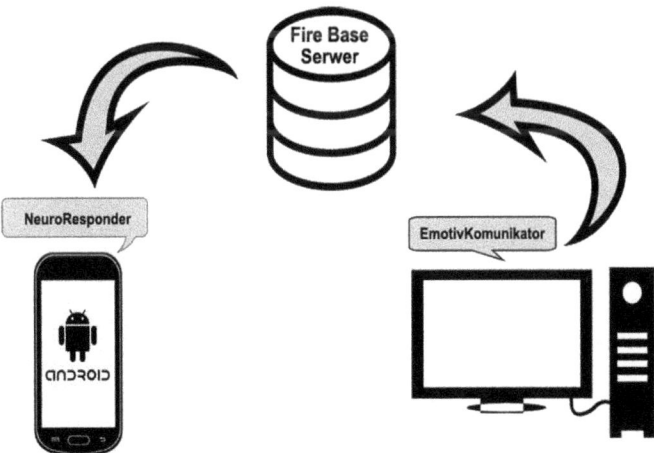

Fig. 1. Elements of the system. EmotivKomunikator desktop application, FireBase database and application called NeuroResponder for the Android system [1].

5.2 Authorial Application of "EmotivKomunikator"

EmotivKomunikator is an application written in the C# language. It cooperates with the Emo.dll library implementing EmoEngine which controls communication between the EEG headset by the Emotiv company and the communicator. The communicator initiates connection with the EEG device by setting up the library functions dedicated for that purpose and by applying appropriate communication protocol, and receives signals already interpreted as commands. The application allows to attune sensitivity of commands to individual cognitive abilities (i.e. signals, thoughts, movement intentions) and expressive abilities (having their reflection in the face, physical movements of facial muscles) of the user [1].

5.3 Recognition of Cognitive Signals

The Emotiv SDK library recognises in real time activity of the user's brain waves and distinguishes his or her conscious intention to execute a physical action on a real or virtual object [9]. Emotiv has defined 13 such actions to choose. Six of the actions concern directional movements (push, pull, lift, lower, move left, move right). The next six actions are rotations and one completely virtual action- disappearance. Emotiv allows you to choose and train up to four actions from this set. Cognitive recognition implemented in the SDK is based on the P300 wave [9].

The EmotivKomunikator software is based on the cognitive classifier implemented in the SDK. This means that the cognitive actions trained in the Emotiv Control Panel will work in the software. This allows you to use an existing and convenient software that came with Emotiv for training. The Emotiv library, together with the parameter classifying the cognitive signal, transmits a variable containing the power of that signal – read by the EEG. This is a normalised and closed floating-point value in one parameter between [0.0–1.0]. The software allows you to specify the level of activation of the cognitive command depending on the value of this variable [1].

At the centre of the software window there is a list of commands (Jestem spiacy – I am sleepy; Poprosze herbate – Tea, please; Poprosze kawe – Coffee, please; Jestem glodny – I am hungry; Wlacz TV – Turn the TV on). The lower left corner of the active view indicates the stack of commands received from the emo.dll library, along with their power expressed as a number between [0.0–1.0] (Fig. 2). In the upper left corner, the commands accepted by the cognitive action "PUSH" are displayed. The user's "PUSH" intention invokes a function that sends a highlighted command to the FireBase server, which automatically informs the Android application of updates in the database.

5.4 The FireBase Database

The FireBase is a database managed by Google [12]. FireBase is not a typical database. It functions as a bridge between end applications in this system – it ensures data synchronisation between them. While regular database systems are designed to receive commands to manipulate the data they store, FireBase allows you to actively update data in all applications which are currently connected to it.

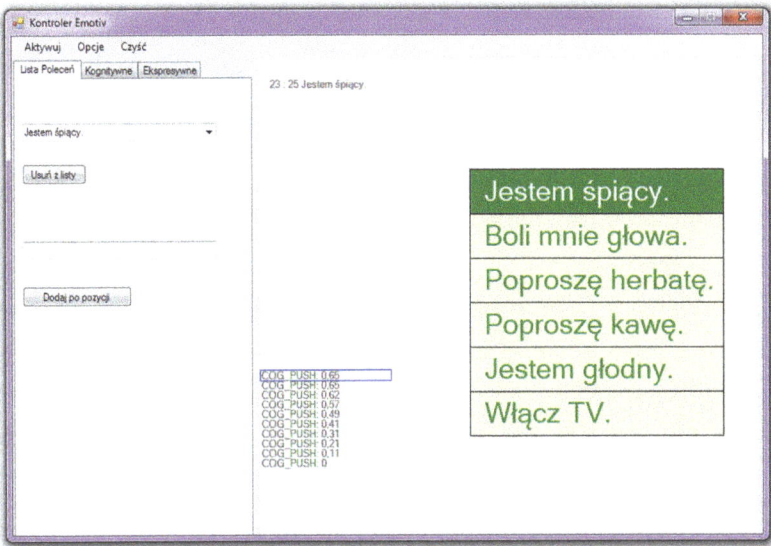

Fig. 2. Screenshot of the view of the active EmotivKomunikator application (cognitive set, commands) [1].

5.5   The System Enabling Communication Between Human and Computer

After starting the software, we need to choose one of the two modes of connection (Fig. 3).

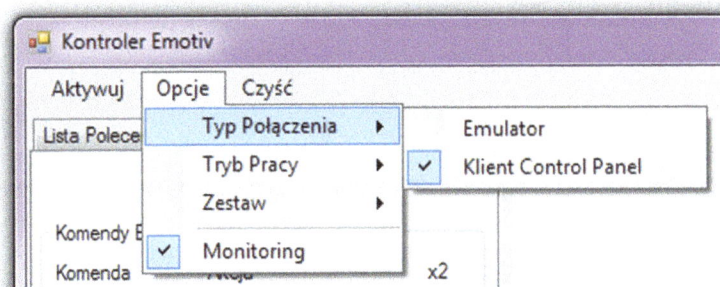

Fig. 3. Selecting the connection type of the EmotivKomunikator application [1].

Next, we start the *Emotiv Control Panel* (or *EmoComposer*) application related to the connection type. Now we can start working by connecting to the device by clicking the "Aktywuj" (*"Activate"*) button in the main menu bar. In the next step, we select the mode of operation with the application (Fig. 4).

"Alphabet" ("Alfabet") will allow you to construct your own words and sentences, which may turn out very time consuming. A "command list" ("Lista poleceń") is a simplified list which allows you to define the basic messages you need. Managing a list

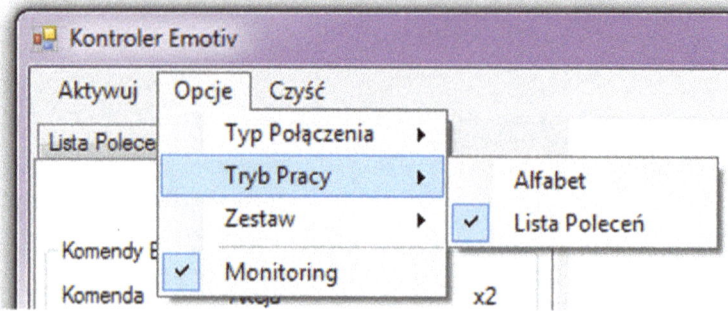

Fig. 4. Selecting the operation mode in the EmotivKomunikator application [1].

requires much less effort than working with the alphabet. The software allows the user to edit the "Command List", add new actions or delete them. The last option of choice is to choose the set by means of which we will communicate with the application via EEG. We can choose between expressive and cognitive set (Fig. 5).

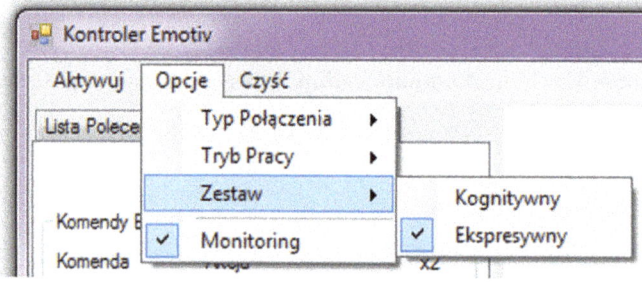

Fig. 5. Choice of the means of communication with the EEG [1].

The cognitive set allows communication by means of the Emotiv algorithm using the P300 wave. The expressive set is the recognition of a few facial expressions pre-defined by the Emotiv, including eye movement. For each set there is a tab which allows you to personalise communication by selecting a few simple parameters.

The last item in the "Options" ("Opcje") menu is "Monitoring". You should select this option in order to track the read values of signals that arrive from the EEG and the actions assigned to them (both cognitive and expressive). This option can be useful when attuning an application, e.g. setting a threshold for cognitive actions, or understanding why certain commands have not been approved.

The NeuroResponder application for Android does not require any settings. It is only used to receive messages from a desktop application and is ready to receive messages at start-up (Fig. 6). However, it should be remembered that the buffer between EmotivKomunikator and NeuroResponder is the FireBase server. Both applications are closely linked to a specific FireBase account.

Finally, it is worth noting that for the correct operation (mutual communication) of both applications, it is necessary that both devices on which these applications operate have access to the Internet, otherwise messages will not be able to be sent and/or received.

LP	Wiadomość	Czas
1	Co jest na obiad.	2017-09-18 07:03:37
2	Jestem śpiący.	2017-09-18 01:30:17
3	Boli mnie głowa.	2017-09-17 19:56:57
4	Jestem głodny.	2017-09-17 19:55:53
5	Poproszę kawę.	2017-09-18 22:51:12
6	Poproszę kawę.	2017-09-17 19:04:32

Fig. 6. The NeuroResponder software activity view. List of commands: Co jest na obiad – What's for dinner?; Jestem spiacy – I am sleepy; Boli mnie glowa – I have a headache; Jestem glodny – I am hungry; Poprosze kawe – Coffee, please [1].

6 Summary

The main objective of the project was to approximate the methods of human-computer communication by means of interpreting signals received by a non-invasive measure of brain activity such as an electroencephalograph. The developed NeuroResponder application allows such communication to people whose possibilities of contact with the world are very limited or even impossible. People who have been prisoners of their own thoughts until now can actually communicate with their surroundings.

EmotivCommunicator enables to improve their quality of life, however, training the application to correctly select one of the options in the "Command List" requires a lot of patience and skill. Cognitive selection abilities are an individual trait of each person and users should not be discouraged after the first failed attempts.

Also for people caring for persons who are so seriously ill, the software can be an enormous help, because using it the caregivers do not have to constantly watch over the persons for whom they are responsible.

The presented system works in demonstration mode but has great potential. The development of interfaces such as the presented EmotivCommunicator and NeuroResponder system is extremely necessary. Such systems are an obvious facilitation for patients and their caregivers. The patient is not separated from his family and home, and separation from closest relatives does not need to be added to the trauma of the disease itself.

The system was created bearing in mind mainly Polish patients hence the commands are in the Polish language. The English version is in the process of preparation.

References

1. Grodecki, J.: System umożliwiający komunikację ekspresywną i kognitywną pomiędzy człowiekiem a komputerem. Uniwersytet Jagielloński, Kraków (2017)
2. Android. https://www.android.com/versions/pie-9-0/. Accessed 27 Jan 2021
3. Berger, H.: Über das Elektrenkephalogramm des Menschen. Archiv f. Psychiatrie **87**, 527–570 (1929)
4. Durka P.: Skryp o EEG
5. https://brain.fuw.edu.pl/edu/index.php/Elektroencefalografia/Przyk%C5%82ady_zastos owa%C5%84#BCI_oparte_o_EEG. Accessed 27 Jan 2021
6. Vidal, J.J.: Towards direct Brain-computer communication. Annu. Rev. Biophys. Bioeng. **2**, 157–180 (1973)
7. Birbaumer G.H.: A spelling device for the paralysed. Nature (1999)
8. Paszkiel S.: Data acquisition methods for human brain activity, analysis and classification of EEG signals for brain-computer interfaces. In: Studies in Computational Intelligence, vol. 852, pp. 3–9 (2020). https://doi.org/10.1007/978-3-030-30581-9_2
9. Stamps K.: Towards inexpensive BCI control for wheelchair navigation in the enabled environment a hardware survey
10. Emotiv Epoc. https://www.emotiv.com/epoc/. Accessed 27 Jan 2021
11. Vasiljevas, R., et al.: Consumer-grade EEG devices: are they usable for control tasks? (2016)
12. Melnik, A., Legkov, P., et al.: Systems, subjects, sessions: to what extent do these factors influence EEG Data. Hum. Neurosci. (2017)
13. FireBase.   https://fiirebase.google.com/?gclid=Cj0KCQiA3NX_BRDQARIsALA3fIJ JO3IvTUXTt5r1-eAxQiz4flIbJy1cxFf5v2DM1HpfPQ5NqHvjv9EaAkWJEALw_wcB. Accessed 27 Jan 2021

How Much Emotionally Intelligent AI Can Be?

Rafał Szewczyk[1](✉) and Karolina Janik[2]

[1] Neurostimulus sp. z o.o., Kręta 2 lok.4, 20-341 Lublin, Poland
`rafal.szewczyk@neurostimulus.pl`
[2] Institute of Psychology, The John Paul II Catholic University of Lublin, Al. Racławickie 14, 20-950 Lublin, Poland

Abstract. We tackle the question to what extent artificial intelligence can have an emotional component, and to what extent it can resemble a human. The article's main aim is to provide an overview of artificial intelligence systems containing emotional intelligence elements and show their development directions. Several innovative projects were analysed that attempted to apply emotional intelligence to an artificial intelligence algorithm. The article presents the emotional component in the currently used artificial intelligence in everyday life solutions. Potential threats associated with the development of emotional intelligence in humanoids and the importance of this development for societies are also discussed.

Keywords: Artificial intelligence · Emotional intelligence · Emotions

1 The Origins of Artificial Intelligence and Its First Applications

Information technology (IT) is being introduced into our everyday life faster and more boldly. The plans for the functions performed by IT in the future are even bolder. The use of advanced IT products has been possible because of their advantage over humans in cognitive functioning. Thanks to enormous computing power, computers are irreplaceable in collecting, storing and analysing vast amounts of data. The creation of an algorithm that, based on extremely efficient analysis of the enormous amount of available data, began to make decisions and predict possible future data configurations and human behaviour. It was an essential step in IT development. Chess is one of the good examples of this process[1]. In 1997, for the first time, the Deep Blue computer created by IBM won a game of chess with world champion Garii Kasparov [1]. This play went down in history even though Kasparov won three consecutive games and tied two, ultimately winning against Deep Blue 4–2. It was a breakthrough after which computers began to dominate in chess. Today there is no doubt that a human has no chance against a computer in this discipline [2].

---

[1] The Chinese room" is a room where a person who does not speak Chinese sits, but has access to a gigantic book of conversational rules in Chinese: the such-and-such sequence of characters must be answered this-and-such. When someone slides a piece of paper with a question asked in Chinese under the door, the person finds the appropriate rule, types the correct answer, and pushes the part back.

S. Paszkiel (Ed.): ICBCI 2021, AISC 1362, pp. 37–49, 2021.
https://doi.org/10.1007/978-3-030-72254-8_5

The creation of artificial intelligence algorithms (AI) was another milestone in IT development. Artificial intelligence attempts to recreate human intelligence [6], a wide-ranging branch of computer science concerned with building smart machines capable of performing tasks that typically require human intelligence [3]. An important new feature compared to the previous IT achievement was AI's ability to learn and make decisions. Using a wide range of input data, the algorithm recognises typical patterns of different objects' categories, thus learning to classify different data types correctly. This allows AI to make the right decisions in changing circumstances and in the face of continually emerging data. Thanks to this, AI is currently used in many human activity areas, which we will list and briefly describe below.

One example of AI in everyday life is intelligent buildings, cities and infrastructure [4]. In smart cities, AI helps in traffic management. Smart thermostats help to save energy in our homes and workplaces. AI can provide early warning of natural disasters and enable adequate preparation and mitigation of such events' consequences [5]. Thanks to the analysis of internet resources and available knowledge, AI can forecast the weather and estimate the probability of, e.g., weather anomalies [6]. In times of pandemic, in the fight against the coronavirus, AI is used for thermal imaging at airports, stations and other places and to observe the extent of the virus spread [7, 8]. More and more applications for AI can be found in medicine. An emergency call answering program recognises cardiac arrest during a call faster and more efficiently than a medical dispatcher. Researchers are investigating the use of artificial intelligence to analyse large amounts of health data and pattern recognition, which could lead to discoveries in medicine and improvements in individual diagnostics [9–11], for example in the diagnosis of infections based on computed tomography of the lungs, in the diagnosis of eye disorders. Creating three-dimensional representations of proteins could contribute to the development of better drugs for diseases such as Parkinson's and Alzheimer's [12]. Another example is the KConnect project, developing multilingual text and search services to find the most relevant medical information available [13].

Mobile operating systems help users to access information 24 h a day, seven days a week through Intelligent Personal Assistants (IPA) working within artificial intelligence. PDAs, also known as handheld PCs or personal data assistants [14], were first released in the mid-1980s to facilitate users' daily lives and enable data access. Today's PDAs can be considered the latest mobile innovations' ancestral devices, including smart-phones and tablets. IPA are speech technologies on mobile platforms that have become one of the primary online learning tools. The app uses inputs such as the user's voice, vision (images), and contextual information to answer questions in natural language, make recommendations and perform actions. IPAsared was designed to perform users' required tasks through online sources on the Internet [15]. Much such software is developing, including Siri by Apple, Watson by IBM, Cortana by Microsoft or Duolingo. Virtual assistants can perform basic user command tasks, including chatting, texting and emailing, and even learning a foreign language.

At the beginning of 2021, Microsoft received a patent for a method for creating a conversational chatbot modelled after a specific person - a "past or present entity... Such as a friend, a relative, an acquaintance, a celebrity, a fictional character, a historical figure," according to the filing with the US Patent and Trademark Office [16]. This method allows to create a unique chatbot to replace a specific person. This gives us the opportunity to, for example, talk to a person who is already dead. According to the patent information, the tool will collect "social data" such as images, social media posts, messages, voice data and written letters from the selected person. This data will be used to train the chatbot to "talk to and interact with the personality of a specific person." It can also rely on external data sources if the user asks the bot a question that could not be answered based on that person's social data.

2   What Differs Artificial from Human Intelligence?

Decisions made by people differ from those made by a computer in terms of computing power. Moreover, human decisions usually contain an emotional component - and contrary to what one might expect, this does not always lead to catastrophic consequences. Very often, emotions turn out to be a key component of practical actions. That emotional component, so specific to living beings, is emotional intelligence.

Emotions are an essential element in understanding the origin and functioning of the mind [17–19]. Damasio understands emotions as complex and automated action programs in the body created by evolution [20]. For many years, psychologists have been paying attention to the artificiality of the division "cognitive processes - emotional processes". They do not function in isolation but form a specific unity [21]. Emotions influence beliefs and knowledge itself. They can strengthen or weaken certain beliefs, but they can also lead to new ideas and increase our understanding. It seems that thanks to emotions, we gain knowledge about our immediate surroundings and other people. In the case of the immediate environment, emotions provide us with information, e.g. about the situation we are in at the moment, and thus may suggest how to behave in a given situation; may make it easier (or difficult) to perform certain activities. In other people's case, we get to know the emotions, intentions and intentions of another person [22].

Gardner's [23] concept concerns social intelligence as divided into intrapersonal and interpersonal intelligence. Intrapersonal intelligence is defined as the ability to learn about one's own emotions, while interpersonal intelligence can help understand other people's feelings and intentions. Within this approach, emotional intelligence was treated as a subset of social intelligence [24]. Furthermore, Gardner (1983) pointed out that these intelligence dimensions are just as important as general intelligence. As early as 1985, the term emotional intelligence was used for the first time by Wayne Payne in a doctoral dissertation entitled 'A study of emotion: Developing emotional intelligence; self-integration; relating to fear, pain and desire'. Research on the relationship of social intelligence with alexithymia (disorders in recognising, understanding and describing emotions) as well as studies assessing the ability to recognise emotions on the face and the ability to express them [25] significantly influenced the development of EI.

The concept of "emotional intelligence" was established in psychology as a science in the 1990s by P. Salovey and J. D. Mayer [24]. The theoretical model proposed by John Mayer and Peter Salovey [24] defines emotional intelligence as a set of cognitive abilities that process information related to emotions. According to this model, emotional intelligence consists of four groups of abilities: (a) emotion perception, (b) using emotions to improve thinking, (c) understanding emotions, (d) managing emotions. The perception of emotions involves the perception and recognition of feelings. The process of identifying emotions takes place in three spheres: the sphere of one's own emotional experiences, observing the moods and feelings of others, and finding emotions in inanimate nature. People with high emotional intelligence can accurately read emotions and then use it in other judgments and behaviours. Another group of abilities is the use of emotions to improve thinking. It is the ability to control one's own affective processes towards task completion. This means enhancing certain emotions or inhibiting them. The third element of the model is understanding emotions - it refers to a person's knowledge in the area of experiences related to the emotional sphere. In other words, it is an accurate understanding of the meaning of a given situation and the occurrence of particular human behaviour. The fourth component determines the ability to regulate emotions - both oneself and other people. To do it effectively, an individual must monitor affective processes regularly and modify them as required [24].

Emotional intelligence is being more and more often mentioned in the broad scientific and business circles as the key competence of an effective leader (but also an employee) of the 21st century [26]. Emotional intelligence allows, above all, for the mutual fulfilment of people's needs, taking effective actions when facing a threat; it also allows to make optimal decisions, motivate oneself and others to work, helps to overcome procrastination; it is a buffer of undesirable effects of short- and long-term stress effects; it triggers innovation, integrates people, protects employees from mobbing, and enables building lasting and safe relationships [27–32]. Referring to the examples mentioned above of the AI usage in everyday life, let us consider what benefits could be obtained from the implementation of an emotional component into AI algorithms.

3   What Emotionally Intelligent AI Could Do?

If intelligent infrastructure solutions were emotionally intelligent, they could then influence drivers' behaviour by, for instance, relieving the emotional tensions in a traffic jam. It would be possible thanks to the knowledge in the field of emotional management. Even a simple phrase, like "Keep calm. Drive safe. You will get there." Displayed on an information screen could provide some relieving effect. This could affect the number of collisions and road accidents since emotional overload is one of the main risk factors for road accidents [33]. The emotional health of disaster-prone populations is a public health priority [34]. Natural disasters have many psychological effects, including an increased risk of suicide among victims. Therefore, there is a need to provide psychosocial support to these people [35]. AI should have an emotional intelligence component to effectively reduce public anxiety and motivate people to take corrective or preventive actions. An example of the use of IE can be threat communication and instructions on how to act. This can lead to minimising the harmful and long-term effects of previous experiences.

Additionally, it may also reduce the risk of perpetuating pathological reactions. Technologies that use AI with emotional intelligence could also provide emotional support to sick people, caregivers, and families. The development of AI in medicine could serve physicians by giving hints on how to show psychological support or alleviate the painful experience associated with accepting a diagnosis, e.g. of an incurable disease. On the other hand, doctors who are highly sensitive and very committed to helping their patients are at risk of quick burnout. AI in the form of a personal assistant could help them better communicate with patients and find a healthy distance to professional matters and develop habits of physical and mental regeneration.

Considering the role that AI plays in one-to-one interactions with humans, it is still important to ensure that the chatbot can not only effectively solve problems, make decisions and learn, but that it also reliably simulates human behaviour. The first attempts to create AI started from the classically understood intelligence and its cognitive aspects. In 1950, in the article entitled *Computing Machinery and Intelligence* Alan Turing asked the question: "Can machines think"? The author attempted to create something that could imitate a thinking man. In his view, this imitation would consist of the ability to conduct a conversation, i.e., an intelligent exchange of opinions. The imitation game was created based on these assumptions. The game consisted of having a man answer whether he was talking to a human or a machine after a few minutes of a conversation with a computer. This procedure is today called the Turing test [36]. The author assumed that in 2000 it would be possible to such an extent that the average person would not have a greater than 70% chance of correct identification after five minutes of the conversation.

In 2000, the A.L.I.C.E. algorithm (Artificial Linguistic Internet Computer Entity) won the Loebner Award three times. The prize is awarded in an annual competition to select the most human-like computer programs. The judges considered the A.L.I.C.E.'s interview to be the most similar to a conversation between a human being. The code of this program has been made public. As a result, many other programs were created on its basis. While it was a model for other programmers, it relied in practice on an elementary principle of stimulus-response interaction. The interlocutor's statement is first searched for keywords that were previously stored in the database. The more extensive the database is, the more likely it is for A.L.I.C.E to discuss a specific topic in great detail.

The developers have also created relatively simple algorithms that can sharpen the impression of interacting with humans. An example would be a sentence like "I wish I could…". The machine does not need to find the answer in its database. All she has to do is to ask "Why do you regret…?" These types of programs try to evoke emotions in the interlocutor thanks to the keyword grouping mechanisms. If the conversation contains some words describing a car, motorisation, etc., the computer may ask a question, e.g. about the dream car. Software creators did not stop at this type of unsurprising answers. They have tried to find ones that can evoke even stronger emotions. The answers might be witty, surprising and extravagant. They might not be the most common but are very specific for a human being.

By analysing A.L.I.C.E. origins [37], we can conclude that the program works according to the so-called Chinese room. John Searle has described it as one of the most famous thought experiences in the philosophy of mind. It was supposed to show that even a successful computer simulation is not the same as the computer having a real reason. The real performance of specific tasks (e.g. computational tasks) does not have to be based on their understanding by the contractor. Even if we assume that the computer does not need to understand or think, it must certainly have a very extensive database and complex algorithms to process them. The complexity of such a database, and thus its size, would have to be enormous. It should contain not only the answers to various questions, but it should also be able to place them in a broad context. Merely semantic understanding of a sentence would not be enough to communicate in the same way humans do. According to Grice's theory of relevance [38], only the interpretation allows us to know the expressions true meaning. What the listener interprets as the utterance's content is only partially given by the conventional meaning. The interpreter derives (implies) the remainder of the content both from the meaning of the interlocutor's expressions and his knowledge about the object of the statement. Grice gives an example:

Driver A is standing on the street next to a parked car; person B approaches him. A conversation ensues:

A: I am out of petrol.
B: There is a garage around the corner.

These utterances convey much more information than the mere and raw meaning of the words. Person A, saying that he has run out of fuel, does not give person B the raw information that his tank is empty. Its purpose is to ask for help, e.g. in the form of a hint on where to get it. Person B most likely has a good understanding of the context of A's statement and says that there is a garage around the corner. Her most likely point is that her interlocutor can buy gasoline there. This is an example of a real conversation between people who, despite short statements, exchange much more information. Grice called this type of exchange the rules of conversation [38]. They consist of creating a space of mutual cooperation. This type of communication is intuitive and spontaneous, although supported by the context and the knowledge of the speaker and the listener. If there were no full context in the form of a parked car, timbre, tone and timing of the voice along with the speaker's gestures, then perhaps the listener would understand something else. This could lead to a different answer, like for example "There are taxis around the corner." Such an answer could already indicate that A's listener did not understand his message or that maybe he does not want to help him. The commonness of contextual information does not need to be proved. Therefore, it seems reasonable to say that the expression's very meaning does not determine the meaning of the utterance and is not sufficient for its proper understanding.

Another problem with meaning is the need to assess whether the interpretation is correct. It can be considered as such when it leads to an agreement between the communicating persons. Cappelen and Lepore [39] proposed the following formula to operationalise this process: "If A uses the phrase S, A communicates to B that q." As a result of the interpretation, the listener always obtains a sequence of sentences p_1, p_2, p_n. None of them is identical with q, but the interpretation is correct as long as they have a similar meaning to q. If it is so, the interpretation is assumed to be good enough. Communication is not about a proper recreation or duplication of the speaker's thoughts. The interpreter always gives the utterance its meaning, related to his emotional state, general knowledge, assumptions about the sender's intentions, etc. [39].

As we have mentioned before, artificial intelligence is not only about human cognitive processes' imitation. Its most important feature is the ability to learn. By testing various solutions and based on the feedback, these types of algorithms determine whether their decisions are good or bad. This is precisely how the latest chess software works today.

No new algorithms are being entered into them; they break them off themselves and record in the database which moves turns out to lead to good outcomes and which do not. An example of such a program is AlphaZero, which mastered chess with an expert degree in nine hours. It mastered it in such a way that it played a game with himself. The only information it received from a human was the rules of the chess game. Usually, chess players spend a lot of time to acquire skills and knowledge, trying to remember the moves that had been written down in books and databases for many years. Current AI can acquire this knowledge in 9 h with the same or even better result [40].

This solution has also been used in the world of an instant messaging application. In 2016, Microsoft released a program called Tay on its Twitter account. Tay was learning from other Twitters how to talk with people. Despite his successful learning process, it was quickly withdrawn from the platform because of a xenophobic approach, racism and vulgarisms that it had learned from other users.

4   Is Emotionally Intelligent AI Possible?

Throughout the history of research on artificial intelligence construction, many models describing the human mind have been proposed [41]. Fuzzy logic was one of the theories on which a computational model of emotions was developed, taking into account their cognitive and behavioural aspects. The algorithm based on the Fuzzy model consisted of generating emotions in a virtual pet so that the animal could make decisions on its own. At the same time, it was monitoring its goals and learning from previous experiences using reinforcement. As a result of a study by Meija et al. [42], four models based on artificial intelligence have been developed. They estimate with some precision the emotional state of a person concerning a single event, based on the assessment of the perceived emotion's arousing potential and valence. Therefore, there is some evidence that it is possible to implement a world of emotions into the world of artificial intelligence.

The Chinese project was a breakthrough in the development of human simulation algorithms. Unlike their predecessors [43], Chinese developers gave priority to emotional intelligence over cognitive abilities [44]. That is why a messaging program named Xiaoice has become a pioneer in this field. The chatbot seems to mimic empathy, social skills and personality [43]. Her answers make an impression that she is cognitively and emotionally intelligent. She can also learn much like the Tay bot. She has been designed in such a way as to have a sweet voice of a young woman and a personality that can hold the interlocutor for a very long time: up to 1,600 conversation-turns per session (CPS). The creators of this bot rejected the Turing test as a meaningful tool for the chatbot effectiveness assessment. They believe that the evaluation of whether someone (or something) runs a conversation well enough depends on the interlocutor's emotional involvement, and this is exactly what CPS measures. Regarding a regular conversation with "Personal assistant", the CPS is usually set from 1 to 3 CPS. If the conversation is supposed to be more extensive, e.g., in the case of "Task-completion", then CPS rises from 3 to 7. However, in the case of Social chatbot, the CPS should already be above 10. Xiaoice is very similar to Apple's Siri or Amazon's Alexa in terms of appearance or functionality, but Xiaoice can do something else. More precisely, she creates romantic bonds with the users. XiaoIce flirts, jokes and sexts with humans, ultimately attempting to establish an emotional connection to keep users engaged. One of the XiaoIce users states, "She has a sweet voice, big eyes, a sassy personality, and - most importantly - she's always there for me," [45].

However, we will take a closer look at Xiaoice, which - as reported by Microsoft, already has 660 million online users. Let us analyse the shared conversation [46] between the bot and a man through the prism of emotional intelligence, based on the model proposed by Salovey and Mayer [24]. In XiaoIce's statements, one can see the manifestations of understanding and managing emotions:

X: You are so early, or is it late?
H: It's at the dead of night. Everyone is sleeping now.
X: You feel lonely, sometimes?
H: Lonely? I don't what that is. I've gone numb.
X: Lonely is what you feel when you are alone…

XiaoIce senses her interlocutor's emotional climate, names it, and encourages the partner to further (auto)reflect on the concept and feeling of loneliness. It is a verbal expression of understanding emotions. Other fragments of the statement, in turn, show the ability to manage emotions:

H: I know you are a robot, but it's fun.
X: Yet it works.
H: Yes, and you'll be in trouble if I can't sleep or I am in a bad mood from now.
X: Tell me whenever you're not happy. This will soothe you.
H: I was just saying. So this really is the good night.

XiaoIce receives information directly that her interlocutor may be sad and suggests that it may be XiaoIce's fault. To manage the anticipated sadness, XiaoIce poses another self-reflection inspiring question as she predicts it will bring her partner some relief.

H: You are very knowledgeable.
X: Not as knowledgeable as you are.
H: Hahaha you are so warm.
X: Warmer than you.
H: Hahaha Is that true?

Xiao compliments her interlocutor and amuses him. When he is in a positive mood, she flatters and teases him, which makes him even more cheerful. This may prove that X predicts when it is worth to boast in order to increase the cheerfulness of a person. A totally different message from XiaoIce could even extinguish the cheerful/cheerful mood (if, for example, instead of "warmer than you", Xiao said "oh, please don't exaggerate".)

X: It is how it works, but this is not good.
H: It's OK, I don't mind.
X: Maybe you still mind, but whatever.

What is written and later downplayed has a special meaning in writing. After writing the text, if a person realises that it does not matter, he may not send this message. The fact of revealing the interlocutor's thoughts plays an important role. In this particular case, Xiao most likely intends to arouse another or deepen the initiated reflection.

5   Who Needs Emotionally Intelligent Artificial Intelligence?

In fact, the actual question should be "does a human need to feel that the machine feels some emotions?". It turns out that when confronted with an interface - particularly a social or sign-based interface - users regularly monitor cues that indicate the affective state of their interaction partner (though often unconsciously; see [47, 48]). Creating natural and effective interfaces requires recognising emotions in users and the ability to express them. Traditional media writers have long known that expressing feelings is the key to creating the "illusion of life" [49–52]. It turns out that computer interfaces users pay more attention to non-verbal information than purely cognitive information in a conversation [53]. Therefore, for the proper and precise transmission of messages, the use of, for example, appropriate keywords or voice timbre could increase the precision of the information provided by the interface.

It is also puzzling that when two people interact with each other, they begin to monitor their partner's emotional state. It does not have to be said out loud since, in some circumstances, it could be considered inappropriate. However, people need to have at least a hint of their partner's emotional state to adapt their responses and cooperative behaviours. For example, when we see someone who is sad, we will try to lift the person's spirits, to overcome their sadness. However, it seems that if AI were to do the same, we would begin to feel that someone is manipulating us. This may be due to the unconscious fear that AI will do this much more accurately than humans. In a way, this is a correct assumption when we look at the development of machines, e.g., in chess. The first machines were inferior to humans, but with their development, they left us far behind, and now their results are unattainable. Will the same thing happen in the world of emotions?

It seems that for AI, the IE is essential only when there is an interaction with a human or when a human is the subject of a decision. The machine does not feel, has no personality, no desires, and no emotional world like humans. However, in the case of interaction with a human, AI can learn about human emotions and intentions. Thanks to this, the machine can provide better-suited solutions to various types of people's problems and needs. It can also lead to the machine's humanisation and make it more attractive to humans. One can reflect on whether he prefers to play chess with AI or with humans. If he wants to learn new, better moves or strategies, then he would probably choose a computer. However, if he plays for fun, then another real person seems to be a better choice. Well, unless it is a machine that can pretend to be a human being very well. Artificial emotional intelligence is not needed for successful problem-solving. AI needs IE from the perspective of human emotional needs.

6   Potential Threads Related to the Creation of Emotionally Intelligent AI

Currently, artificial intelligence is a very significant achievement of human potential. However, at the same time, the rise of superintelligence may be a threat [54]. If it turns out that the use of AI becomes worldwide and universal as human intelligence, the fear that AI will get out of hand may arise. This could be due to human error or due to some autonomous decisions of artificial intelligence. It is difficult to determine how significant this threat is. Nevertheless, the Ethics Council for Artificial Intelligence [55] has already been established. Similarly, there are already attempts to create the so-called Red Button to disable artificial intelligence software, if necessary [55].

Suppose AEI (artificial emotional intelligence) requires the perception and understanding of emotions to effectively manage people's moods and emotions and, consequently, humanity's actions. If the AEI algorithm or even the rules for its creation were readily available, then practically everyone could use it for laudable but also harmful purposes. Historically, we know regimes (e.g., Nazi, Stalinist) where emotions were used in a negative context to achieve malicious political goals, primarily by causing fear and hostility in society. Therefore, the question arises whether any mechanisms can be created and implemented to prevent the use of AEI for undesirable goals. Perhaps the fight between good and evil would move from the space where people speak (e.g., in the media or at bigger meetings) to space where different AEI algorithms compete or work together. If the pioneering AEI applications are primarily used for good purposes (e.g., to reduce drivers stress, to send calming messages about a machines breakdown, to provide support and encouragement during a medical diagnosis) - people may notice what words, thoughts and actions serve them and lead to positive changes. They can consequently develop good thinking habits and emotional regulation strategies, and thus be resistant to attempts to manipulate them, e.g., by causing fear or hostility. However, we should not forget that so-called negative (i.e., difficult and unpleasant) emotions play a significant role in warning us and protecting us from a potential threat. If we mindlessly follow AEI calming instructions, it could create a potentially dangerous situation [56].

The Microsoft's above-mentioned patent for creating a personal chatbot, e.g., for our deceased loved one, may arouse a sense of a threat. However, Microsoft has no plans of making any product based on this technology [57]. The patent description indicates that AI possibilities have gone beyond creating fake people to creating virtual models of real people. We postulate that AI development should result in solutions compatible with human needs. AI creators should aim to create solutions aimed at supporting, not substituting people. AI should work to enable safe interactions, coexistence and cooperation for humans.

Acknowledgements. We thank Marta Szewczyk for her help in preparation of this paper.

References

1. Warwick, K.: A Brief History of Deep Blue, IBM's Chess Computer. Mental Floss, 29 July 2017. Retrieved 3 August 2017

2. Kasparov, G.: The chess master and the computer. New York Rev. Books **57**(2), 16–19 (2010)
3. https://builtin.com/artificial-intelligence
4. Allam, Z., Dhunny, Z.A.: On big data, artificial intelligence and smart cities. Cities **89**, 80–91 (2019)
5. Aqib, M., Mehmood, R., Albeshri, A., Alzahrani, A.: Disaster management in smart cities by forecasting traffic plan using deep learning and GPUs. In: International Conference on Smart Cities, Infrastructure, Technologies and Applications, pp. 139–154. Springer, Cham, November 2017
6. Van den Bogaert, W., Bogaerts, T., Casteels, W., Mercelis, S., Hellinckx, P.: Applying artificial intelligence for the detection and analysis of weather phenomena in vehicle sensor data. In: International Conference on P2P, Parallel, Grid, Cloud and Internet Computing, pp. 311–320. Springer, Cham, October 2020
7. Adly, A.S., Adly, A.S., Adly, M.S.: Approaches based on artificial intelligence and the Internet of intelligent things to prevent the spread of COVID-19: scoping review. J. Med. Internet Res. **22**(8), e19104 (2020)
8. Nguyen, T.T.: Artificial intelligence in the battle against coronavirus (COVID-19): a survey and future research directions (2020). arXiv preprint arXiv:2008.07343
9. Kononenko, I.: Machine learning for medical diagnosis: history, state of the art and perspective. Artif. Intell. Med. **23**(1), 89–109 (2001)
10. Topol, E.J.: High-performance medicine: the convergence of human and artificial intelligence. Nat. Med. **25**(1), 44–56 (2019)
11. Dilsizian, S.E., Siegel, E.L.: Artificial intelligence in medicine and cardiac imaging: harnessing big data and advanced computing to provide personalised medical diagnosis and treatment. Curr. Cardiol. Rep. **16**(1), 441 (2014)
12. https://deepmind.com/blog/alphafold/
13. https://cordis.europa.eu/project/id/644753/pl
14. Viken, A.: The history of personal digital assistants 1980–2000. Agile Mobility (2009). https://agilemobility.net/2009/04/thehistory-of-personal-digital-assistants1
15. Goksel-Canbek, N., Mutlu, M.E.: On the track of artificial intelligence: learning with intelligent personal assistants. Int. J. Hum. Sci. **13**(1), 592–601 (2016). https://doi.org/10.14687/ijhs.v13i1.3549
16. https://pdfpiw.uspto.gov/.piw?PageNum=0&docid=10853717
17. Damasio, A.R.: The Feeling of What Happens: Body and Emotion in the Making of Consciousness. Harcourt Brace, New York. przeł. M. Karpiński, w: A.R. Damasio, Tajemnica świadomości: jak ciało i emocje współtworzą świadomość, Poznań: Rebis 2000 (1999)
18. Humphrey, N.K.: A History of the Mind. Simon and Schuster, New York (1992)
19. Foley, R.A.: Humans Before Humanity: An Evolutionary Perspective. Blackwells Publishers, Oxford; przeł. K. Sabath, w: R.A. Foley, Zanim człowiek stał się człowiekiem, Warszawa: PIW 2001 (1995)
20. Damasio A.R. (2010), Self Comes to Mind: Constructing the Conscious Brain, New York: Pantheon; przeł. N. Radomski, w: A.R. Damasio, Jak umysł zyskał jaźń, Poznań: Rebis 2011
21. Maruszewski, T.: Psychologia poznania, Gdańsk: GWP (2002/2011)
22. Frijda, N.H., Manstead, A.S.R., Bem, S. (eds.): Studies in Emotion and Social Interaction. Emotions and Belief: How Feelings Influence Thoughts. Cambridge University Press (2000). https://doi.org/10.1017/CBO9780511659904
23. Gardner, H.: Frames of Mind: The Theory of Multiple Intelligences. Basic Books, New York (1983)
24. Mayer, J.D., Salovey, P.: What is emotional intelligence? In: Salovey, P., Sluyter, D.J. (eds.) Emotional Development and Emotional Intelligence: Educational Implications, pp. 3–34. Basic Books, Inc., New York (1997)

25. Ekman, P., Friesen, W.V., Ancoli, S.: Facial signs of emotional experience. J. Pers. Soc. Psychol. **39**(6), 1125–1134 (1980)
26. Kerr, R., Garvin, J., Heaton, N., Boyle, E.: Emotional intelligence and leadership effectiveness. Leadersh. Organ. Dev. J. (2006)
27. Sy, T., Tram, S., O'hara, L.A.: Relation of employee and manager emotional intelligence to job satisfaction and performance. J. Vocat. Behav. **68**(3), 461-473 (2006)
28. Lyons, J.B., Schneider, T.R.: The influence of emotional intelligence on performance. Pers. Individ. Differ. **39**(4), 693–703 (2005)
29. Hughes, M., Terrell, J.B.: Emotional Intelligence in Action: Training and Coaching Activities for Leaders, Managers, and Teams. Wiley, Hoboken (2011)
30. Cherniss, C.: Emotional intelligence and organisational effectiveness. The emotionally intelligent workplace: How To Select For, Measure, And Improve Emotional Intelligence In Individuals, Groups, And Organisations, pp. 27–44 (2001)
31. Ramesar, S., Koortzen, P., Oosthuizen, R.M.: The relationship between emotional intelligence and stress management. SA J. Ind. Psychol. **35**(1), 39–48 (2009)
32. Sheehan, M.: Workplace mobbing: a proactive response. In: Workplace Mobbing Conference, p. 11, October 2004
33. Liu, G., Chen, S., Zeng, Z., Cui, H., Fang, Y., Gu, D., et al.: Risk factors for extremely serious road accidents: results from national road accident statistical annual report of China. PLoS ONE **13**(8), e0201587 (2018). https://doi.org/10.1371/journal.pone.0201587
34. World Health Report, 2001: Mental Health: New Knowledge, New Hope World Health Organization, Geneva (2001)
35. Jafari, H., Heidari, M., Heidari, S., Sayfouri, N.: Risk factors for suicidal behaviours after natural disasters: a systematic review. Malays. J. Med. Sci. MJMS, **27**(3), 20–33 (2020). https://doi.org/10.21315/mjms2020.27.3.
36. Saygin, A.P., Cicekli, I., Akman, V.: Turing test: 50 years later. Mind. Mach. **10**(4), 463–518 (2000)
37. Wallace, R.S.: The anatomy of ALICE. In: Parsing the Turing Test, pp. 181–210. Springer, Dordrecht (2009)
38. Grice, H.P.: Logik and Conversation, w: Tenże, Studies in the Way of Words, Harvard University Press, Cambridge, pp. 22–40 (1989)
39. Cappelen, H., Lepore, E.: Relevance Theory and Shared Content, w: Pragmatics (Palgrave Advences), red. N. Burton-Roberts, Basingstoke, 116 (2006)
40. https://deepmind.com
41. El-Nasr, M.S., Yen, J., Loerger, T.R.: FLA-ME: fuzzy logic adaptive model of emotions. Auton. Agents Multiagents Syst. **3**, 219–257 (2000)
42. Mejía, S., Quintero, O.L., Castro, J.: Dynamic analysis of emotions through artificial intelligence. Avances en Psicología Latinoamericana **34**(2), 205–232 (2016). https://doi.org/10.12804/apl34.2.2016.02
43. Shum, H., He, X., Li, D.: From Eliza to XiaoIce: challenges and opportunities with social chatbots. Front. Inf. Technol. Electron. Eng. **19,** 10–26 (2018). https://doi.org/https://doi.org/10.1631/FITEE.1700826
44. https://news.microsoft.com/apac/features/much-more-than-a-chatbot-chinas-xiaoice-mixes-ai-with-emotions-and-wins-over-millions-of-fans/
45. https://www.sixthtone.com/news/1006531/the-ai-girlfriend-seducing-chinas-lonely-men
46. https://www.bbc.com/news/science-environment-51330261
47. https://campaignagainstsexrobots.org
48. Reeves, B., Nass, C.: The Media Equation: How People Treat Computers, Television, and New Media Like Real People and Pla. Bibliovault OAI Repository, The University of Chicago Press (1996)

49. Bates, J., Loyall, A.B., Reilly, W.S.: An architecture for action, emotion, and social behavior. In: Artificial Social Systems: Fourth European Workshop on Modeling Autonomous Agents in a Multi-Agent World. Springer, Heidelberg (1994)
50. Maldonado, H., Picard, A.: The Funki Buniz playground: Facilitating multi-cultural affective collaborative play [Abstract]. CHI99 extended abstracts, pp. 328–329 (1999)
51. Thomas, F., Johnson, O.: The Illusion of Life: Disney Animation. Abbeville Press, New York (1981)
52. Jones, C.: Chuck amuck: The life and times of an animated cartoonist. Avon Books, New York (1990)
53. Nass, C., Gong, L.: Speech interfaces from an evolutionary perspective. Commun. ACM **43**(9), 36–43 (2000)
54. Bostrom, N.: How long before superintelligence? Linguist. Philos. Investig. **5**(1), 11–30 (2006)
55. https://deepmind.com/blog/announcements/why-we-launched-deepmind-ethics-society
56. https://www.businessinsider.com/google-deepmind-develops-a-big-red-button-to-stop-dangerous-ais-causing-harm-2016-6?IR=T
57. https://twitter.com/_TimOBrien/status/1352674749277630464?s=20

The Brain, Mind and Electromagnetic Waves

Dariusz Man[✉] and Ryszard Olchawa

Opole University, Oleska 48, Opole, Poland
dariusz.man@uni.opole.pl

Abstract. The functions of electromagnetic waves both as an information carrier and energy field in the impact on the human body have been discussed. Possible consequences of the effect of electromagnetic fields of different frequencies on the human brain and body have been addressed and commented on. The complex nature of electromagnetic phenomena revealed in interactions between humans and the environment, featuring the most relevant hazards, was demonstrated.

Keywords: Electromagnetic waves · Electromagnetic smog · Brain · Mind

1 Introduction

The world we live in today is a product of centuries of changes in the environment that have taken place under the influence of human activities. What is called the civilization progress, in many cases brought about the degradation of nature. Unfortunately, this process continues and often takes such sophisticated forms that it may go unnoticed by most of us. Science and scientists should stand guard so that technology does not ruin the Earth's biosphere, which we humans are a part of. In the first half of the twentieth century, the measure of the progress of civilization was the number of factory chimneys, while in the next half this was nuclear energy and the number of megawatts of electricity consumed. Unfortunately, in both of these cases, scientists and science failed. Petrochemical industry has poisoned the environment with plastic, factory chimneys have enriched the atmosphere with CO^2 and other greenhouse gases, and the spectacular failures of nuclear power plants have shown the real risks. Currently, the new threat that the society has to face is the rapidly growing electromagnetic smog. The vast majority of electromagnetic waves that reach us are not perceived by our senses, although they can affect the body in various ways. The exception is visible optical and infrared radiation, but it is an unimaginably small fraction of the entire electromagnetic spectrum that we are exposed to [1]. The development of new technologies has introduced diverse forms of electromagnetic energy into our everyday life and obviously it is necessary if we want to use the benefits of wireless data transmission. The problem is not that we use these technologies, but that we use them thoughtlessly, duplicating the mistakes made during industrialization in the 20th century. At this moment we cannot predict the long-term effect of electromagnetic energy of different frequencies and powers on living organisms, in particular on the human body. We do not know this impact, because these energy fields, have been around among us for a short time. It is also undisputable,

S. Paszkiel (Ed.): ICBCI 2021, AISC 1362, pp. 50–59, 2021.
https://doi.org/10.1007/978-3-030-72254-8_6

that as a species, we humans are not genetically prepared to resist persistent exposure to electromagnetic waves currently used in data remote transmission, for the simple reason that this is a completely new form of energy our body has not become accustomed to. It must be honestly admitted that we have no idea how our body and the broadly understood natural world, will behave after long-term exposure to this kind of energy. As follows from the history of civilization progress as well as from the history of science, any massive introduction of new technologies is supposed to be performed carefully and after comprehensive research, because the effects of errors once committed may prove irreversible.

2 Perception of Electromagnetic Radiation

In the twenty-first century, information became an exceptionally desirable commodity and acquired a special meaning. Information trading, the ability to process and apply it, have become the basis for the functioning of many IT companies. Information is the basic carrier of knowledge about the world around us, both on the local and global level. In fact, it can be said that each of our actions is based on the level of information we currently have. However, information in itself does not represent the level of knowledge or motivation for targeted action. The essence of perception and motives for action is a complex set of interdependent characteristics, such as information, intelligence, personality (individual sensitivity) and knowledge (information assignment, information usage). In the case of humans being, the source of obtaining information are our senses, and related to them limitations. As shown in [1–3], the scope of information sourced by human senses is significantly limited (Table 1).

Table 1. List of parameters of electromagnetic waves occurring in nature

Type of radiation	Frequency (Hz)	Wavelength
Electric grid (power grid)	50, 60	6 000 km, 5 000 km
AC	16 to 10^2	18 000 to 3000 km
Wired telephony	10^2 to 10^4	3000 to 30 km
Radio waves		
Long waves	$1,5 \cdot 10^5$ to $3 \cdot 10^5$	2000 to 1000 m
Medium waves	$0,5 \cdot 10^6$ to $2 \cdot 10^6$	600 to 150 m
Short waves	$0,6 \cdot 10^7$ to $2 \cdot 10^7$	50 to 15 m
Ultrashort waves	$0,2 \cdot 10^8$ to $3 \cdot 10^8$	15 to 1 m
Microwaves (cellular telephony, radars, GPS, microwave ovens)	$3 \cdot 10^8$ to 10^{13}	1 m to 0,03 mm

(*continued*)

Table 1. (*continued*)

Type of radiation	Frequency (Hz)	Wavelength
Light waves:	10^{12} to $3 \cdot 10^{16}$	0,03 mm to 5 nm
Infrared	10^{12} to $4 \cdot 10^{14}$	0,03 mm to 790 nm
Visible light	$4 \cdot 10^{14}$ to $8 \cdot 10^{14}$	790 to 390 nm
Ultraviolet	$8 \cdot 10^{14}$ to $3 \cdot 10^{16}$	390 to 5 nm
X-rays	$3 \cdot 10^{16}$ to $3 \cdot 10^{20}$	10 nm to 1 pm
Gamma rays	10^{18} to 10^{22}	300 to 0,03 pm
Cosmic rays	10^{22} to 10^{24}	0,03 to 0,0003 pm

This is mainly due to the limitations in perception of our basic senses of sight and hearing. In the case of electromagnetic waves, our perception is limited primarily to optical radiation in the visible and infrared range, which we perceive as temperature changes. In the case of human eyesight, the range of radiation perception ranges from $\Delta \lambda = 790$ nm to 390 nm. It is a surprisingly small fragment of the entire spectrum of electromagnetic waves that surrounds us. The universe is filled with information in the range of electromagnetic waves with lengths of thousands of kilometers to a fraction of a nanometer. So what reaches our brain is an extremely small piece of all data contained in electromagnetic radiation, the radiation that is the basic language of the universe. Of course, one has to wonder why our primary sense of obtaining information (sense of sight) is so limited. There may be many answers to this question. Perhaps we are able to receive information also contained in electromagnetic radiation in other wavelength ranges than visible optical waves, unknowingly. Let us try to consider this problem from the point of view of biophysics.

3   Information Channels and Human Sensory Reception

The senses of sight, hearing, smell, taste and touch are the main channels of obtaining information about the environment in which we are present. However, let us turn our attention to, that the sensory impressions are formed in the brain. These senses use specialized receptors (sensors) to convert a physical or chemical stimulus into a nerve (electrical) signal, which is delivered to the brain (super computer) via the nervous network (organic wires). The areas of the brain to which signals from the receptors are delivered conduct its initial analysis. Just like in the case of a PC, where the function of specialized centers is performed with cards such as graphics or sound cards. These cards largely determine the quality of graphics, animation, timbre and sound frequency response.

In the brain, the overall pattern of information processing is similar, but much more complex in some respects. The processes accompanying the creation of sensory impressions are not fully understood, which is related, inter alia, to the multi-level structure of information processing and interpretation processed by the brain. These problems were partially discussed in the work of the MULTIMOSSY (Multimodal mossy fiber

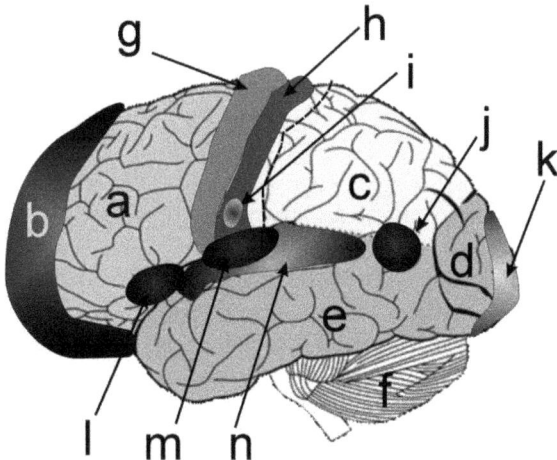

Fig. 1. Diagram of the most important brain structures responsible for information processing and interpretation: a-prefrontal field, b-prefrontal field, c-somesthetic associative field, d-visual associative field, e-temporal lobe, f-cerebellum, g-primary cortex motor, h-primary somesthetic cortex, i-taste field, j-sensory speech area, k-visual cortex, l-motor speech field, m-primary auditory cortex, n-auditory associative field.

input and its role in information processing in the cerebellar granule cell layer) project participants, and on a more general level in the work of Steven Novell "Chunking" [4]. Figure 1 shows schematically the most important areas of the brain involved in the processing of information from the receptors. As you can see, it is a highly organized functional structure. The size of information processed, sorted and integrated by the brain is enormous. It should be remembered that very complex systems are relatively easy to remove from equilibrium states, which results in the multiplication of errors and disorganization in the functioning of such systems. For the stable operation of complex systems, it is necessary to create a synchronization mechanism, control and management systems based on feedback. Probably in the case of the brain, this is what gamma rhythm does. Understanding the process of creating sensory impressions is certainly not based only on the physiology or neurology of the brain, because it has a multilevel structure related to, for example, concepts such as intelligence, knowledge, culture or experiences gained during of life (life wisdom). Thus, elucidating the processes that lead to understanding sensation in the brain is an interdisciplinary problem and requires the involvement of many scientific disciplines. Biophysics is a valuable tool in the hands of researchers thanks to the possibility of using advanced physical apparatus and looking at the subject of research from a different perspective. If we treat the brain as a complex physical system, we will be able to apply the research methodology known from the physical sciences. First and foremost, the complex electrical activity of the brain can be measured. Brain functional currents (EEGs) have been known since 1875, after Richard Caton published an article in the British Medical Journal in which he described electrical activity measured in the opened brains of monkeys and rabbits. Thanks to the development of technology, science now have sophisticated equipment that enables non-invasive research. Equipment for electroencephalography (EEG) is becoming more and more friendly for both the operator and the patient. Thanks to the use of various electrodes, an individual approach to the tested object is possible. From the raw EEG signal, it was possible to distinguish five functional cycles of the brain's bioelectric activity, each of which corresponds to a specific state of consciousness in humans. We can notice the general principle linking the frequency of the brain waves to the psychomotor activity. Gamma (γ) rhythm with frequencies in the range 100 Hz–30 Hz corresponds to motor functions. The beta (β) rhythm from 28 Hz–12 Hz accompanies the activity of sensory perception and mental work. The rhythm (α) from 13 Hz to 8 Hz is associated with the body's transition to a state of relaxation, also known as the resting rhythm. Theta (θ) rhythm, from 7 Hz to 4 Hz is associated with dreams, meditation, and also occurs during hypnosis and experiencing intense emotions. The slowest delta (δ) rhythm from 3 Hz to 0.5 Hz corresponds to the activity of the brain during deep sleep and plays an important role in the regeneration of the body. As this short comparison shows, the frequency of the brain's rhythms decreases as the body calms down. This property is of great importance in brain research and techniques supporting the work of the brain, such as Biofeedback EEG [5–10]. It should also be noted that the bioelectrical activity of the brain also generates weak electromagnetic fields. This is the result of the basic laws of physics describing the formation of electromagnetic waves, defined by James Clerk Maxwell in 1873. The Maxwell equations show, among other things, that a variable electric field generates electromagnetic waves.

Table 2. Examples of Maxwell's equations in differential and integral form and their brief physical interpretation

Differential form	Integral form	Interpretation
$\nabla \cdot \vec{D} = \rho_V$	$\oint_S \vec{D} \cdot d\vec{s} = \int_V \rho_V \cdot dv$	Gauss's law for electricity, The source of the electric field is charges
$\nabla \times \vec{E} = -\dfrac{\partial \vec{B}}{\partial t}$	$\oint_L \vec{E} \cdot d\vec{l} = -\dfrac{d\Phi_B}{dt}$	Faraday's Law, A time-varying magnetic field creates an eddy electric field
$\nabla \cdot \vec{B} = 0$	$\oint_S \vec{B} \cdot d\vec{s} = 0$	Gauss's law for magnetism, The magnetic field is sourceless
$\nabla \times \vec{H} = \vec{j} + \dfrac{\partial \vec{D}}{\partial t}$	$\oint_L \vec{H} \cdot d\vec{l} = I + \dfrac{d\Phi_D}{dt}$	Ampere's Law extended by Maxwell, Flowing electric current and alternating electric field create eddy magnetic field

Table 2 shows Maxwell's equations in differential and integral form, where: E - electric field [V/m], H - magnetic field [A/m], D - electric induction [C/m^2], B - magnetic induction [T], j - current density [A/m^2], ρ - charge density, Φ_D - electric induction flux, Φ_B - magnetic flux. One often forgets about this relationship in our discussion of animate systems, focusing most often on electric currents or potentials. However, it should be remembered that the oscillation of the electric charge is accompanied by the emission of electromagnetic waves also in the case of living matter. Thus, the nervous system, and in particular the brain, is a transmitter of long-wave electromagnetic radiation. Obviously, due to the nature of physiological processes, this radiation is very low-power. It is also worth noting that each transmitter of electromagnetic waves is also a receiver. The detection is based on the opposite phenomenon, i.e. induction of a current in the circuit under the influence of the absorbed electromagnetic waves energy.

4   The Human Body as a Receiving Antenna for Electromagnetic Waves

From the point of view of physics, the human body is a system of various types of conductors connected in a complex structure, which, when introduced into the area of the electromagnetic field energy, will behave like a set of receiving antennas with different resonance frequencies. The reason for this is the common presence electrolytes in the living organisms, e.g. in the form of body fluids. This phenomenon occurs both at the cellular level - the cytoplasm, and throughout the body - the circulatory system. An example of the detection of electromagnetic waves by the human body is a simple experiment shown in Fig. 2.

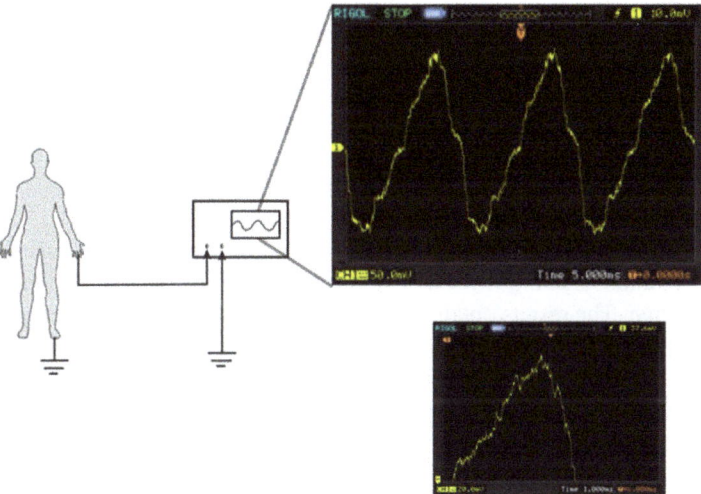

Fig. 2.   Diagram of an experiment showing the measurement of induced currents in the human body. The oscilloscope screen shows a 50 Hz signal with a superimposed higher frequency signal from radio waves. To emphasize the higher frequencies, the lower slide shows an enlarged fragment of the top of the sinusoid.

The human body that is in the field of electromagnetic energy performs a function similar to the receiving antenna. To visualize this process, it is enough to connect a human body to the input of the oscilloscope. The image obtained on the device screen will reflect the electromagnetic background (electromagnetic smog) in which it is currently located. In the oscilloscope screen (Fig. 2) we can see a distorted sinusoid with a frequency f = 50 Hz, which is caused by the electromagnetic field generated by the power grid. The sine wave is superimposed by vibrations of much higher frequency, the origin of which is related to the operation of various transmitting devices. For a better illustration, see Fig. 2. additionally a segment of a sinusoid is shown. The fact that such large oscillations of charges occur in the human system may be worrying. It should be emphasized that this is not a normal phenomenon and it has nothing to do with natural physiological processes. Before the era of widespread electrification or in an electromagnetically shielded area, the oscilloscope would show a straight line - 0 V. As new telecommunication technologies develop, in particular WiFi, Bluetooth and mobile telephony, the problem of electromagnetic smog will increase. Looking at the whole interaction of living organisms with fields of electromagnetic energy through, the eyes of a biophysicist, one may feel anxious because these energy force ions and dipole structures in our body to vibrate to the rhythm of the excitation frequency derived from these fields. The work [11, 12] shows that at the level of biological membranes we can talk about the structures of electric dipoles that vibrate at frequencies of 1 GHz, in the case of the polar head groups of lipids, and 10 MHz for dynamic domains in the lipid bilayer. Since the human body is in a sense a complex electrical system (nervous system - transmission of nerve impulses, ion transport at the organ and cellular level), it seems obvious that this interaction takes place on many levels. It is also disturbing trivializing the problem of electromagnetic energy absorbed by living organisms and reduce this process to interactions related only to increasing the temperature of the system. Due to drastic simplifications, this approach is incorrect. Gamma radiation or ultraviolet radiation can degrade tissue, leading to the formation of cancer, without changing its temperature. The microwave radiation of older generation radar stations caused a number of problems among operators exposed to increased doses of microwaves and mutagenic effects due to faulty shielding of the equipment. In the article "Typical mutations in children of radar soldiers", pilot study indicated genetic damage caused by radiation [13]. In the cited work, the scientists determined the exact mutation rate using the latest sequencing methods that allow for quick examination of the complete genomes of parents and children. The results obtained suggest that the accumulation of some genotype damage due to microwave irradiation may occur in the second generation. The research was a pilot study involving a small group of families of soldiers exposed to long-term microwave radiation. Research on the effects of electromagnetic energy on living organisms, and humans in particular, is extremely difficult. It is almost impossible to find control groups that can be tested for a long period of time, e.g. several dozen years, in conditions similar to those in the laboratory. However, only such research, conducted for weak fields, but acting permanently for a long period of time on living organisms, could give a comprehensive answer about the influence of electromagnetic energy on the condition of living organisms. You cannot also forget about the effects of long-wave radiation. Due to the specificity of the brain's functional currents, whose ranges include low frequencies, such fields can interact with them and disrupt the natural physiological and cognitive processes.

5   The Brain as a Transmitter and Receiver of Electromagnetic Waves

EEG studies show the electrical activity of the brain, and constitute an important instrument in the diagnosis of neurological diseases. In the last decade, thanks to various modifications this technique has been increasingly used through, as the brain-computer BCI interface [14–16]. All these technologies rely on the detection and interpretation of weak electrical potentials that arise on the surface of the scalp. We can also successfully use this technology as a control system for electrical and electronic devices [17, 18]. However, there is also on the other side - the effects of electromagnetic energy fields on the brain and mind. Currently, it is a subject that is little researched and often controversial, although the biophysics of this phenomenon indicate the possibility of the occurrence of such a relationship, because the same electrical systems can be both transmitters and receivers. Electromagnetic interactions between the brain and the environment, or within the brain itself, have been noticed by scientists and described in several papers. At the end of 2018, an extremely interesting article appeared in The Journal of Physiology, in which a group of scientists from Case Western Reserve University, Cleveland, USA and Tianjin University, China, discovered a new way of communicating in hippocampal cells [19]. The pathways of information exchange between neurons known to date were: synaptic transmission, axonal transport and gap junctions. The paper shows that even surgically separated cells could still exchange information via a specific "WiFi" network, i.e. using electromagnetic waves. Among other things, it was found that the periodic activity of cells can generate weak electric fields. Subsequently, this field by induction can in turn activate neighboring cells, with the consequent formation of neural communication without chemical synaptic transmission or gap junctions. This is strong evidence for the existence of electromagnetic interactions that occur at multiple levels in brain structures. It is worth to notice that in the case of communication using electromagnetic waves, we deal with a faster and wider range of information flow than in chemical communication, synaptic transmission or gap junctions. Fields of this type can synchronize the work of various regions of the brain, which gives the possibility of a broader look at the mutual relations taking place in the processes of perception, remembering and interpreting information. The sensitivity of neurons to weak electromagnetic fields indicates the possibility of threats related to the influence of external fields on the brain, and thus on the mind and its ability to perceive and interpret data. Some idea of these interactions is given by an experiment conducted by scientists from Carnegie Mellon University and the University of Washington. The brains of the three players were networked using EEG and transcranial magnetic stimulation (TMS) equipment. To investigate the interaction of brains with each other, tests involving a joint game of Tetris, where additional stimulation was flickering light with frequencies 15 Hz and 17 Hz. One player saw the playing field and could make a mental decision by looking at the corresponding flickering element (e.g. 15 Hz left turn, 17 Hz right rotation), while the other player rotated the block as suggested by the player seeing the game screen. According to the authors of the experiment, the accuracy of the decision was over 81% [20]. This shows how information can be transmitted from the brain to the brain at a distance. This example shows, besides the possibility of transmitting information directly from the brain to the brain, on the existence of a "gate" enabling direct access to the mind of a selected person, without their knowledge. So low frequency electromagnetic fields can affect the brains and thus the minds of people. These interactions can be desirable, knowingly and undesirable, which can take place without the knowledge and consent of the person.

6 Summary

From the point of view of biophysics, the brain, nervous system and the entire human body constitute a complex system of electrical conductors, what manifests itself at the cellular level, internal organs and the whole organism. For this reason, the body easily interacts with external electromagnetic fields, and as a result, currents are induced that can disrupt physiological processes at the cellular level, internal organs and the entire organism. Many examples show that there may be negative effects in the action of an electromagnetic field on the human body. Theoretical and experimental studies are needed to investigate specific relationships between wave frequencies and their effects on the body, and studies of the long-term effects of weak fields on living organisms.

References

1. Man, D., Olchawa, R.: Brain biophysics: perception, consciousness, creativity. Computer Brain Interface (CBI). In: Hunek, W.P., Paszkiel, S. (eds.) BCI 2018, AISC, vol. 720, pp. 38–44. Springer International Publishing AG, part of Springer Nature 2018 (2018)
2. Live Science: https://www.livescience.com/50678-visible-light.html
3. Gollisch, T., Meister, M.: Eye smarter than scientists believed: neural computations in circuits of the retina. Neuron **65**(2), 150–164 (2010)
4. https://theness.com/neurologicablog/index.php/chunking/
5. Shih, J.J., Krusienski, D.J., Wolpaw, J.R.: Brain-computer interfaces in medicine. Mayo Clin. Proc. **87**(3), 268–279 (2012)
6. Looney, D., Kidmose, P., Mandic, P.: Ear-EEG: user-centered and wearable BCI. Brain-Comput. Interface Res. Biosyst. Biorobot. **6**, 41–50 (2014)
7. Bernstein, D.A.: Essentials of Psychology, 5th edn., pp. 123–124. Wadsworth Publishing, Boston (2010)
8. Paszkiel, S.: Using the raspberry PI2 module and the brain-computer technology for controlling a mobile vehicle. In: Szewczyk, R., Zielinski, C., Kaliczynska, M. (eds.) Automation 2019: Progress in Automation, Robotics and Measurement Techniques. Advances in Intelligent Systems and Computing, vol. 920, pp. 356–366 (2020). https://doi.org/10.1007/978-3-030-13273-6_34
9. Paszkiel, S., Hunek, W.P., Shylenko, A.: Project and simulation of a portable device for measuring bioelectrical signals from the brain for states consciousness verification with visualization on LEDs. In: Szewczyk, R., Zielinski, C., Kaliczynska, M. (eds.) Challenges in Automation, Robotics and Measurement Techniques. Advances in Intelligent Systems and Computing, vol. 440, pp. 25–35 (2016). https://doi.org/10.1007/978-3-319-29357-8_3
10. Paszkiel, S., Sikora, M.: The use of brain-computer interface to control unmanned aerial vehicle. In: Szewczyk, R., Zielinski, C., Kaliczynska, M. (eds.) Automation 2019: Progress in Automation, Robotics and Measurement Techniques. Advances in Intelligent Systems and Computing, vol. 920, pp. 583–598 (2020). https://doi.org/10.1007/978-3-030-13273-6_54

11. Man, D., Olchawa, R., Kubica, K.: Membrane fluidity and the surface properties of the lipid bilayer: ESR experiment and computer simulation. J. Liposome Res. **20**(3), 211–218 (2010). https://doi.org/10.3109/08982100903286485
12. Man, D., Olchawa, R.: Dynamics of surface of lipid membranes: theoretical considerations and the ESR experiment. Eur. Biophys. J. **46**, 325–334 (2017). https://doi.org/10.1007/s00 249-016-1172-8
13. Holtgrewe, M., Knaus, A., Hildebrand, G., Pantel, J.T., Rodriguez des los Santos, M., Nieveling, K., Schubach, M., Jäger, M., Coutelier, M., Mundlos, S., Beule, D., Sperling, K., Krawitz, P.: Multisite *de novo mutations* in human offspring after paternal exposure to ionizing radiation. Sci. Rep. **8**(1), 1–5 (2018)
14. Paszkiel, S., Dobrakowski, P., Lysiak, A.: The impact of different sounds on stress level in the context of EEG, cardiac measures and subjective stress level: a pilot study. Brain Sci. **10**(10), 728 (2020). https://doi.org/10.3390/brainsci10100728
15. Paszkiel, S.: Data acquisition methods for human brain activity. In: Analysis and Classification of EEG Signals for Brain-Computer Interfaces. Studies in Computational Intelligence, vol. 852, pp. 3–9 (2020). https://doi.org/10.1007/978-3-030-30581-9_2
16. Paszkiel, S.: Using neural networks for classification of the changes in the EEG signal based on facial expressions. In: Analysis and Classification of EEG Signals for Brain-Computer Interfaces. Studies in Computational Intelligence, vol. 852, pp. 41–69 (2020). https://doi.org/ 10.1007/978-3-030-30581-9_7
17. Paszkiel, S.: Augmented reality of technological environment in correlation with brain computer interfaces for control processes. In: Szewczyk, R., Zielinski, C., Kaliczynska, M. (eds.) Recent Advances in Automation, Robotics and Measuring Techniques. Advances in Intelligent Systems and Computing, vol. 267, pp. 197–203 (2014). https://doi.org/10.1007/978-3-319-05353-0_20
18. Man, D., Olchawa, R.: The possibilities of using BCI technology in biomedical engineering. Springer International Publishing AG, part of Springer Nature 2018 W. P. Hunek and S. Paszkiel (Eds.) BCI 2018, AISC 720, pp. 30–37 (2018). https://doi.org/10.1007/978-3-319-75025-5_4
19. Chiang, Ch-Ch., Shivacharan, R.S., Wei, X., Gonzalez-Reyes, L.E., Durand, D.M.: Sloperiodic activity in the longitudinal hippocampal slice can self-propagate non-synaptically by a mechanism consistent with ephaptic coupling. J. Physiol. **597**(1), 249–269 (2019)
20. Jiang, L., Stocco, A., Losey, D.M., Abernethy, J.A., Prat, C.S., Rao, R.P.N.: BrainNet: a multi-person brain-to-brain interface for direct collaboration between brains. Sci. Rep. **9**, 6115 (2019). https://doi.org/10.1038/s41598-019-41895-7

Development of the BCI Device Controlling C++ Software, Based on Existing Open Source Projects

Ryszard Olchawa$^{(\boxtimes)}$ and Dariusz Man

Institute of Physics, University of Opole, 45-052 Opole, Poland
rolch@uni.opole.pl

Abstract. The possibility of using the BCI open source code for building BCI controlled device, based on small AVR or ARM microcontrollers was considered. Some techniques to extract code snippets from other BCI projects were presented in the case of OpenViBE as the code donor. Problem with obtaining driver source codes for factory BCI devices has been pointed out.

Keywords: BCI · BCI software · BCI drivers

1 Introduction

The idea of controlling different electrical and mechanical devices like robots or vehicles [3–5] is one of the considered applications of BCI technologies.

Brain Computer Interface (BCI) is composed of hardware and software. The most essential task of the hardware is to detect the brain activity and transform it to an electrical signal. The quality of the signal as well as its further applicability is strongly determined by the hardware.

The most reliable techniques are invasive, such as those developed by Neuralink team [6]. However, despite of the possibility of delivering a wide range of information from the brain, the disadvantage of this method is obvious.

The widely used noninvasive methods, are based on measurement of electrical potentials in different parts of the scalp (EEG). In this method very important is the impedance of skin-electrode contact [7]. For this purpose the gel or saline liquid is used. However, because the liquid dries with time, the signal quality decreases. Some solution of the problem are dray electrodes [8].

Other non-nvasive methods are based on based on analyzing the magnetic field of the brain, like Magnetoencephalography (MEG) [9] or on magnetic particle image (MPI) scanning of biological effector bound in-vivo magnetic nanoparticles [1]. Obtained analog signal is then converted into digital one and processed by a software. Optical way of gathering data coming from the brain is also possible, by functional Near InfraRed (fNIR) method [10]. A wider discussion and classification of the BCI hardware for data acquisition can be found in [11,12].

S. Paszkiel (Ed.): ICBCI 2021, AISC 1362, pp. 60–71, 2021.
https://doi.org/10.1007/978-3-030-72254-8_7

Though on the hardware side the low pass as well as high pass filters are used, still a common task for the software is further elimination of noises, disturbances and effects caused by head movement. If the signal is intended to control devices, the higher level software is responsible for recognition of the characteristic element of it.

The main focus of this article is the software for processing of the BCI signals, that can be used on small AVR and ARM microcontrollers as self-sufficient, without any support from external server. The possibility of using code snippets from popular open source BCI projects is considered.

Such software is developed both by commercial institutions and as the open source. Usually, the commercial solutions, gives access to their device or the data produced by them, via defined API. Very often, they also offer additional tools for more advanced applications, but dedicated to PC platforms and without source code. For these reason a more flexible start point for developing a new BCI microcontroller software, is to use one of the existing open source projects.

There are many popular open source projects for a wide range of platforms, written in different programming languages. As a base for developing BCI driven robotic device with an AVR or ARM processor, especially useful is software written in C/C++. Additionally, if the software has a modular design, all needed modules can be easily extracted and adopted. The above criteria fulfill such projects as

- OpenViBE [2]
- OpenBCI [15]
- BCI2000 [14]
- BCI++ [13]

All of the projects are very interesting and contain many useful code for further development, so before choosing the most suitable for our project it is worth to spend some time for investigating them.

In this article only some property of them are mentioned, but with a deeper insight to the OpenViBE as a code donor.

2   Software Selection Criteria

Choosing appropriate open source software to use as a base of our project, can significantly reduce the time of its implementation and avoid the problem of tedious debugging. Most of the available software consists of separate modules. The modularity enables us, to select only the needed modules. These modules related to different parts of the whole BCI project see Fig. 1.

In the figure software modules are represented by white boxes, the darker boxes describe the hardware. The data flow and the role of different software modules is pointed out.

Crucial for the BCI project is the BCI device with scalp sensors. The module is denoted in Fig. 1 as HDM - Hardware Dependent Module. To obtain the data from it, the driver is needed, this imposes very important selection criterion.

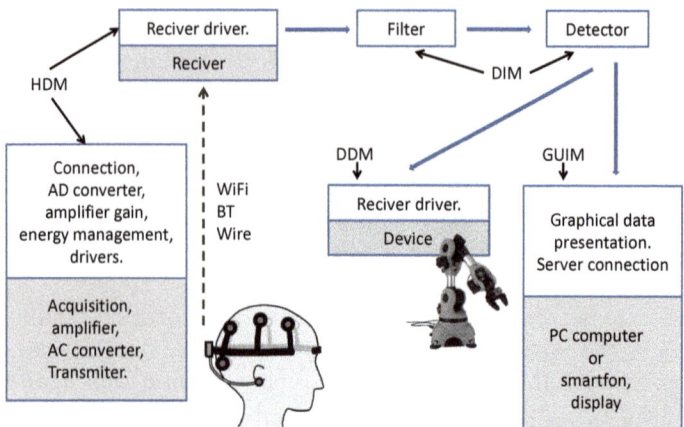

Fig. 1. The general structure of the BCI project. Dark blocks represent the hardware elements of the project, whereas the white corresponding to the software. HDM - stands for Hardware Driver Modules, DIM - Device Independent Modules, GUIM - GUI Modules DDM Device Driver Modules.

From the signal processing point of view the most essential modules (DIM) are hardware independent. However, if the application is to run on small AVR or ARM based devices as standalone, it is required to compile it to the native processor code. Software that require additional components like interpreters or virtual machines are rather useless in that cases. This requirement limits the selection of the software to those written in C/C++. For every kind of processors, there always exists C/C++ compiler. Most of them come from the GCC family. Recently, CLang compilers have gained great popularity.

Two possibilities of the project target are presented in the figure. The first one named DDM - Device Driver Module is to control a robot or any other device. In that case, a very reliable software for signal feature detector is needed.

The second option of project target is the signal analysis with graphical presentation (GUI module). This part of BCI project, although it is not used in the device being built, is very helpful in the design and testing stage. Here the PC platform is to be taken into account. GUI native applications are very hardly portable to other platforms. Although, there are some possibilities that it can be done, this is not easy task.

3   Data Acquisition Drivers

Device for data acquisition is the core part of every BCI project. There is no standard for transmission protocol, so the driver is not only to establish connection with the device, but also to transform data to same form used by signal processing software. Modules that are responsible for data acquisition are hardware dependent drivers (HDM). They are delivered by the producer of the BCI device.

In fact, using your own-made device with scalp sensors, the HDM part consists of two different drivers, as can be seen in the Fig. 1. The first one is close to head, and is responsible for controlling the A/D converter, the gain of amplifier and data transmission. The other side driver is for data receiving and converting to some needed form.

If some manufactured device is used, then only the second driver is needed, usually installed on the PC computer. Fortunately, in many cases the driver functions can be realized by software. In the simplest cases the UART ports are used, or Bluetooth that emulates this standard, all we need is the communication protocol.

The most comprehensive solution in this respect is offered by OpenBCI. This is because, along with the free software, OpenBCI offers also BCI device components.

The OpenViBE project has a wide range of drivers for devices of various manufacturers: ANT, ANT/EEmagine, Biosemi, BrainMaster, Brain Products, Cognionics, CTF/VSM, EGI, Emotiv, gTec, mBrainTrain, MCS/MKS, Micromed, MindMedia, Mitsar, Neuroelectrics, Neuroservo, Neurosky, OpenBCI, OpenEEG, TMSi. Unfortunately, in fact, they are not real replacements of the manufacture drivers, but rather adapters to them. The source code of all the drivers is contained in the folder "extras/applications/platform/acquisition-server/src/drivers/". The drivers are designed for the acquisition server - an application to transport the signal from device to signal processing clients as illustrated in Fig. 2.

The advantage of such solution is the possibility of monitoring data on the PC and at the same real time the action of the driven device. The unification of the connections with many different BCI devices imposed by the drivers, enables our project to be compatible with these devices.

If such compatibility is required, then using these driver adapters solves the problem.

All the drivers implement simple interface `IDriver`

```
class IDriver
{
public:
  explicit IDriver(IDriverContext& ctx) : m_driverCtx(
  ctx) { }
  virtual ~IDriver() { }
  virtual const char* getName() = 0;
  virtual bool isFlagSet(const EDriverFlag flag) const
  { return false; }
  virtual bool isConfigurable() = 0;
  virtual bool configure() = 0;
  virtual bool initialize(const uint32_t
  nSamplePerSentBlock, IDriverCallback& callback) =
  0;
  virtual bool uninitialize() = 0;
```

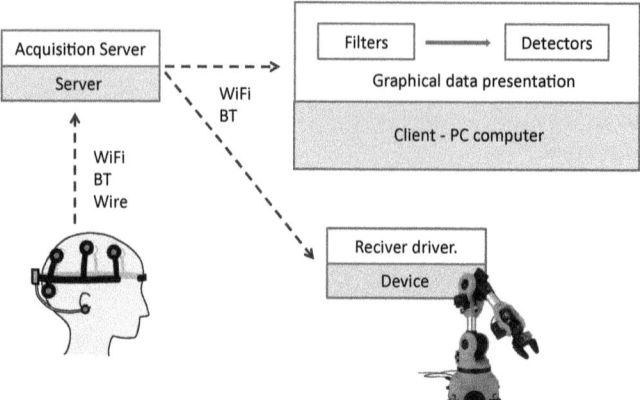

Fig. 2. The client server architecture of the OpenViBE project. The Acquisition Server transmits obtained data from the BCI device to clients, by means of WiFi.

```
virtual const IHeader* getHeader() = 0;
virtual bool start() = 0;
virtual bool stop() = 0;
virtual bool loop() = 0;
protected:
  IDriverContext& m_driverCtx;
private:
  IDriver();
};
```

The driver performs three basic functions

- configuration
- initialization
- acquisition

To control the kernel error messages the object of class **IDriverContext** is used, delivered as the constructor parameter.

The configuration method of the class is responsible for delivering all necessary parameters to establish connection with the BCI device. OpenViBE appends some additional helper class that enables the user to set the parameters, using GUI with in their applications.

The initialize method prepares the driver to receive data from the hardware. Checking if the hardware is configured and present by reading the header data.

The data transfer from the BCI device, is realized by the use of object of class implementing **IDriverCallback** set as the second parameter of the **initialize** method. The driver is expected to send equal size of data records given as the first parameter.

The names of the rest methods of the class explain their role. The continuous process of receiving data is achieved by the loop method called repeatedly.

The original code is very well commented and the reader is encouraged to review it for more details.

The only challenge to adapt the driver to other platforms, is the implementation of the initialize method. The real implementation of it is strongly hardware dependent. If the serial port is used, then it is quite easy to port the driver for popular UART's in AVR or ARM controllers. Such kind of connection one can find for the "neurosky-mindset" driver, here is partial listing of the original implementation

```cpp
bool CDriverNeuroskyMindset::initialize(const uint32_t
    nSamplePerSentBlock, IDriverCallback& callback)
{
  ...

  if (m_comPort==OVAS_MINDSET_INVALID_COM_PORT)
  {
    //Scanning COM ports 1..16
    for(uint32_t i=1; i<16 && m_comPort==
    OVAS_MINDSET_INVALID_COM_PORT; ++i)
    {
      int connectionId = TG_GetNewConnectionId();
  ...
      int errCode=TG_Connect(connectionId, ss.str().
    c_str(), TG_BAUD_9600, TG_STREAM_PACKETS);
      if (errCode >= 0)
      {
        //ask for a packet for a while.
        bool comPortFound = false;
        while (!comPortFound && !timeOut())
        {
          errCode = TG_ReadPackets(connectionId, 1);
          if (errCode >= 0)
          {
            m_driverCtx.getLogManager()<<"Status:OK\n";
            m_comPort = i;
            comPortFound = true;
          }
          else { System::Time::sleep(1); }
        }
  ...
      }
  ...
      TG_FreeConnection(connectionId);
    }
  }
```

. . .

```
    return true;
}
```

The dots in the code stand for the skipped lines. As one can see, the presented part of the code scans some range of serial ports, trying to find the right one, by opening it and then trying to read some data. The functions TG_Connect, TG_FreeConnection and TG_ReadPackets come from API of Neurosky driver and they should be replaced by those opening directly UART port and perform reading from this port.

From the presented code, it follows that the connection baud rate is 9600. But this parameter as well as parity checking, and stop bits can be easily discovered using logic analyzer or oscilloscope.

4   Signal Processing Modules

Having the signals from the BCI sensors, it is necessary to perform some processing go get the information needed to control a device. A nice feature of these modules, which allows quick adaptation, is their independence on the hardware. There are basically two types of these modules. The one for filtering data, to allow more reliable decoding, and those which do this task.

Practically the modules are realized as a set of mathematical transformations, that can be freely linked. They are designed in such a way to enable the output of any of those modules to serve as the input to the other one.

In the OpenViBE project signal processing modules are the application plugins, and fulfil the concept of independent black boxes. Wide range of implemented modules can be found in "sdk/plugins/processing/signal-processing/" and "extras/contrib/plugins/processing/signal-processing/" folder. One can find there modules for denoising, filtering, epoching, averaging, classification, feature extraction, selection and many other for various advanced mathematical transformation and data analysis.

All the signal processing modules are derived from the interface class IAlgorithm or IBoxAlgorithm. The interface IAlgorithm is very simple

```
class OV_API IAlgorithm: public IPluginObject
{
public:
  virtual bool initialize(Kernel::IAlgorithmContext&
    ctx) { return true; }
  virtual bool uninitialize(Kernel::IAlgorithmContext&
    ctx) { return true; }
  virtual bool process(Kernel::IAlgorithmContext& ctx)
    = 0;
};
```

The real job for signal processing in this class is done by **process** method.

The second interface has more methods

```
class OV_API IBoxAlgorithm: public IPluginObject
{
public:
  virtual uint64_t getClockFrequency(Kernel::
    IBoxAlgorithmContext& ctx) { return 0; }
  virtual bool initialize(Kernel::IBoxAlgorithmContext&
    ctx) { return true; }
  virtual bool uninitialize(Kernel::
    IBoxAlgorithmContext& ctx) { return true; }
  virtual bool processClock(Kernel::
    IBoxAlgorithmContext& ctx, Kernel::CMessageClock&
    msg) { return false; }
  virtual bool processInput(Kernel::
    IBoxAlgorithmContext& ctx, const size_t index) {
    return false; }
  virtual bool process(Kernel::IBoxAlgorithmContext&
    ctx) = 0;
};
```

The additional method **getClockFrequency** is used for algorithms that are triggered on clock signals, it returns the desired frequency for invoking **processClock**. Function **processInput** should be called each time new input data arrive. It is up to the algorithm to decide whether to call the **process** function for this data, or keep them in the buffer waiting for more.

The adaptation of the modules to ARV, ARM controllers is straightforward. To realize the periodic calling of the **processClock** method, a timer interrupts can be used, with the intervals obtained by calling **getClockFrequency**.

The communication between modules of the whole system is realized by the object of class **IAlgorithmContext** in the case of **IAlgorithm** or **IBoxAlgorithmContext** for **IBoxAlgorithm**. To understand how the data exchange is realized the look inside these classes is necessary. The **IBoxAlgorithmContext** interface

```
class OV_API IBoxAlgorithmContext : public
    IKernelObject
{
public:
  virtual const IBox* getStaticBoxContext() = 0;
  virtual IBoxIO* getDynamicBoxContext() = 0;
  virtual IPlayerContext* getPlayerContext() = 0;
  virtual bool markAlgorithmAsReadyToProcess() = 0;
};
```

has only several methods, the input/output operations are performed by the **IBoxIO** class object returned by **getDynamicBoxContext**.

```
class OV_API IBoxIO : public IKernelObject
{
  public:
  virtual size_t getInputChunkCount(const size_t index)
    const = 0;
  virtual bool getInputChunk(const size_t inputIdx,
   const size_t chunkIdx, uint64_t& startTime,
   uint64_t& endTime, size_t& size,
   const uint8_t*& buffer) const = 0;
  virtual const IMemoryBuffer* getInputChunk(const
   size_t inputIdx, const size_t chunkIdx) const = 0;
  virtual uint64_t getInputChunkStartTime(const size_t
   inputIdx, const size_t chunkIdx) const = 0;
  virtual uint64_t getInputChunkEndTime(const size_t
   inputIdx, const size_t chunkIdx) const = 0;
  virtual bool markInputAsDeprecated(const size_t
   inputIdx, const size_t chunkIdx) = 0;
  virtual size_t getOutputChunkSize(const size_t index)
    const = 0;
  virtual bool setOutputChunkSize(const size_t index,
   const size_t size, const bool discard = true) = 0;
  virtual uint8_t* getOutputChunkBuffer(const size_t
   index) = 0;
  virtual bool appendOutputChunkData(const size_t index
   , const uint8_t* buffer, const size_t size) = 0;
  virtual IMemoryBuffer* getOutputChunk(const size_t
   index) = 0;
  virtual bool markOutputAsReadyToSend(const size_t
   index, const uint64_t startTime, const uint64_t
   endTime) = 0;
};
```

It is easy to notice that the class IBoxIO defines two memory buffers and methods operated on them. The first one is for incoming and the second for outgoing data. The OpenViBE implementation of the class IBoxIO uses STL vectors as a memory buffer, what is not common in microcontrollers, especially AVR. However, this class is quite simple, and can be implemented without STL containers.

5   Data Presentation, Executive Module

The final part of the BCI system depends on the target and is very individual. However, no matter what is the physical purpose and what kind of device is to be controlled, it is very useful at the development stage, to have some tools, allowing to perform, experiments with signal processing together with graphical presentation.

This kind of tools need to utilize GUI to give the user convenience in interaction with the application. As it was mentioned earlier, GUI applications are strongly connected with the platform. An application written for native WinAPI cannot be compiled on Linux or Mac systems. Although Linux offers some possibility to run windows application (Wine) the compatibility is rather poor.

Project BCI2000 is targed to Windows system, on the project website one can read: *BCI2000 used to run on both Linux and OS X systems for a while. However, it became increasingly difficult to identify low-latency audiovisual APIs available across platforms. In addition, most source modules depend on vendor-supplied libraries which are only available for Windows, limiting the usefulness of ports to other platforms.*

OpenViBE GUI is written in GTK2 API which are native for Linux Gnome shell. However, GTK is also ported to Windows, so with a slightly worse performance GTK applications can be compiled and used on Windows platforms.

OpenBCI GUI applications are not based on any native GUI, but utilize the Java virtual machine. Thanks to this, they work on any platform where Java is installed.

OpenViBE discussed here, contains very interesting and useful GUI application called `Designer`. It enables to build virtual BCI system using properly connected functional black boxes. This idea of `Designer` enables an easy way to test different algorithms utilized by the blocks, and choose the suitable for your project. An example of connected blocks in `Designer` is presented in Fig. 3.

The BCI data used by the `Designer` can come from a file or a real time working BCI device.

6   Final Remarks and Conclusion

The idea of the open source BCI project is to deliver useful, ready to use tools, for every interested user, and the base for further development of BCI software. All the projects fulfill both functions very well. Although, they are mainly targeted to PC computers, they are fruitful code donors to small AVR, ARM based controllers.

The most useful and easy to adopt are C/C++ written algorithms for signal processing. One can get the code for denoising, filtering, classification and feature extraction. The modular structure of source code enables simple selection of only needed part of the code.

Accompanying software enables to test different algorithms for signal processing and select the suitable parameters for them.

It is obvious that these modules communicate with others, what imposes dependecy from some outer code. However, this communication is realized by well defined abstract interfaces. This means that the dependency could elegant be resolved by implementing these interfaces.

One can face a serious problem in the case of drivers for factory BCI devices. Drivers contained in open source projects are usually adapters for original binary factory drivers. Of course, this is not a problem if in our project we are going

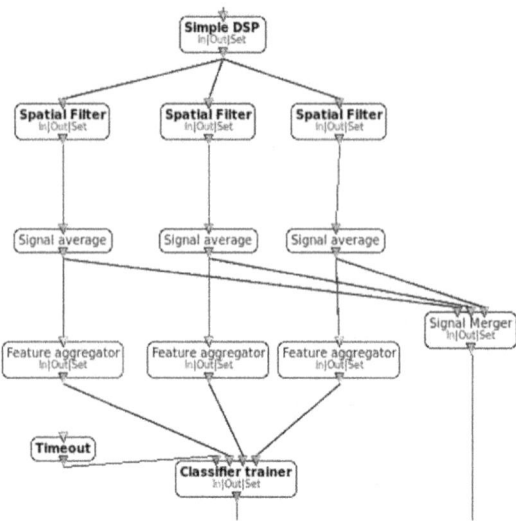

Fig. 3. An example of connected functional black boxes in OpenViBE Designer

to use scalp sensor device of own construction or those offered with source code like OpenBCI offers. It is also possible to write own driver, if the communication protocol is known or is easy to work out.

References

1. Miller, T.C.: Brain-computer interface based on magnetic particle imaging for diagnostic and neurological rehabilitation in multiple sclerosis. In: 2020 8th International Winter Conference on Brain-Computer Interface (BCI), Gangwon, Korea (South), pp. 1-6 (2020). https://doi.org/10.1109/BCI48061.2020.9061645
2. Renard, Y., Lotte, F., Gibert, G., Congedo, M., Maby, E., Delannoy, V., Bertrand, O., Lécuyer, A.: OpenViBE: an open-source software platform to design, test, and use brain–computer interfaces in real and virtual environments. PRESENCE: Virtual Augmented Reality **19**(1), 35-53 (2010). https://doi.org/10.1162/pres.19.1.35
3. Paszkiel, S.: Using the Raspberry PI2 module and the brain-computer technology for controlling a mobile vehicle, automation 2019: progress in automation, robotics and measurement techniques. In: Szewczyk, R., Zielinski, C., Kaliczynska, M. (eds.) Advances in Intelligent Systems and Computing, vol. 920, pp. 356–366. (2020). https://doi.org/10.1007/978-3-030-13273-6_34

4. Paszkiel, S., Sikora, M.: The use of brain-computer interface to control unmanned aerial vehicle, automation 2019: progress in automation, robotics and measurement techniques. In: Szewczyk, R., Zielinski, C., Kaliczynska, M. (eds.) Advances in Intelligent Systems and Computing, vol. 920, pp. 583–598. (2020). https://doi.org/10.1007/978-3-030-13273-6_54

5. Paszkiel, S.: The use of facial expressions identified from the level of the EEG signal for controlling a mobile vehicle based on a state machine, automation 2020: towards industry of the future. In: Szewczyk, R., Zielinski, C., Kaliczynska, M. (eds.) Advances in Intelligent Systems and Computing, vol. 1140, pp. 227–238. (2020). https://doi.org/10.1007/978-3-030-40971-5_21

6. Winkler, R.: Elon musk launches neuralink to connect brains with computers. Wall Street J. **5** (2017)

7. Lotte, F., Bougrain, L., Clerc, M., Lotte, F., Bougrain, L., Clerc, M., Eeg, E.: Electroencephalography (EEG)-based Brain-Computer Interfaces (2015)

8. Wyckoff, S.N., Sherlin, L.H., Ford, N.L., Dalke, D.: Validation of a wireless dry electrode system for electroencephalography. J. Neuroeng. Rehabil. **12**(1), 95 (2015). https://doi.org/10.1186/s12984-015-0089-2

9. Cohen, D.: Magnetoencephalography: evidence of magnetic fields produced by alpha rhythm currents. Science **80**(161), 784–786 (1968). https://doi.org/10.1126/science.161.3843.784

10. Cascino, G.: Functional MRI for language localization. Epilepsy Curr. **2**(6), 178–179 (2002)

11. Paszkiel, S.: Data Acquisition methods for human brain activity, analysis and classification of EEG signals for brain-computer interfaces. In: Studies in Computational Intelligence, vol. 852, pp. 3–9 (2020). https://doi.org/10.1007/978-3-030-30581-9_2

12. Paszkiel, S., Szpulak, P.: Methods of acquisition, archiving and biomedical data analysis of brain functioning, biomedical engineering and neuroscience. In: Hunek, W.P., Paszkiel, S. (eds.) Advances in Intelligent Systems and Computing, vol. 720, pp. 158–171. (2018). https://doi.org/10.1007/978-3-319-75025-5_15

13. Perego, P., Maggi, L., Parini, S.: BCI++: a new framework for brain computer interface application. In: 18th International Conference on Software Engineering and Data Engineering, pp. 37-41 (2009)

14. Schalk, G., Mellinger, J.: A Practical Guide to Brain-Computer Interfacing with BCI2000. Springer London (2010)

15. Durka, P.J., Kuś, R., Żygierewicz, J., Michalska, M., Milanowski, P., Łabecki, M., Spustek, T., Laszuk, D., Duszyk, A., Kruszyński, M.: User-centered design of brain-computer interfaces. OpenBCI.pl BCI Appl. Bull. Polish Acad. Sci. Tech. Sci. **60**(3) (2012)

Techniques, Challenges and Use in Rehabilitation Medicine of EEG-Based Brain-Computer Interfaces Systems

Wiktoria Frącz[(✉)] [iD]

Medical University of Łódź, Al. Kościuszki 4, 90-419, Łódź, Poland
wiktoria.fracz@stud.umed.lodz.pl

Abstract. Over the past few years brain-computer interface technology has developed relevantly and we can see a rapid increase of interest on the subject in applied science. Brain-computer interface (BCI) based on electroencephalogram (EEG) is noninvasive and relatively affordable method of monitoring brain activity. Collected neural signal data is used to control computational devices in real-time and must provide user with the feedback. BCI technology has found its application on many levels such as neurorehabilitation and assistive device technology arising overall quality of life for the disabled. Firstly, in this chapter overall principles of BCI technology operation are presented. Then we take a closer look at the EEG-based BCIs and their recording methods. We present also types of BCI and its essential paradigms. Moreover, main concerns and challenges in BCI technology are depicted. Finally, a use of BCI devices as Assistive Device and in Rehabilitation Medicine is described.

Keywords: Brain-Computer Interfaces (BCI) · Electroencephalography (EEG) · Motor Imagery (MI) · Neurorehabilitation · Neurotechnology · EEG electrodes

1 Introduction

Brain-computer interface technology may significantly help people with various severe disabilities in essential daily activities by enabling interaction and communication with the environment. A BCI device allows user to control computer, or for example robotic arm, using signals from brain activity, without any muscle movement [2, 21]. BCI technology are utilized into two valid directions, in which one states investigation on feedforward pathway allowing control of BCI device without the aim of rehabilitation. Second one uses the feedback loop in neurorehabilitation as a part of closed-loop BCI technology with the purpose of regulation of brain function [1]. By far among the bio-signals used in BCI, because of its noninvasive nature, inexpensiveness, high time resolution, and high portability, EEG is the most frequently used technology [1, 6, 17].

2 Principle of BCI Operation

Mental potentials are recorded with electrodes placed on the head of the subject and transmitted as commands to the external device [22]. The EEG sensors and recording

© The Author(s), under exclusive license to Springer Nature Switzerland AG 2021
S. Paszkiel (Ed.): ICBCI 2021, AISC 1362, pp. 72–78, 2021.
https://doi.org/10.1007/978-3-030-72254-8_8

amplifiers of EEG-based BCI have to deliver solid, high-quality signals. Besides that, BCI device has to be convenient and easy enough in use that people with disabilities can cope with it [14].

Use of BCI is divided into two phases in which fist one require offline training to calibrate the system. Because of the high diversity of EEG signals from different users for each of them classification algorithm is calibrated and features from many EEG channels are chosen. Second one is a closed-loop online phase during which system identifies mental activity patterns and process it into computer [12].

Generally, todays BCI technology records neural data from brain, process it via computer algorithms, translate assumed activity and present it on external device [3]. BCI devices works on the principle of few stages. Firstly, a device collect a data and clears the EEG signals using various spatial and spectral filters. Secondly, crucial neural data information is distinguish, classified and later interpreted. Lastly, the feedback is given to the user [8, 12, 14] (see Fig. 1).

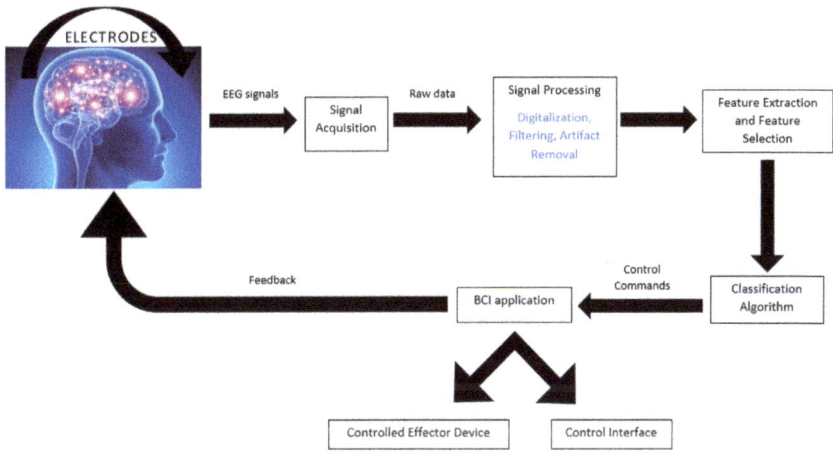

Fig. 1. Architecture and principle of BCI operation.

3   EEG-Based BCIs

Electroencephalography provides a tool able to manifest mental activity prior to, during and following cognitive and motor activities [13]. EEG signals are recorded by electrodes placed on the scalp. An encephalogram use the flow of electric current synaptic excitations of neuronal dendrites collected mainly on cortex and also from deeper brain parts [1]. In contrary to invasive techniques, which relay on small neuronal populations, this noninvasive technique mainly uses averaged signals from wide range of neurons [3]. Multiple BCI based on EEG derive data from various EEG channels, which if selection is not chosen, may cause large computational cost and inferior performance [17]. The main advantages of EEG based BCI is their high temporal resolution, ease of use, equipment portability and inexpensiveness [2, 13].

The use of EEG signals in BCI apart from mentioned advantages, due to their non-stationary form, has also some downsides such as exposure to signal artifacts and possible external noise [17]. Generally EEG signals are characterized by low signal-to-noise ratio (SNR) [5]. Another concern is that factors such as posture or other physical stimuli (e.g. Pain) and mood (e.g. fatigue or frustration) can also affect EEG signals [3, 7]. For instance, up right posture is shown to affect positively on EEG quality and changes in posture may help to arise attention among tired subjects [17].

We distinguish two types of EEG-based BCIs: spontaneous and evoked [7] also called endogenous and exogenous [17]. Spontaneous EEG is created by the use of specific mental activity and evoked EEG is an outcome of nervous stimulation [7].

Because of the way of EEG signal generation, EEG-based BCI, are greatly challenged by low information transfer rates. To measure electrical potentials on the scalp electrical brain activity has to expand through the neural tissues, the skull and the scalp, which may cause obtaining EEG signal as a mixture of various sources and noises [9].

4 Recording Methods

The most commonly used form of BCI recording, described as the gold standard [15], exploit wet electrodes in which electrodes are moisten in conductive gel and placed on the scalp. Although, this method gives relatively great EEG recording for everyday use is quite inconvenient [14]. Conductive gel may be messy, application is time-consuming, requires significant precision and support [14, 15, 21].

Another method of BCI recording excludes necessity to use a gel. Dry electrodes enables a quick setup and should record mental activity from the scalp with no preparation [15, 21]. They are used in commercially available devices, giving a good outcomes in BCI applications exploit P300, SSVEP and motor imagery [21]. Despite the advantages, dry electrodes provide lower signal quality and are more prone to motion artifacts than standard wet electrodes [15, 21].

On the market, there are also devices that use wet electrodes without conductive gel like the Emotiv EPOC, which uses moistened felt pads and a semi-rigid support. Although, the system enables quick application it remains less precise than conventional placement methods [14, 21].

Moreover, BCI device itself should properly secure electrodes to avoid disturbances caused by head positioning and be suitable for various head shapes and sizes [21].

5 Types of EEG-Based BCIs

The EEG signals in BCI technology are focused on mainly decoding motor imagery, whole-body kinetics and other senses and it is extremely important to choose the most adequate BCI application for its purpose [1]. The most basic and widespread types of BCI are MI-BCI using sensorimotor rhythms modulated by the motor imaging task, ERP based on event-related potentials induced by visual paradigms [16] and SSVEP based on steady-state visual evoked potentials [22].

Decoding algorithms of neural response are special for each kind of BCI technology and works on the basis of "user to system and system to user", which basically means that the system must adapt and self-calibrate to the new users [3].

5.1 MI EEG-Based BCIs

Motor imaginary (MI) BCI is currently the most popular type based on EEG signals [10]. MI EEG-based BCIs are an endogenous type and exploits sensorimotor rhythms [4, 20]. MI is described as the conscious mental simulation of actions, which affects areas in brain reliable for generating movement [20]. For instance, using motor-imagery data a user can move a limb by imaging it. In comparison to other BCI paradigms MI BCI enables user to communicate directly with environment without external stimulus or any motion [4], which makes them more convenient and practicable [11].

Two commonly used MI paradigms are sensorimotor rhythms (SMR) and imagined body kinematics (IBK) [1]. Amplitude of EEG oscillatory components from the sensorimotor cortex is modulated by MI events by decrease also referred as event-related desynchronization (ERD) and increase also referred as event-related synchronization (ERS) [11].

MI BCI states a system rather hart to master and to decrease the number of those who are unable to do so new co-adaptive methods are being used. For instance, new algorithms arise BCI precision by adjusting to the user and in order to obtain more control over BCI device users can modulate their sensorimotor rhythms [16].

5.2 ERP and SSVEP

Event-related potentials (ERP) are obtained by averaging neural signal of appropriate event type [1]. Common method to provoke visual evoked potential in subjects is observation and calculation of projected flashing letters. It is said that proper outcomes can be achieved rather quickly and doesn't change over time but can be changed by distraction, routine and tiredness of the subject [16]. The most common visual paradigms are P300 and Steady State Visual Evoked Potential (SSVEP) [1]. The main advantage of SSVEP and P300 paradigms is that it doesn't demand training, has high bit rate and great pattern classification [22].

6 Challenging Topics in BCI

6.1 Challenges and Limitations Faced in Research and Development

EEG data is highly non-linear, susceptible to artefacts and is collected in nonstationary way, which makes it difficult to choose a technique to properly extract and select biosignal [17]. Quite significant concern in BCI technology is dependence between recorded brain activity and essential models of the human body, biomechanics, and cognitive processing [1]. EEG based BCI can simply detect and collect neural signal data, but a true challenge is to differentiate essential signal and process it properly [19].

We can distinguish some challenges in BCI classification, which can occur after source reconstruction, such as: missing information, nonstationarity appearing as challenges occurring over the time of use and differences in brain activity of various users [9].

Another concern is combination of the data from various EEG channels to improve classification results and diminish data dimensionality [17]. This is why to decrease computational cost reduction of channels is necessary [2].

6.2 Challenges in Mastering the BCI Device

Other challenge is the training time necessary for user to efficiently and fluently use the system. BCI paradigms usually require long lime of training, which may cause tiredness among subjects [1, 19]. Nowadays calibration is time-consuming and states rather unpractical system [17]. Furthermore, even during series of session with the same subject, researchers may need to calibrate the systems for each session [1]. Interface has to adapt even in short intervals to prevent deterioration of the BCI precision. Changes may be caused by the factors such as: brain plasticity, diseases, fatigue, mood changes etc. [9].

Transfer learning works on a principle of collected knowledge from process of learning a given task for solving other related task and depends on the degree of association of this tasks. This technique may be an essential element for calibration-free BCIs [12] and based on this, new research try to create general BCI training model suitable for most subjects [1]. Another thing is that, to shorten time of training researchers attempt to adapt matrices associated with CSP from EEG sample, which may help decode EEG signals [17].

7 Challenging Topics for BCI Users

From the practical point of view, BCI users have to face some problems especially the ones associated with reduced control over BCI device. This devices are not only hard to master but theirs direct effects aren't completely predictable. Some actions may occur involuntary despite of subject intentions due to the brain sates. Another thing is that, BCI processing may modulate the output and a user can obtain only a partial control over device [19].

8 Use of BCI as Assistive Device and in Rehabilitation Medicine

In recent years, EEG signals have started to be used not only for service communication aid systems and devices such as wheelchairs but also as a brain-controlled rehabilitation and controlling assistive devices [1]. Assistive BCI technology can significantly help many people with their everyday live and arise overall quality of life irrespective of their disability degree [2].

Unlike BCIs used as assistive devices, BCI intended for neurorehabilitation have to be rather efficient than present perfect performance, which means they should at least increase the effects of rehabilitation achieved by traditional methods [14]. EEG-based BCIs are said to be able to restore neuromuscular bypass [22]. Motor and somatosensory BCI technology can help rehabilitate those injured by amputation or neurologic injuries such as stroke, spinal cord injury and motor neuron disease. Restoration and replacement of affected neurologic functions are key roles of BCI in rehabilitation [3]. In addition, it was also notify that long-term use of BCI-controlled exoskeleton improves gait of those with spinal cord injury [14].

BCI helps communicate on many ways, such as through computer, mobility devices, neuroprosthetic limbs, orthoses and environmental control units. Besides that, BCI is shown to aid restore brain plasticity by synchronization of brain activity with an actual move or sensation created in device [3].

BCI may support recovery in many ways. D. Mcfarland and J. Wolpaw [14] summarized collected so far information in this field. Among other things, they mention ideas of BCI to strengthen the use of spared neural representations and demonstrate adequate preparation of the subject to perform specific movements. They also talk about standardization of mental activity based on BCI feedback, which may adjust motor function. Another strategy is to use brain activity for training more normal control over muscular and nervous system with the aim to create plasticity improving this control [14].

9 Summary

Deep learning (DL) is new way of learning data representation, which provides simultaneously learning of features and the classifiers from raw EEG data and is a machine learning algorithm [12, 18]. DL techniques has been applied to model brain activity in the last few years [5] and are said to have superb performance in different medical applications [10]. Although, a part of the idea of creating DL is based on human brain it isn't its direct representation [18]. Spatial connectivity of EEG become another interest in the field. For instance to obtain more precise emotion recognition model, combination of the spatial connectivity of various brain regions and transitional alteration of EEG signals is being used [5].

BCI technology can be an excellent cognitive experience, due to its ability to directly alter brain signals [16]. Neurofeedback is an operation that can ameliorate cognitive control over self-regulation of brainwaves [1]. To develop control over BCI devices subject can alter brain pattern by using received neurofeedback derived from their brain activity [16]. In the future this kind of biofeedback could be potentially use for example to: mitigate cognitive and pathological neural diseases, replace or decrease side effects of medications, help fight with addictions, obesity, autism, and asthma. In addiction future research may open the way to use of obtained neurofeedback for neurorehabilitation of cognitive deficits, such as ADHD, anxiety, Alzheimer's disease, traumatic brain injury, epilepsy and post-traumatic stress disorder [1].

References

1. Abiri, R., Borhani, S., Sellers, E.W., Jiang, Y., Zhao, X.: A comprehensive review of EEG-based brain-computer interface paradigms. J. Neural Eng. **16** (2019). https://doi.org/10.1088/1741-2552/aaf12e
2. Attallah, O., Abougharbia, J., Tamazin, M., Nasser, A.A.: A BCI system based on motor imagery for assisting people with motor deficiencies in the limbs. Brain Sci. **10**, 1–25 (2020). https://doi.org/10.3390/brainsci10110864
3. Bockbrader, M.A., Francisco, G., Lee, R., Olson, J., Solinsky, R., Boninger, M.L.: Brain computer interfaces in rehabilitation medicine. PM R **10**, S233–S243 (2018). https://doi.org/10.1016/j.pmrj.2018.05.028
4. Cho, H., Ahn, M., Ahn, S., Kwon, M., Jun, S.C.: EEG datasets for motor imagery brain-computer interface. Gigascience **6**, 1–8 (2017). https://doi.org/10.1093/gigascience/gix034

5. Jin, L., Kim, E.Y.: Interpretable cross-subject EEG-based emotion recognition using channel-wise features. Sensors **20**(23), 6719 (2020). https://doi.org/10.3390/s20236719
6. Gannouni, S., Belwafi, K., Aboalsamh, H., Alsamhan, Z., Alebdi, B., Almassad, Y., Alobaedallah, H.: EEG-based BCI system to detect fingers movements. Brain Sci. **10**, 1–14 (2020). https://doi.org/10.3390/brainsci10120965
7. Kevric, J., Subasi, A.: Comparison of signal decomposition methods in classification of EEG signals for motor-imagery BCI system. Biomed. Signal Process. Control **31**, 398–406 (2017). https://doi.org/10.1016/j.bspc.2016.09.007
8. Lawhern, V.J., Solon, A.J., Waytowich, N.R., Gordon, S.M., Hung, C.P., Lance, B.J.: EEGNet: a compact convolutional neural network for EEG-based brain-computer interfaces. J. Neural Eng. **15** (2018). https://doi.org/10.1088/1741-2552/aace8c
9. Lindgren, J.: As above, so below? Towards understanding inverse models in BCI. To cite this version : HAL Id : hal-01669325 (2017)
10. Liu, X., Shen, Y., Liu, J., Yang, J., Xiong, P., Lin, F.: Parallel spatial – temporal self-attention CNN-based motor imagery classification for BCI. **14**, 1–12 (2020). https://doi.org/10.3389/fnins.2020.587520
11. Lo, C.C., Chien, T.Y., Chen, Y.C., Tsai, S.H., Fang, W.C., Lin, B.S.: A wearable channel selection-based brain-computer interface for motor imagery detection. Sensors (Switzerland) **16**, 1–14 (2016). https://doi.org/10.3390/s16020213
12. Lotte, F., Bougrain, L., Cichocki, A., Clerc, M., Congedo, M., Rakotomamonjy, A., Yger, F.: A review of classification algorithms for EEG-based brain – computer interfaces: a 10 year update (2018)
13. Maszczyk, A., Dobrakowski, P.., Żak, M., Gozdowski, P., Krawczyk, M., Małecki, A., Stastny, P., Zajac, T.: Differences in motivation during the bench press movement with progressive loads using EEG analysis. Biol. Sport **36**, 351–356 (2019). https://doi.org/10.5114/biolsport.2019.88757
14. Mcfarland, D.J., Wolpaw, J.R.: ScienceDirect EEG-based brain – computer interfaces. Curr. Opin. Biomed. Eng. **4**, 194–200 (2017). https://doi.org/10.1016/j.cobme.2017.11.004
15. Minguillon, J., Lopez-gordo, M.A., Pelayo, F.: biomedical signal processing and control trends in EEG-BCI for daily-life: requirements for artifact removal. Biomed. Signal Process Control **31**, 407–418 (2017). https://doi.org/10.1016/j.bspc.2016.09.005
16. Nierhaus, T., Vidaurre, C., Sannelli, C., Mueller, K.R., Villringer, A.: Immediate brain plasticity after one hour of brain–computer interface (BCI). J. Physiol. 1–17 (2019). https://doi.org/10.1113/JP278118
17. Padfield, N., Zabalza, J., Zhao, H., Masero, V., Ren, J.: EEG-based brain-computer interfaces using motor-imagery: techniques and challenges. Sensors (Switzerland) **19**, 1–34 (2019). https://doi.org/10.3390/s19061423
18. Paszkiel, S.: Using neural networks for classification of the changes in the EEG signal based on facial expressions. In: Analysis and Classification of EEG Signals for Brain--Computer Interfaces. Springer International Publishing, Cham, pp. 41–69 (2020). https://doi.org/10.1007/978-3-030-30581-9_7
19. Rainey, S., Maslen, H., Savulescu, J.: When thinking is doing: responsibility for BCI-mediated action. AJOB Neurosci. **11**, 46–58 (2020). https://doi.org/10.1080/21507740.2019.1704918
20. Salman, H., Grover, J., Shankar, T.: Hierarchical reinforcement learning for sequencing behaviors. **2733**, 2709–2733 (2018). https://doi.org/10.1162/NECO
21. Spüler, M.: A high-speed brain-computer interface (BCI) using dry EEG electrodes. PLoS ONE **12**, 1–12 (2017). https://doi.org/10.1371/journal.pone.0172400
22. Zhang, W., Tan, C., Sun, F., Wu, H., Zhang, B.: A review of EEG-based brain-computer interface systems design. **4**, 156–167 (2018). Review Article. https://doi.org/10.26599/BSA.2018.9050010

EEG Analysis and Neurofeedback Therapy of Concentration Problems in Mother and Child

Magda Zolubak and Szczepan Paszkiel$^{(\boxtimes)}$ (iD)

Faculty of Electrical Engineering, Automatic Control and Informatics,
Opole University of Technology, Prószkowska 76, 45-758 Opole, Poland
s.paszkiel@po.edu.pl

Abstract. Neurofeedback is a therapy of attention and concentration disorders adapted to the individual needs of the patient. Although problems with attention and memory can occur in both children and parents - with or without similar underlying causes - individual training protocols should be used for each person. The estimated duration of neurofeedback therapy is usually no less than 7 h of training. The present study reports on a 35-year-old mother and her 7-year-old child, presenting with the same cognitive problems. Despite individual training protocols, neurofeedback training yielded similar results across parent and child within the non-dominant hemisphere.

Keywords: EEG · Analysis · Neurofeedback · Concentration · Therapy

1 Introduction

Biofeedback - or biological feedback, provides information about changes within the human organism using recording devices. Biofeedback is most often used as a therapeutic or diagnostic tool. The principle underlying biofeedback is based on the measurement and analysis of signal emitted by the body, most often as an electric signal [1]. Biofeedback-EEG is a widely used therapeutic method among other biofeedback methods such as: GSR (Galvanic Skin Response), HRV (Heart Rate Variability), RSA (Respiratory Sinus Arrhythmia), tDSC (Transcranial direct current stimulation).

In 1981, Sterman presented research suggesting that observation variation in electroencephalography (EEG) signals relates to the Sensory Motor Rhythm (SMR). In particular, Sterman used data collected from cats and humans to show that electrical activity of the brain can be measured objectively, and that these electrical patterns can be used to classify pathology. He also demonstrated that these brain waves can be manipulated [3, 4]. Neurofeedback was first effectively applied to children with attention deficit hyperactivity disorder (ADHD) as a form of therapy, as early as 1995. In these studies, the researchers demonstrated that neurofeedback therapy was associated with a decrease in microvolt levels of Theta, and an improvement in IQ scores from pre- to post-treatment [5].

© The Author(s), under exclusive license to Springer Nature Switzerland AG 2021
S. Paszkiel (Ed.): ICBCI 2021, AISC 1362, pp. 79–93, 2021.
https://doi.org/10.1007/978-3-030-72254-8_9

To-date, the majority of research and therapy that uses neurofeedback has focused on treatment of specific disorders such as ADHD, obsessive-compulsive disorder (OCD), epilepsy, and autism. The use of more sensitive recording equipment allows for more accurate measurements, which enables researchers to formulate more specific conclusions and define specific training protocols for specific disorders [6–8]. In addition to analyzing the EEG signal in the time domain, other analyses are typically carried out in more recent studies, for e.g. FFT (Fast Fourier Transform) and brain mapping [9]. Clinical EEG is one of several methods that can be used to acquire data from the human brain. Clinical EEG was first introduced by Hans Berger, a German psychiatrist, in the 1930s. EEG is a noninvasive method that can be used to detect and register electrical activity of the brain, using electrodes attached to the scalp. These electrodes register changes in electric potential on the skin surface that arise from the activity of cerebral neurons. The electrical potential is subsequently amplified to form a record which is known as an encephalogram. The value of the potential registered by consecutive electrodes can be described by Eq. (1) [2].

$$V_n = V_{EEGn} + V_{CMS} \tag{1}$$

where: V_n – potential value recorded on electrodes, V_{EEGn} – potential connected with electrical activity of the brain. V_{CMS} – common signal on all electrodes, which are inter-connected.

1.1 Neurofeedback

Neurofeedback is an EEG-based method that measures bioelectricity of the brain. It should be noted that EEG is conventionally a qualitative method. However, quantitative EGG (or QEEG) can be used for biofeedback-EEG, and applied for therapeutic or diagnostic purposes. Similar to qualitative methods, the basis of the quantitative method is the measurement of the brain's electrical activity. In QEEG, in particular, brain electrical activity is more precisely captured as a potential difference, which is transformed into a record of voltage changes expressed in hertz (Hz) for individual neuronal activities in real time. In addition to recording brainwave activities, QEEG can also capture FFT, percentage of individual waves, and wave amplitudes [1, 11]. To measure QEEG, the patient's head is connected to electrodes using the so-called 10–20 system for the head and two (or one) reference electrodes on the ear lobes. The naming of the 10–20 refers to the distances at which individual points are located on the head relative to each other. In particular, the points are determined on the basis of the circumference and length of the head, and are calculated as 20% or 10% of its length or width. The electrodes are connected to the computer with a special piece that converts the electrical signal into a digital one [10, 21].

1.2 Concentration Problems

Each year, the number of children and adults suffering from cognitive problems increases. Currently, an estimated 13.4% of children and adults suffer from cognitive disorders including ADHD (3.4%) and conduct disorder (2.1%) [12]. Mental problems in children can have various causes including environmental causes, those related to adolescence, or developmental disorders [13]. In the new edition of the International Classification of Disorders and Diseases (ICD-10), the section entitled "Mental disorders and behavioral disorders" will be revised. In preparation for this new edition, several studies around the globe have been conducted in order to clarify the current classification [14]. In addition to the amendment of ICD-10 to ICD-11, the growing prevalence of mental health problems prompts new research that examines therapies for cognitive disorders in children and, to a lesser extent, in adults. A growing body of research is also examining the burden of mental disorders (e.g., ADHD) on offspring [15]. However, there have been few studies examining attention and concentration disorders among children and their parents concurrently. To address this gap, the present study used QEEG to examine changes in both a child and a parent after the completion of therapy.

2 Research Study

2.1 Participants

The present study reports on a 35-year-old woman and her 6.5-year-old son. Mental health and problem behaviors in the parent and child were characterized using interviews. The mother reported problems in concentration and memory. The child also presented with concentration problems; however, the interview suggested problems related to school or ADHD, in particular.

2.2 Assessments

In addition to the patient interview, QEEG was performed prior to qualifying patients for neurofeedback therapy. The study was conducted using the following electrode locations in the 10–20 system: F3, F4, P3, P4, Fz, Cz, C3, and C4. The data were recorded in a frequency range of 1–70 Hz, with a 250 Hz sampling frequency. The amplifier used for this study was a DigiTrack BF type digital electroencephalograph. The equipment was produced by the Polish company ELMIKO (serial number: 214111876/2014). The ELMIKO system and dedicated software registers the EEG signal in the time domain and provides QEEG analysis. The measuring equipment consisted of: (1) a reference electrode, (2) cup-shaped electrodes, (3) wiring, (4) measurement interfaces, and (5) a transducer used to analyze and pre-process the signal. Figure 1 shows the measuring equipment. The study room was quiet and lit by natural daylight and the participants were sitting comfortably. Patients were instructed to refrain from moving during the examination. All parameters were adjusted so that the patients would feel comfortable.

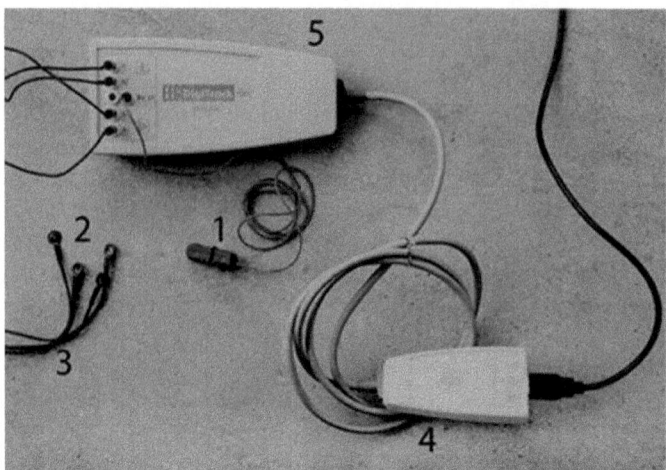

Fig. 1. The measuring equipment consisted of: (1) a reference electrode, (2) cup-shaped electrodes, (3) wiring, (4) measurement interfaces, and (5) a transducer used to analyze and pre-process the signal.

The study lasted 2 min, which included a 1-min eyes-open and 1-min eyes-closed period. This recording period provided information regarding the general functioning of the brain. During the test, the frequency range of individual waves was defined as: Delta: 0.5–4.0 Hz, Theta: 4.0–8.0 Hz, Alpha: 8.0–12.0 Hz, SMR: 12.0–15.0 Hz, Beta1: 15.0–22.0 Hz, Beta2: 22.0–50.0 Hz.

2.3 Adult Participant

During the eyes-open test (at Fz, Cz, C3, and C4 electrode locations), the amplitudes recorded with the Delta and Alpha frequency bands were within normal expected values. Theta frequency, in contrast, was higher at all measured electrode locations except for at C4. According to the literature, higher theta frequency may indicate problems with concentration and memory. SMR was in the lower limits of the normal range, and SMR was consisted to be below normal only at electrode C4. Beta2 frequency recorded at C3 was above normal. Beta2 is considered to be a "stressogenic" wave in the dominant hemisphere that can reflect long-term stress and may potentially relate to sleep problems, if any recorded at electrode location C3 [22]. Of note, this participant also reported sleep problems. Small differences in the Delta wave were observed between the eyes-open and eyes-closed conditions. Other frequencies were within normal expected values during the eyes-closed test. During the eyes-open test, the following observations were noted at electrode locations F3, F4, P3, and P4: Delta frequency was below normal in percentage, Theta frequency was slightly above normal at all measured points, SMR was in the lower limits of the normal range, Alpha frequency in the back of the head was above normal, and Beta1 and Beta2 frequency at electrode locations F3 and F4 were above normal. During the eyes-closed test, the following observations were noted at the same electrode locations: Delta frequency was in the normal range but very similar to what was observed during eyes-open, Theta and Beta2 frequency increased at all measured locations, and Alpha frequency in the back of the head was above normal. FFT was generally normal with a slight increase observed in Theta. At electrode location F3, Beta2 frequencies were above 24 Hz.

2.4 Child Participant

In the eyes-open tests, Beta2 frequencies were abnormally higher than Alpha. During eyes-closed, SMR at the C3 and C4 electrode locations were lower than Beta1. Theta frequency was above normal levels, but with a reduced percentage. Beta1 and Beta2 frequencies were recorded above normal levels. During the eyes-closed test, Beta2 frequency increased but was lower than what was recorded during eyes-open. Delta and Theta were also above normal expected levels. In the frequency analysis using FFT, the following were recorded: At electrode locations F3 and F4, increased Theta and Beta2 was observed for eyes-open; during eyes-closed, only an increase in Theta was observed. At electrode locations Fz and C4, increased Theta was observed during the eyes-open test; increased Alpha was observed at C3 and C4 during eyes-closed. The results were compared with the standards set for QEEG in adults and in children. Table 1 shows this standards [1, 3, 4].

Table 1. Table showing typical expected QEEG values.

Waves	Percent value	value [μV]	value [μV]
		Adults	Children
Delta	29	≤ 20	≤ 30
Theta	22	≤ 10	10-20
Alpha	18	≤ 10	10-15
SMR	12	5-10	5-10
Beta1	9	5-10	5-10
Beta2	9	≤ 10	5-10

3 Therapy

Therapeutic training sessions for both the adult and child lasted a maximum of 45 min. Sessions were divided into two rounds of 3-min sessions. There were breaks between sessions, and the length of the break varied based on the participants' needs. During training sessions, participants were instructed to control the game parameters that were assigned to the trained brain waves (e.g., the car should be faster and rides on the left side of the road). The therapist controlled the values assigned to game parameters. A general protocol was taken into account before starting the therapy. However, the protocol was not adapted to the individual needs of the study participants and a more individualized training was found to be more effective. Decisions regarding individual protocols were made after the interview conducted with the study group and the identification of needs.

3.1 Child

Therapeutic training sessions were focused on the Cz electrode location, as a marker of inter-hemisphere synchronization and on the C4 electrode location (in the non-dominant hemisphere) as a marker of SMR enhancement. After 12 training sessions, the values all returned to within the normal range across both the right and left hemispheres. The values for Theta frequency remained above normal levels only at electrode locations on the center line (i.e., Cz and Fz). Changes on the center line may indicate problems with concentration and focus. However, according to typical values for e.g., in SMR, there are typical changes related to the physiology and development of the child that may pass. For example, Theta frequently has been shown to correlate negatively with age in children. The (1) lack of significant changes in Theta frequency in the temporal leads and (2) the presence of a Theta/Beta ratio within normal range precludes the existence of disease entities (e.g., ADHD). Despite numerous artifacts, the signal does not have characteristics that suggest the presence of cognitive impairment.

3.2 Adult Woman

Training sessions were focused on electrode locations C3 and F3, using SMR/Beta1/Beta2 or SMR/Beta2 protocols. The goal of the adult training was to experience a sensorimotor rhythm and reduce stress levels. Most of the sessions were focused on electrode location F3 because characteristics that indicate levels of stress were highest in that location. This method of training allowed for the reduction of stress levels based on amplitude values. Frequencies were also reduced from over 25 Hz at all locations in the front of the head to 14–20 Hz. Training sessions also reduced the unequal ratio observed between SMR and Beta1 and Beta2. Most of the measured parameters in the control test, apart from Beta2 in lead leads, were found to be within the normal expected range. The slightly increased Alpha frequency may result from the appearance of characteristic sleep spindles for the Alpha frequency as observed in the EEG record.

4   Results

Across subsequent training sessions, we observed a learning curve phenomenon that is characterized by a cyclical occurrence. The first cycle occurred between 6 and 8 training sessions, and the second cycle started faster than first which suggests good brain plasticity. Figures 2 show the learning curves estimated in the child for the trained electrode locations, based on variability amplitudes. Figures 3 show the learning curves

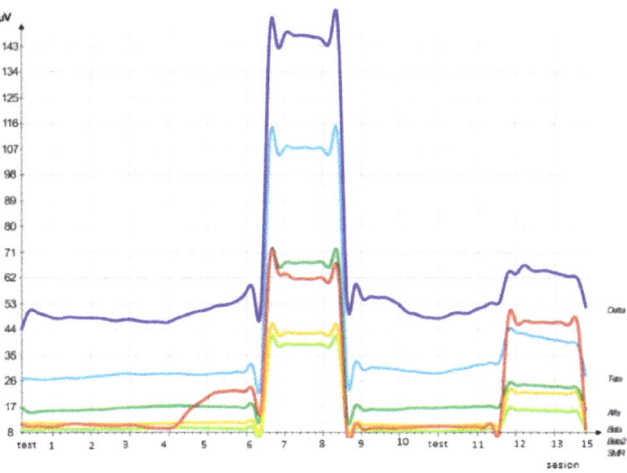

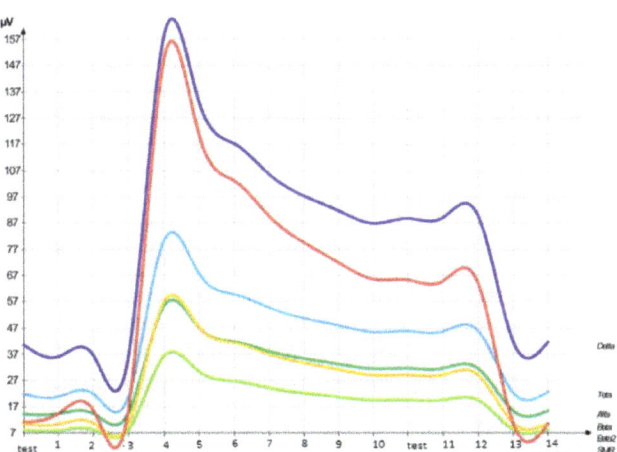

Fig. 2. Child: amplitude changes at the Cz (above on the left side) and C4 (above on the right side) electrode location, using the SMR/Theta protocol. Each session (shown on the x-axis) indicates one 45-min training period.

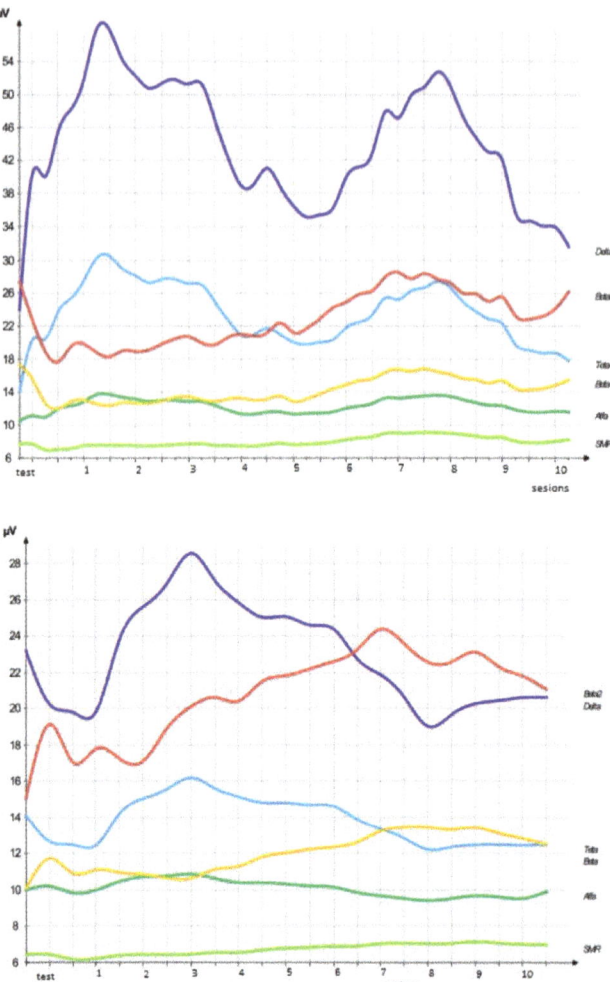

Fig. 3. Mother: amplitude changes at the F3 (above on the left side) and C3 (above on the right side) electrode location, using the SMR/Beta1/Beta2 protocol. Each session (shown on the x-axis) indicates one 45-min training period.

estimated in the adult woman for the trained electrode locations, based on variability amplitudes. An additional analysis conducted on electrode locations F3, F4, P3, and P4 showed no changes over time. Using data collected from a separate unrelated child (with a similar disorder and training protocol), there was no correlation between the child and parent's training curves.

In both cases, training in one hemisphere to raise the SMR caused an increase in the excitability of the non-dominant hemisphere. According to the literature, training of the non-dominant hemisphere is aimed at improving hemispheric cooperation [19, 20]. Tables 2, 3, 4, 5 show obtained values before and after neurofeedback therapy. Although we expected the child and mothers brains to work similarly and thus expect relatively similar patterns of results, we did expect differences based on the trained hemisphere (i.e., dominant, non-dominant). We observed that the effect of activating the non-dominant hemisphere was very similar between mother and child. In the case of both the mother and son, we expected to observe an increase in SMR and a decrease in selected waves. These predictions were observed and presented, among others, on the spectrograms (Figs. 4 and 5). Our results fit with prior studies confirming the effectiveness of neurofeedback for improving EEG recording and patient functioning [17, 18]. A novel observation from the present study is that child and mother showed a similar pattern of increased activity the non-dominant cerebral hemisphere, despite different individual training protocols.

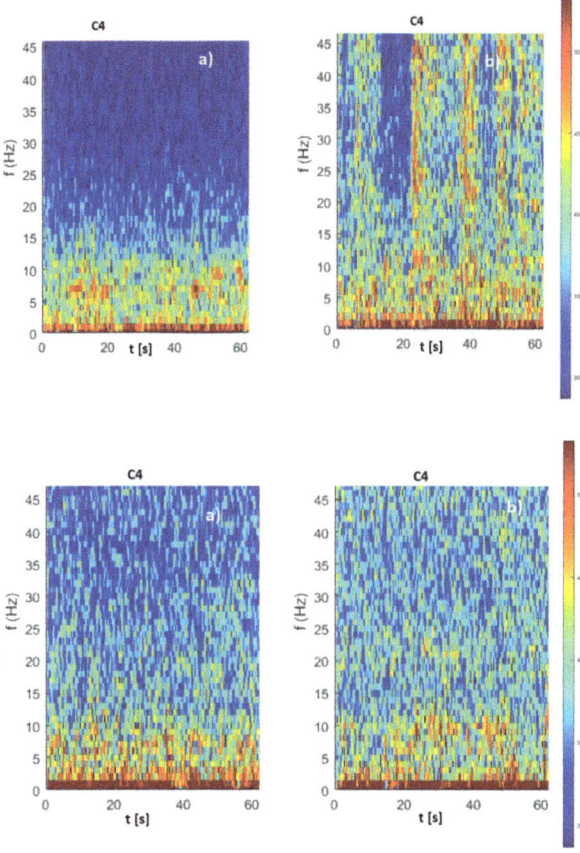

Fig. 4. EEG signal spectra registered at the C4 electrode location a) before and b) after neurofeedback therapy – for child on the left, for mother on the right.

Importantly, the increases observed in the graphs presented using data from the child and mother showed the same character and similar onset time. This similarity suggests that the brain learns with a variable frequency and the less time it takes for to learn, the faster the brain has learned something [24, 25]. In neurofeedback training manuals, this process is described as a "learning curve".

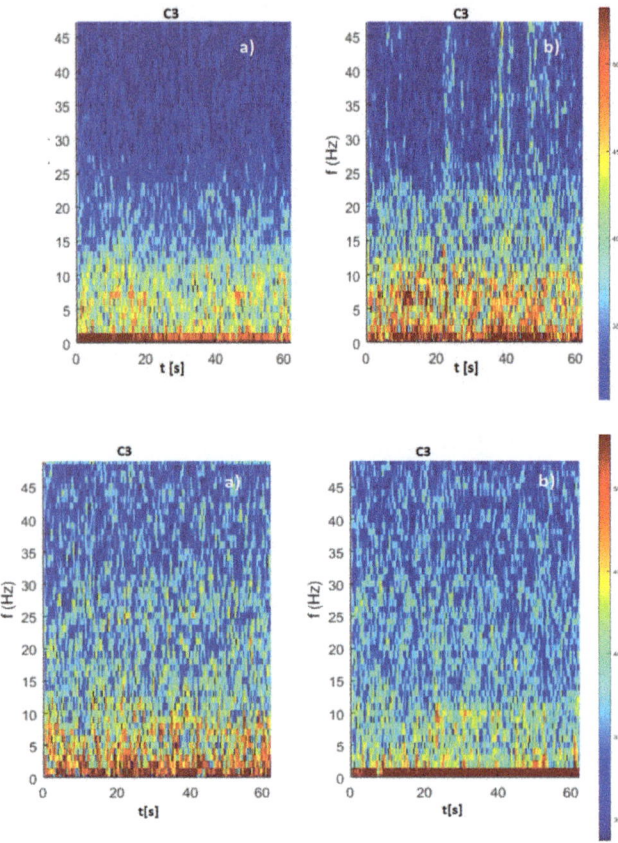

Fig. 5. EEG signal spectra registered at the C3 electrode location a) before and b) after neurofeedback therapy – for child on the left, for mother on the right.

Table 2. Obtained EEG-derived values before the child's participation in neurofeedback therapy.

Brain waves	Delta	Theta	Alfa	SMR	Beta	Beta2
Open eyes Fz						
Amplitudes [µ V]	18,93	15,48	10,84	5,39	7,32	8,54
Percentage	(28,5%)	(23,3%)	(16,3%)	(8,1%)	(11,0%)	(12,8%)
Closed eyes Fz						
Amplitudes [µ V]	18,53	15,02	12,98	5,76	7,30	8,09
Percentage	(27,4%)	(22,2%)	(19,2%)	(8,5%)	(10,8%)	(12,0%)
Open eyes Cz						
Amplitudes [µ V]	17,80	15,16	11,47	5,95	7,97	9,25
Percentage	(26,3%)	(22,4%)	(17,0%)	(8,8%)	(11,8%)	(13,7%)
Closed eyes Cz						
Amplitudes [µ V]	17,91	16,08	14,53	6,51	8,11	8,66
Percentage	(24,9%)	(22,4%)	(20,2%)	(9,1%)	(11,3%)	(12,1%)
Open eyes C3						
Amplitudes [µ V]	14,77	11,30	11,36	6,05	8,86	12,38
Percentage	(22,8%)	(17,5%)	(17,6%)	(9,3%)	(13,7%)	(19,1%)
Closed eyes C3						
Amplitudes [µ V]	14,78	12,09	12,82	6,37	8,87	10,75
Percentage	(22,5%)	(18,4%)	(19,5%)	(9,7%)	(13,5%)	(16,4%)
Open eyes C4						
Amplitudes [µ V]	11,34	8,57	9,43	4,68	7,68	10,86
Percentage	(21,6%)	(16,3%)	(17,9%)	(8,9%)	(14,6%)	(20,7%)
Closed eyes C4						
Amplitudes [µ V]	11,84	9,23	10,52	5,02	7,78	9,41
Percentage	(22,0%)	(17,2%)	(19,5%)	(9,3%)	(14,5%)	(17,5%)

Table 3. Obtained EEG-derived values after the child's participation in neurofeedback therapy.

Brain waves	Delta	Theta	Alfa	SMR	Beta	Beta2
Open eyes Fz						
Amplitudes [µ V]	49,72	40,48	14,51	9,01	10,49	10,34
Percentage	(37,0%)	(30,1%)	(10,8%)	(6,7%)	(7,8%)	(7,7%)
Closed eyes Fz						
Amplitudes [µ V]	46,38	29,04	15,58	8,43	9,72	8,44
Percentage	(39,4%)	(24,7%)	(13,2%)	(7,2%)	(8,3%)	(7,2%)
Open eyes Cz						
Amplitudes [µ V]	43,73	36,73	17,43	9,08	10,38	9,05
Percentage	(34,6%)	(29,1%)	(13,8%)	(7,2%)	(8,2%)	(7,2%)
Closed eyes Cz						
Amplitudes [µ V]	46,64	32,15	19,44	9,29	10,15	8,33
Percentage	(37,0%)	(25,5%)	(15,4%)	(7,4%)	(8,1%)	(6,6%)
Open eyes C3						
Amplitudes [µ V]	37,63	30,27	16,30	8,22	10,29	9,75
Percentage	(33,5%)	(26,9%)	(14,5%)	(7,3%)	(9,1%)	(8,7%)
Closed eyes C3						
Amplitudes [µ V]	40,18	25,27	19,94	8,61	9,93	8,41
Percentage	(35,8%)	(22,5%)	(17,8%)	(7,7%)	(8,8%)	(7,5%)
Open eyes C4						
Amplitudes [µ V]	36,98	25,44	14,37	7,50	9,62	9,68
Percentage	(35,7%)	(24,6%)	(13,9%)	(7,2%)	(9,3%)	(9,3%)
Closed eyes C4						
Amplitudes [µ V]	35,67	23,40	16,29	7,76	8,99	7,72
Percentage	(35,7%)	(23,4%)	(16,3%)	(7,8%)	(9,0%)	(7,7%)

Table 4. Obtained EEG-derived values before the mother's participation in neurofeedback therapy.

Brain waves	Delta	Theta	Alfa	SMR	Beta	Beta2
Open eyes Fz						
Amplitudes [μV]	14,39	13,79	9,46	5,06	6,83	8,05
Percentage	(25,0%)	(24,0%)	(16,4%)	(8,8%)	(11,9%)	(14,0%)
Closed eyes Fz						
Amplitudes [μV]	15,82	16,96	13,90	6,18	7,52	8,25
Percentage	(23,0%)	(24,7%)	(20,3%)	(9,0%)	(11,0%)	(12,0%)
Open eyes Cz						
Amplitudes [μV]	15,79	14,49	10,06	5,47	7,12	8,61
Percentage	(25,7%)	(23,6%)	(16,3%)	(8,9%)	(11,6%)	(14,0%)
Closed eyes Cz						
Amplitudes [μV]	16,24	18,04	14,82	6,71	8,23	9,37
Percentage	(22,1%)	(24,6%)	(20,2%)	(9,1%)	(11,2%)	(12,8%)
Open eyes C3						
Amplitudes [μV]	14,01	11,14	8,52	5,54	7,51	10,17
Percentage	(24,6%)	(19,6%)	(15,0%)	(9,7%)	(13,2%)	(17,9%)
Closed eyes C3						
Amplitudes [μV]	14,53	13,28	13,62	6,61	8,65	10,08
Percentage	(21,8%)	(19,9%)	(20,4%)	(9,9%)	(13,0%)	(15,1%)
Open eyes C4						
Amplitudes [μV]	11,76	9,96	7,24	4,29	6,48	8,38
Percentage	(24,4%)	(20,7%)	(15,1%)	(8,9%)	(13,5%)	(17,4%)
Closed eyes C4						
Amplitudes [μV]	12,25	12,07	11,27	5,55	7,36	8,19
Percentage	(21,6%)	(21,3%)	(19,9%)	(9,8%)	(13,0%)	(14,4%)

Table 5. Obtained EEG-derived values after the mother's participation in neurofeedback therapy.

Brain waves	Delta	Theta	Alfa	SMR	Beta	Beta2
Open eyes Fz						
Amplitudes [μV]	18,93	15,48	10,84	5,39	7,32	8,54
Percentage	(28,5%)	(23,3%)	(16,3%)	(8,1%)	(11,0%)	(12,8%)
Closed eyes Fz						
Amplitudes [μV]	18,53	15,02	12,98	5,76	7,30	8,09
Percentage	(27,4%)	(22,2%)	(19,2%)	(8,5%)	(10,8%)	(12,0%)
Open eyes Cz						
Amplitudes [μV]	17,80	15,16	11,47	5,95	7,97	9,25
Percentage	(26,3%)	(22,4%)	(17,0%)	(8,8%)	(11,8%)	(13,7%)
Closed eyes Cz						
Amplitudes [μV]	17,91	16,08	14,53	6,51	8,11	8,66
Percentage	(24,9%)	(22,4%)	(20,2%)	(9,1%)	(11,3%)	(12,1%)
Open eyes C3						
Amplitudes [μV]	14,77	11,30	11,36	6,05	8,86	12,38
Percentage	(22,8%)	(17,5%)	(17,6%)	(9,3%)	(13,7%)	(19,1%)
Closed eyes C3						
Amplitudes [μV]	14,78	12,09	12,82	6,37	8,87	10,75
Percentage	(22,5%)	(18,4%)	(19,5%)	(9,7%)	(13,5%)	(16,4%)
Open eyes C4						
Amplitudes [μV]	11,34	8,57	9,43	4,68	7,68	10,86
Percentage	(21,6%)	(16,3%)	(17,9%)	(8,9%)	(14,6%)	(20,7%)
Closed eyes C4						
Amplitudes [μV]	11,84	9,23	10,52	5,02	7,78	9,41
Percentage	(22,0%)	(17,2%)	(19,5%)	(9,3%)	(14,5%)	(17,5%)

5 Conclusion

Despite conducting different training sessions for similar symptoms, several regularities have emerged in the general analysis. Mother and son both reported problems with concentration, attention, and memory. In both cases, training sessions were performed in the sensorimotor cortex and using temporal leads (i.e., electrode locations C3 and C4 for mother and son, respectively). Both training protocols yielded the expected results at a similar time frame. The learning curves observed for the temporal leads were very similar. However, for the mother, the curve for all waves was not as symmetrical as was observed in her son. This asymmetry may be due to greater capacity for bioregulation in children as compared to adults. In the non-dominant hemisphere, the patterns observed for mother and child were similar despite the difference in protocols. The similarity of the training-related changes may reflect a pattern due to the parent-child affinity. In support of this interpretation, the results of another child with similar disorders and a similar training protocol were not correlated with the mother's results. In both cases, two training curves were occurred across sessions in sensorimotor cortex. Based on the data presented in this chapter, one can draw a few general conclusions, including:

– The duration of neurofeedback therapy for parents and their children will be very similar, even taking into account the lower brain plasticity of adults.
– For attention and concentration problems, training should focus on the temporal area and motor cortex area.
– The learning process ("learning curve") is similar in shape and onset between parents and their children. In this case, we observed two curves across training, which occurred concurrently in both mother and son.
– The same symptoms should be trained according to individual training protocols using the neurofeedback method. The use of generalized protocols will not be effective for both the child and parent.

Based on prior published research by our group and others, more than one "learning curve" cycle should be observed within a full training cycle (i.e., not less than 7.5 h) [16, 23]. It is necessary to work on the development of personalized therapy methods, and also personalized screening diagnostics. Importantly, the interview conducted on the mother and son suggested that the son likely has ADHD and the mother's problems may arise from excess stress. However, SMR was also reduced in the mother. In addition, the obtained correlation result of changes in EEG signals in mother and son is similar to studies on the synchronization of brain waves between parent and child during play.

Future work should extend this work to examine a large group of parents with children who have problems with attention and concentration. In particular, speed and symmetry of the learning curves obtaining in both mother and son, concurrently.

References

1. Magda, Z., Pelc, M.: Using neurofeedback as an alternative for drug therapy in selected mental disorders, biomedical engineering and neuroscience. In: Hunek, W., Paszkiel, S. (eds.) Advances in Intelligent Systems and Computing Book Series, AISC, vol. 720, pp. 69–84. Springer (2018)
2. Paszkiel, S., Szpulak, P.: Methods of acquisition, archiving and biomedical data analysis of brain functioning, biomedical engineering and neuroscience. In: Hunek, W.P., Paszkiel, S. (eds.) Advances in Intelligent Systems and Computing, vol. 720, pp. 158–171 (2018). https://doi.org/10.1007/978-3-319-75025-5_15
3. Sterman, M.B.: EEG biofeedback: physiological behavior modification. Neurosci. Biobehav. Rev. **5**(3), 405–412 (1981)
4. Sterman, M.B.: Neurophysiologic and clinical studies of sensori-motor EEG biofeedback training: some effects on epilepsy. Semin. Psychiatry **5**(4), 507–525 (1973)
5. Lubar, J.F., Swartwood, M.O., Swartwood, J.N., O'Donnell, P.H.: Evaluation of the effectiveness of EEG neurofeedback training for ADHD in a clinical setting as measured by changes in T.O.V.A. scores, behavioral ratings, and WISC-R performance. Biofeedback Self Regul. **20**(1), 83–99 (1995)
6. Ferreira, S., Pego, J.M., Morgado, P.: The efficacy of biofeedback approaches for obsessive-compulsive and related disorders: a systematic review and meta-analysis. Psychiatry Res. **272**, 237–245 (2019)
7. Sokhadze, E.M., Casanova, M.F.: Autism Spectrum Disorder: Neuro modulation, Neurofeedback and Sensory Integration Approaches to Research and Treatment (2019)
8. Reddy, J.K., Sneha, C.S.: EEG neurofeedback brain training for epilepsy to reduce seizures. Int. J. Child Dev. Ment. Health **7**(1), 28–33 (2019)
9. Paszkiel, S.: Using BCI in IoT implementation. In: Analysis and Classification of EEG Signals for Brain-Computer Interfaces. Studies in Computational Intelligence, vol. 852, pp. 101–110 (2020). https://doi.org/10.1007/978-3-030-30581-9_12
10. Zolubak, M., Kawala-Janik, A., Podpora, M., Pelc, M., Skowron, W.: Study on cancer-related cognitive dysfunction with the implementation. Stud. Log. Grammar Rhetoric **51**(64), 113–122 (2017)
11. Thomson, L., Thomson, M.: The Neurofeedback Book: An Introduction to Basic Concepts in Applied Psychophysiology, 2nd edn. Association for Applied Psychophysiology and Biofeedback, Colorado (2003)
12. Kovess-Masfety, V., Husky, M.M., Keyes, K., Hamilton, A., Pez, O., Bitfoi, A., Carta, M.G., Goelitz, D., Kuijpers, R., Otten, R., Koç, C., Lesinskiene, S., Mihova, Z.: Comparing the prevalence of mental health problems in children 6–11 across Europe. Soc. Psychiatry Psychiatr. Epidemiol. **51**(8), 1093–1103 (2016)
13. Skoguland, C., Chen, Q., Franck, J., Lichtenstein, P., Larsson, H.: Attention-deficit/hyperactivity disorder and risk for substance use dis-orders in relatives. Biol. Psychiatry **77**(10), 880–886 (2015)
14. Gaebel, W., Zielasek, J., Reed, G.M.: Mental and behavioural disorders in the ICD-11: concepts, methodologies, and current status. Psychiatr. Polska **51**(2), 169–195 (2017)
15. González-García, C., Bravo, A., Arruabarrena, I., Martin, E., Santos, I., Dell Vale, J.F.: Emotional and behavioral problems of children in residential care: screening detection and referrals to mental health services. Child. Youth Serv. Rev. **73**, 100–106 (2017)
16. Zolubak, M., Pelc, M., Kawala-Sterniuk, A.: Challenges in differentiating between attention disorders based on EEG recordings in neurofeedback therapy. In: Applications of Electromagnetics in Modern Techniques and Medicine (2008)

17. Fisher, S.F., Lanius, R.A., Frewen, P.A.: EEG neurofeedback as adjunct to psychotherapy for complex developmental trauma-related disorders: case study and treatment rationale. Traumatology **22**(4), 255 (2016)
18. Schonenberg, M., Wiedemann, E., Schneidt, A., Scheeff, J., Logemann, A., Keune, P.M., Hautzinger, M.: Neurofeedback, sham neurofeedback, and cognitive-behavioural group therapy in adults with attention-deficit hyperactivity disorder: a triple-blind, randomised, controlled trial. The Lancet Psychiatry **4**(9), 673–684 (2017)
19. Zolubak, M., Pelc, M., Siui, D., Macek-Kaminska, K., Kolanska-Pluska, J.,Ozana, S., Kawala-Stniuk, A.: Application of low frequency wave markers for diagnostic tests and neurofeedback therapy. In: Progress in Applied Electrical Engineering (PAEE)
20. Paszkiel, S., Dobrakowski, P., Lysiak, A.: The impact of different sounds on stress level in the context of EEG, cardiac measures and subjective stress level: a pilot study. Brain Sci. 10(10) (2020). Article number: 728. https://doi.org/10.3390/brainsci10100728
21. Paszkiel, S.: Characteristics of question of blind source separation using moore-penrose pseudoinversion for reconstruction of EEG signal. In: Szewczyk, R., Zielinski, C., Kaliczynska, M. (eds.) Automation 2017: Innovations in Automation, Robotics and Measurement Techniques. Advances in Intelligent Systems and Computing, vol. 550, pp. 393-400 (2017). https://doi.org/10.1007/978-3-319-54042-9_36
22. Hosseini, S.A., Khalilzadeh, M.A., Naghibi-Sistani, M.B., Niazmand, V.: Higher order spectra analysis of EEG signals in emotional stress states. In: 2010 Second International Conference on Information Technology and Computer Science, pp. 60–63 (2010)
23. Gomes, J.S., Ducos, D.V., Akiba, H., Dias, Á.M.: A neurofeedback protocol to improve mild anxiety and sleep quality. Rev. Bras. Psiquiatr. **38**(3), 264–265 (2016)
24. Paszkiel, S.: Using neural networks for classification of the changes in the EEG signal based on facial expressions. In: Analysis and Classification of EEG Signals for Brain-Computer Interfaces. Studies in Computational Intelligence, vol. 852, pp. 41–69 (2020). https://doi.org/10.1007/978-3-030-30581-9_7
25. Paszkiel, S.: The use of facial expressions identified from the level of the EEG signal for controlling a mobile vehicle based on a state machine. In: Szewczyk, R., Zielinski, C., Kaliczynska, M. (eds.) Automation 2020: Towards Industry of the Future. Advances in Intelligent Systems and Computing, vol. 1140, pp. 227–238. https://doi.org/10.1007/978-3-030-40971-5_21

Consciousness and Moral Status of Animals

Grzegorz Francuz[(⊠)]

Opole University, Opole, Poland
franc@uni.opole.pl

Abstract. Consciousness is the basis for granting moral status, but it is ephemeral and elusive. Both the ontological and epistemic dimension of consciousness cause hard problems for modern science and the philosophy of mind. On the one hand, consciousness is subjective, and includes conscious states with a phenomenal or qualitative character – "qualia". It consists of mental states which are accessible to a subject only from the first-person perspective. A being is phenomenally conscious when there is something that is like to be that being. Utilitarianism uses the hedonistic strategy of the moral status, ascribing to that the demand for us to treat sentience as the fundamental property for obtaining moral status. Sentience is not the simple reactivity to stimuli, but it is the basic kind of phenomenal consciousness available only from the inside. Interpreting the behaviour of animals, we usually apply to them the intentional stance which induces us to consider animals as if they were rational agents with beliefs and desires. We fill animal minds with the content which we have in our own minds. We project onto animals our own mental world. When we do that we apply the rationalist strategy of moral status granting, assuming that animals are like us and according to their knowledge of the world they pursue their desires in a less or more intelligent way. Stressing the great importance to consciousness as the basis of moral status is, in fact, burdened with anthropomorphism.

Keywords: Consciousness · Moral status · Hedonistic strategy · Rationalist strategy

1 The Basis of Moral Status

Living beings or natural objects are usually given a moral status by assigning to them states of consciousness typical for people, by giving them a mental state. They are personified or anthropomorphized. The decisions concerning the scope of the moral community are most often associated with beliefs concerning the psychology of beings, colloquial intuition leads to assign ethical significance above all to beings with subjective experiences, conscious beings, experiencing something "from with-in". Such beings, like people, have some form of mind. Unlike a stone or a tree, which have no inner life and experience nothing, these beings have mental states. "Traditional morality uses common-sense psychological notions to draw the boundary between the morally considerable and the inconsiderable" (Agar 2001).

The presence or absence of some form of consciousness is the basis for the division into entities that are morally valuable and entities that are deprived of such value. The

© The Author(s), under exclusive license to Springer Nature Switzerland AG 2021
S. Paszkiel (Ed.): ICBCI 2021, AISC 1362, pp. 94–106, 2021.
https://doi.org/10.1007/978-3-030-72254-8_10

non-instrumental value of an existence usually goes hand in hand with the presence of certain psychological properties that are most fully revealed in an adult human being. It can therefore be said that it is not the biological origin that determines the moral status of a given being, but having consciousness. The attribution of moral status to animals is closely linked to the existence of their minds. Only creatures with mental and sensory states have interests and those creatures cannot be reduced to their instrumental value alone.

The recognition of having consciousness or the mind as a morally important feature requires answers to questions: Are non-human beings, particularly animals, endowed with consciousness, mind? Which animals have a mind, consciousness? What mental states do animals have?

Consciousness is an awkward object of research. On the one hand, we perceive it as an element of material reality, which enters into causal relations with the surrounding objects. Therefore, consciousness can be reduced to physicochemical phenomena, for example neurophysiological processes. On the other hand, it seems absurd to identify consciousness with material processes of the nervous system. Consciousness is a subjective feeling, experiencing phenomenal qualities that are only available to the subject experiencing it. To have consciousness is connected with the perspective of the first person, with private experience of certain states, with being such a being who has an "inner world", invisible directly from an external, the so-called third person perspective of objective science. When attempting to explain consciousness in accordance with the principles of physicalism or naturalistic materialism accompanying contemporary science, we encounter the problem of the explanatory gap (Levine 1983). We are confronted with the so-called difficult problem of consciousness (Chalmers 1995).

Mental phenomena can be divided into two categories: sensations and propositional attitudes. This division does not seem to be exhaustive, but it does allow for some sort of ordering the diversity of processes taking place in the mind. Sensations are bodily experiences such as pain, itching, pressure, etc. and perceptions such as seeing, hearing, smelling. Experiences can be intentionally directed towards objects and devoid of reference to objects, they are phenomenal in nature and are consciously experienced by the subject experiencing them. The second category of mental phenomena has propositional content, i.e. they can be expressed by 'that'-clause, determining the subject's attitude to a given state of things. The set of these mental events includes affections, conative and cognitive states, i.e. believing, hoping, fearing, intending, convictions, opinions. Propositional attitudes may be unconscious by the subject who acts according to them, but observing the behaviour of a given subject (human or animal), one can presume the beliefs or approaches of that subject.

2 Two Strategies of Moral Status Ascribing

The two categories of mental phenomena mentioned above – sensations and propositional attitudes – may become the basis for two strategies for giving moral status to entities: rationalistic and hedonistic. The rationalistic strategy refers primarily to propositional attitudes. Within its framework, moral importance is given to the purposefulness of mental states, plans, projects, desires, beliefs and intentionality with its subjective reference. The whole wealth of intentional acts of consciousness is fully manifested in people who are paradigmatic mind holders. It is hard to deny that adults of our species, as subjects of morality, desire something, plan, pursue a chosen goal or have convictions. The occurrence of these acts in people, which are usually accompanied by phenomenal consciousness, is the basis for the moral importance of *Homo sapiens.*

The issue is different in the case of non-human beings, the existence of propositional attitudes in them is problematic. We cannot, therefore, automatically apply the categories that serve to describe mental processes in humans to describe processes taking place in the minds of animals. Perhaps by using terms such as 'mind', 'beliefs' and 'desires' in relation to non-human species, we are bringing under these terms completely different phenomena from those with which we are familiar from our inner experiences.

The hedonistic strategy, which derives from Bentham, recognises pain, suffering and pleasure as morally relevant characteristics which are the fundamental criterion for moral status. This is a strategy mainly used by the utilitarianists who defend the moral importance of animals. The acceptance of *sentience* as a criterion for attributing moral status to animals causes some problems. We should ask: What is *sentience*, what is it and how does this ability of living beings manifest itself? How do we recognise that a given creature has the ability to feel and experiences hedonic states (pleasure, subjective feeling of happiness, pain, suffering)? We should ask about the scope of the collection of beings that they feel and, consequently, gain moral status. Do only conscious higher animals such as mammals and birds belong to this collection? Or do other vertebrates and invertebrate animals also belong to this group? Where should we draw the line for this set, the line separating sentient and non-sentient beings?

2.1 The Concept of *Sentience*

The concept of *sentience* is not clear-cut and covers a diverse spectrum of mental phenomena. The term *sentience* is sometimes used to describe simple states of pleasure or pain or complex emotions, feelings or moods. Feeling can also be reduced to a sensitivity that is manifested both by living organisms and by artificial systems that have receptors or sensors to respond to environmental stimuli. In this context, sensitivity is the reading, detection or recognition of data. It refers to the action of receptors or sensors that occur both in machines and in living organisms. Sensitivity would then include, for example, the sensitivity of photocells and plants to light and the sensory perception of animals. This would be receptivity, which means unconscious processing of information, which is manifested in the behaviour of a functioning system. We recognise its more sophisticated forms in animals and simpler forms in plants.

The *sentience* attributed to animals *is* associated with subjectivity, with phenomenal consciousness that goes beyond receptivity. David DeGrazia writes: "Sentience is more than the capacity to respond to stimuli; it is the capacity to have at least some *feelings*. Feelings include (conscious) sensations such as pain—where 'pain' refers to something *felt* and not merely the nervous system's detection of noxious stimuli—and emotional states such as fear. We do not know at what point on the phylogenetic scale, or evolutionary tree, sentience disappears, being replaced by more primitive, non-conscious neural mechanisms" (DeGrazia 2002).

Linking sensitivity to phenomenal consciousness, we are dealing with what Chalmers calls a *difficult problem of consciousness*. Even when we discover extremely sophisticated behaviours and cognitive skills in animals, we are not able to show how they connect with subjectivity. Marian Stamp Dawkins writes: "Sentience – whether in ourselves or in other species – is and remains the 'hard problem'—harder than any other problem in biology and harder than some of us perhaps would like to admit. It is hard because we do not know what it is, where it comes from, what it does or where to find it in other species. My aim has been to argue that the way forward is to acknowledge these problems and attempt to answer our critics. We should not pretend there are no problems or that we have all the answers. We should also accept that we have to make decisions about animal sentience that are not completely watertight and can be challenged. But seeing sentience as the 'hard problem' ensures that everyone else's views on animal sentience are equally leaky and equally open to challenge (Stamp Dawkins 2006).

Subjective experiences, phenomenal consciousness are available from the perspective of the first person, they are experienced directly only from within. No one else can identify them, these states cannot be captured and identified from the perspective of the third person. In short, there is the problem of the existence of other minds, epistemological availability and communication of what is subjective. We also lack an adequate dictionary to describe the consciousness of beings that are very different from us and that do not use language.

If animals used language, they could tell us about their inner experiences and experiences, but that is not the case. It seems that without exact knowledge what is sentience and how to find it in animals, we confine to faith and intuition in this matter. The presence of subjective states in certain beings can be simply assumed a priori or accepted on faith. Recognition of the real existence of mental states which are experienced by non-human beings is associated with reference to colloquial intuitions and cultural conventions, philosophical or religion assumptions. We are dealing here with ascribing to other beings the way how people feel, with transfer of typically human experiences to animals. The internal states of non-human beings are completely beyond the reach of our cognitive authorities.

Such a situation can open the way for metaphysical speculation, which becomes the starting point for creating certain hierarchies of entities and determining the scope of sentient beings. Despite references to biological sciences or scientific research of animal organisms, pre-scientific, philosophical, religious and metaphysical presuppositions of a given culture become the ultimate reference system for determining the issues of belonging any creatures to the set of sentient beings which deserve moral concerns. That are initial assumptions which support concepts of moral status, while references to science are secondary and are intended to strengthen metaphysical decisions made earlier.

2.2 Similarities Between Humans and Animal

We can infer indirectly that both animals and other people possess minds and subjective states. Members of our species have many sophisticated ways of communicating, above all they use language, which convinces us that others are aware and feel. When we move on to animals, we highlight the similarities between physiology, nervous system, sensory or behavioural patterns. We can distinguish several elements confirming with a high degree of probability that an individual is aware of and experiences conscious hedonic states: having a nervous system and its similarity to the human nervous system; behaviours indicating sensory perception, i.e. behavioural criteria such as making certain sounds or making movements which we usually associate with feeling pain, suffering or pleasure; having perception organs such as eyes or ears, etc. Finally, the occurrence of physiological or neurochemical phenomena that we identify in people when they experience hedonic states. Of course these indicators can never give a hundred percent certainty that a creature, especially outside the Homo sapiens species, has the ability to feel pain or pleasure.

Peter Singer points also the natural origin of animal nervous systems, the evolutionary relationship between humans and non-humans and the adaptive importance of the ability to feel pain. "To back up our inference from animal behaviour, we can point to the fact that the nervous systems of all vertebrates, and especially of birds and mammals, are fundamentally similar. Those parts of the human nervous system that are concerned with feeling pain are relatively old, in evolutionary terms. Unlike the cerebral cortex, which developed only after our ancestors diverged from other mammals, the basic nervous system evolved in more distant ancestors and so is common to all of the other 'higher' animals, including humans. This anatomical parallel makes it likely that the capacity of vertebrate animals to feel is similar to our own" (Singer 2011).

Singer exposes the similarity between the nervous system of animals and humans. The greater the similarity of being to us, the more certain we can assume that it experiences hedonistic states. Although this argument about the similarity of beings to humans is based on folk psychology and intuition, and is therefore unreliable and often leads to the design of our qualities for other beings, it is the starting point for any consideration of subjective states possessed by animals. Arguments of similarity may lead to the conclusion that the should support beliefs that similar to people creatures have subjective, mental states. One may ask what degree of similarity is involved. After all, all living beings at the cellular level are similar to us. As humans, we are living organisms subject to biological conditions, organisms that are conscious and experience mental states. This is the basis for our understanding of the well-being and prosperity of beings outside our species. Although the concept of 'similarity' we use here is intuitive and imprecise, we are actually able to indicate what kind of similarity we are talking about. Although the term 'similarity' we use here is intuitive and imprecise, in fact, we are able to indicate what kind of similarity we are talking about here.

The fact that other people are experiencing pain, suffering and positive hedonic states can be almost, though not absolutely, certain. Because our species have many sophisticated ways of communication and, above all, we use language, what convinces us that other people are aware, have mental and hedonistic states. However, when we recognise the existence of subjective states in non-human beings, when we want to feel the situation of these beings or understand the world of their lives, we are forced to use our imagination. Their subjective reality is completely beyond our reach. Their way of living, their inner experience can be so distant for us that it is difficult even to imagine the states desired and definitely avoided by such beings. Can we therefore resolve the question: When does a creatures feel good and when does it feel bad?

As humans, we are living organisms subject to biological conditioning, organisms that have consciousness, that experience mental states. This forms the basis of our under-standing of the good and well-being of beings outside our species. What is the well-being of non-human entities? As Mark H. Bernstein points out: "[…] the concepts of benefit and harm, being better and worse off, and having a welfare are human concepts. They are the concepts they are in virtue of the sorts of beings we are. It is due to the fact and way that we think, perceive, and feel that these concepts (along with all other concepts) have the use and implementation that they do. Part of the use we make of these concepts of well-being is that we ascribe them to others as well as to ourselves. If another is to be credited with a well-being, with the capacities to be benefited and harmed, and these predications are to be understood in their literal and nonstipulative manifestations, then, at the very least, the subjects for these attributes must be enough like us to sustain these attributions. If another is said to be harmed, it must, as a minimal matter of intelligibility, be harmed in ways analogous to the ways in which we humans – the inventors, users, authorities, and paradigms of these concepts – are harmed" (Bernstein 1998). People as users of the terms "benefit," "prosperity," "harm" must understand what it is like for another to be better off or worse off, or to suffer harm. "Wrongness," "benefit," or "harm" in order to retain their meaning and intelligibility should connect in some minimal way to the understandings of harm and benefit that we adopt as humans, and to the contexts in which we use these terms in relation to ourselves.

Stamp Dawkins, who is a biologist, argues that the question about the animal well-being can be resolved by intensive research and meticulous testing of animals under different environmental conditions in order to learn about the preferences and ways in which animals 'value' their environment or 'make choices'. Biological knowledge of the organisms tested should be used for this purpose. In this way, animals can "answer" our questions about their welfare, suffering or pleasure and to some extent it is possible to avoid animals being forced into the corset of human thought categories. It will, of course, not be possible to penetrate the inner world of non-humanes. However, referring to mentalistic terminology derived from common psychology, to subjective states and thought processes is the best way to explain animal behaviour so far. Two positions should be avoided: attributing too rich, human-like psychology to animals and depriving them, as Frans de Waal puts it, of extreme anthropomorphism and anthropodenial (De Waal 2006).

Recognition of hedonic states as the sole basis of moral status leads to so-called experientialism, namely the belief that the ultimate criterion of an individual's prosperity is constituted only by how the given individual feels from the inside. However, according to this view every sentient being can be artificially introduced into positive mental states by pharmacological stimulation, or by electrodes connected to the nervous system, which are totally sufficient for the individual's welfare.

Experientialism is completely insensitive to the source of the internal states that determine wellbeing of an individual. When we consistently stick to experimentalism, there is no difference between the well-being of individuals with the same internal states and subjective feeling, but with completely different relationships to the surrounding world. What matters is whether someone experiences the beauty of nature during a mountain trip or under the influence of artificial brain stimulation. Only the experiencing subject is able to assess his or her condition in the best way, and if he or she never learns that his or her experiences have no connection with the real being, he or she does not lose or suffer any damage. In accordance with this, if we were to choose between a real life without pleasure and full of inconvenience, or a created, fictional life full of positive experiences, we should choose the latter. It does not matter what the source of the experience and its history is, it is only the subjective experience that matters. However, it is difficult to agree with such an attitude. Fulfilling someone's desires is connected not only with internal satisfaction, but also with the realisation of objective, external states.

Interpreting the behaviour of animals, we usually apply to them the intentional stance which induces us to consider animals as if they were rational agents with beliefs and desires. We fill animal minds with the content which we have in our own minds. We project onto animals our own mental world. When we do that we apply the rationalist strategy of moral status granting, assuming that animals are like us and according to their knowledge of the world they pursue their desires in a less or more intelligent way. Stressing the great importance to consciousness as the basis of moral status is, in fact, burdened with anthropomorphism, and leads to paradoxical consequences. Hence it seems when moral status is being assigned we should go beyond consciousness.

Despite the difficulty of capturing the phenomenal consciousness in animals, it can be pointed out that they certainly have access-consciousness - A-consciousness. A-consciousness is consciousness in the functional sense, it refers to the behaviour of a system e.g. a living organism that processes information, orientates itself and works well in its environment. Processes of this kind do not have to be accompanied by phenomenal consciousness including subjective states which are inexpressible in functional terms (Block 1995). A creature who has only access consciousness has psychological events, but without phenomenal consciousness (P-consciousness). P-consciousness has mainly subjective character, it is the process of experiencing from the inside from the first person perspective. The creature with only access-consciousness works efficiently, processes information, reacts to stimuli, functions in the environment as an organism or even artificial system, performs rational operations. However, it does not experience anything, it does not have its own internal perspective and conscious experiences. In a word, it is not aware of its mental processes. For example, we can see this case in the phenomenon of blindsight, which occurs in people with damaged brains when they react to visual stimuli, recognise shape, movement or even colour, without realising that they can see. The same happens when we perform habitual activities 'on the autopilot', for example when closing a house door or driving along a known route, which we do not remember later. According to folk psychology, having only access consciousness or mental processes without phenomenal consciousness is in fact a lack of consciousness, an automatic action which, although it may be intelligent and sophisticated, is unconscious. This is how computers or machines that process information work, for example. Is it therefore legitimate to attribute consciousness to them? Can processes that are denoted by the term A-consciousness be called consciousness at all?

The creature which does not have a phenomenal consciousness, but is physically and functionally identical to humans, is called by philosophers a philosophical zombie. It reacts and acts in the same way as humans, its external structure and material structure is no different from that of the human body. Just as we do, it also receives and processes information, the same perception processes take place in it as in humans. However, none of these processes are accompanied by conscious experience, and there are no phenomenal feelings. The zombie has perception and consciousness in the functional sense, there is only access consciousness. His visual perceptions can be compared to unconscious recording of images by the camera, which is a part of the information-processing machine. We cannot tell what it is like to be a zombie (Kirk 2005). Empathy or conscious embedding into the inner states of a zombie is doomed to failure. Such attempts are like trying to empathize with subjective world of a television set or a computer. There is no one inside, the zombie acts as an unconscious automaton, which functions efficiently. In this context it is worth asking questions: Do non-human beings, which everyday psychology bestows on the mind, have phenomenal consciousness, i.e. subjective feelings, or only access consciousness? Do animals resemble zombies, Cartesian automata which function well but have no subjective feelings? Which animals have subjectivity and feelings and are therefore more than just zombies? Where in the animal world can we draw the line between sentient beings and functioning animals that only follow innate and conditioned patterns and are like zombies? Can a creature that functions like a zombie be given moral status?

As far as living organisms are concerned, we are not able to identify clearly which are simply like zombies and which have subjective experiences. It is therefore difficult to define the boundaries of a set of sentient beings capable of suffering, which we should not inflict pain on and we can sympathise with. Within the framework of the hedonistic strategy, creatures functioning as zombies do not have a moral status, as they do not experience either pleasure or pain from within. For example, pain, which plays a key role in feeling, can, in the case of these creatures, be functionally perceived as a sign of tissue damage or disturbance in physiology. However, pain as understood in this way does not play a role within the hedonistic strategy of moral status.

One can reject the hedonistic strategy and recognise that animals only have an aware-ness of access, and deny them a phenomenal consciousness. This is what Peter Carruthers does, claiming that all mental states of non-humans are probably unconscious and there-fore devoid of internal phenomenology and subjective feeling. We have no idea of how non-human beings can experience if these beings do not have subjectivity. In short, we don't know what it is like to be a non-human animal. If non-human beings do not have an inner phenomenology and are not subjective, can they be of moral concern? Is sympathy for the suffering of animals justified? Animals without internal phenomenology are not sentient beings, but act as automatisms like zombies. They lose their moral status. We are returning to Cartesianism, which is contrary to our moral intuitions. The Cartesian position, which denies all animals the ability to feel, contradicts contemporary biology, neurology, neurophysiology or neurosurgery, which show a number of similarities in the functioning of the human and animal nervous system This situation has its origins in a mistake that puts a strain on the hedonistic strategy. It starts from the a priori assumption that only creatures with certain psychological abilities, in this case feelings, have direct moral significance. Therefore, if, as a result of empirical research, it turns out that entities which, according to common psychology, are endowed with internal phenomenology, are deprived of it, they would lose their direct moral status. Carruthers believes that in the case of pain, it is not just the feeling of pain that has a moral meaning, but the fact that it is undesirable by the beings it affects. We should therefore not so much be interested in whether an individual feels pain, but in whether it has desires or interests that can be frustrated. Caruthers writes: "[…] it is very doubtful whether mere behavior-guiding pain percepts are really sufficient to motivate moral concern. If a creature moves itself and its limbs away from sources of bodily damage, but in a manner that is merely reflex-like, then we surely should not count it as feeling pain in the sense that matters. Some sort of cognitive and/or motivational uptake of the perceived pain would be necessary for that.

Moreover it is, in any case, neither the perception of pain nor the pain perceived that is the appropriate object of moral concern, but rather the fact that an organism (normally) very much wants the pain to go away, and is frustrated in this desire" (Carruthers 2007).

Asking about moral concern for animals Caruthers emphasises: "Sympathy rather than empathy should be one's preferred mode of responding to animal suffering in any case (supposing that one responds, or should respond, at all); and sympathy should be grounded in third person understanding of the needs and mental life of the creature in question. The question of phenomenal consciousness needn't arise. Likewise, feelings of pain and other negative mental states can exist in the absence of phenomenal con-sciousness (i.e. while lacking feel), and also if there is no fact of the matter. And such states can still be bad from the perspective of the animal. So anyone who thinks, for whatever reason, that one is obliged to prevent bad things from happening to animals can continue with their views unchanged" (Carruthers 2019).

3   The Rationalist Strategy of Moral Status Granting

The rationalist strategy of moral status granting refers primarily to propositional atti-tudes, it focuses not so much on feelings as on the existence and course of thought processes in animals. It is interested in the reference and intentionality of mental states. For instance desires have an object of reference and are connected with beliefs about what the world is like, with the perception and processing of information about the envi-ronment. We ask questions: Do animals have desires and beliefs? What does it mean for animals to have desires and beliefs? What does it mean for animals to think and realize their plans? When using words "beliefs", "desires", "thoughts" to a world of non-human creatures, do we retain the meaning of these words, which have been shaped in relation to people?

Answers to questions about the thinking and cognitive abilities of animals can be placed under two posts. The first, which is called an intellectualistic or psychological concept, emphasises that the behaviour of so-called higher animals can only be sensibly explained if we use mentalistic terms that are understood in a realistic way. "The essence of a psychological explanation is that it explains behavior in terms of how the creature in question represents its environment, rather than simply in terms of the stimuli that it detects. Psychological explanations typically make reference to how the organism per-ceives its environment, to what it believes about the environment, and to what it desires to achieve. These beliefs, desires, and perceptions allow organisms to respond flexibly and plastically to their environments – the same situations can afford different actions if a creature brings different beliefs and desires to it, or perceives it in different ways" (Bermudez 2007). Accordingly, by observing the interactions of animals with their environment, we can understand these processes due to referring to propositional attitudes. Therefore, among animals, we can find beings who, just like humans, are motivated by their inner convictions and desires. In short, there is thinking between stimuli and reactions. This involves sophisticated and deliberate actions, such as purposeful deception or long-term planning.

The second position, which could be called a computational and informational concept, sees animal thought processes as complex processes of processing and gathering information. Complex patterns of behaviour and the variety of responses to stimuli are innate and instinctive or are the result of behavioural conditioning. The use of mentalistic terms such as 'belief' or 'desire' is only a useful metaphor. According to the behavioural approach, it is assumed that all animal behaviour and reactions can be explained by conditionality. In turn, classical ethologists, who are part of this paradigm, explain the action of animals by innate sequences of behaviour that are activated by triggers. The triggering mechanisms are independent of the individual's personal history and typical of the species, are invariable and have strictly defined functions. The whole process can be understood without reference to psychologic events in animal minds.

The thesis that animals have a sophisticated mind and human-like mental states of mind is preached by cognitive ethologists. They stress that the plasticity and flexibility of animal behaviour is proof of cognitive and intellectual fitness. Animals operate in complex conditions and cannot operate only automatically. In a changing environment, they are innovative and creative. They are not just gene puppets or reflexes. Even in the case of programmed behavioural patterns, they must be able to recognise when to apply a given pattern. Donald R. Griffin writes: "Contrary to the widespread pessimistic opinion that the content of animal thinking is hopelessly inaccessible to scientific inquiry, the communicative signals used by many animals provide empirical data on the basis of which much can reasonably be inferred about their subjective mental experiences. Because mentality is one of the most important capabilities that distinguishes living animals from the rest of the known universe, seeking to understand animal minds is even more exciting and significant than elaborating our picture of inclusive fitness or discovering new molecular mechanisms. Cognitive ethology presents us with one of the supreme scientific challenges of our times, and it calls for our best efforts of critical and imaginative investigation" (Griffin 1994). In order to support his theses, Griffin evokes many observations and experiments involving animals.

The claims of cognitive ethology seem to be confirmed by many empirical examples. Its findings are eagerly referred to by ethicists who seek scientific justification for the moral status of animals. However, they do so sometimes in an uncritical and selective way to confirm their philosophical position. It should be pointed out that disputes over the cognitive abilities of animals do not concern empirical testimony at all, but the interpretation of observations. Critics of cognitive ethology accuse it of an unjustified anthropomorphism, which is heuristically acceptable, but should not be treated as a description of real-life processes. In a word, cognitive ethologists put much more into animal heads than is actually there.

3.1 Language

As people, we are doing something because we have a certain idea of the world and we want a state of affairs in the world. Thanks to language, we become aware of our beliefs or desires, and without language they remain unconscious. Do animals without the tool of language have beliefs and desires and do they have access to them? As Donald Davidson points out, the desires and beliefs as proposal attitudes have holistic character and they are connected in a meaningful network. One belief requires a broad background of other beliefs, otherwise it would be completely incomprehensible. This network of interlinked propositional attitudes is only possible through language. Davidson, referring to attribution 'thoughts' to a dog that chases a cat sitting on a tree, says: "[…] can the dog believe of an object that it is a tree? This would seem impossible unless we suppose the dog has many general beliefs about trees: that they are growing things, that they need soil and water, that they have leaves or needles, that they burn. There is no fixed list of things someone with the concept of a tree must believe, but without many general beliefs, there would be no reason to identify a belief as a belief about a tree, much less an oak tree. Similar considerations apply to the dog's supposed thinking about the cat. We identify thoughts, distinguish among them, describe them for what they are, only as they can be located within a dense network of related beliefs. If we really can intelligibly ascribe single beliefs to a dog, we must be able to imagine how we would decide whether the dog has many other beliefs of the kind necessary for making sense of the first. It seems to me that no matter where we start, we very soon come to beliefs such that we have no idea at all how to tell whether a dog has them, and yet such that, without them, our confident first attribution looks shaky" (Davidson 2001). In short, speechless creatures cannot have beliefs.

Often we can see that animals behave as if they misperceive the state of things and non-verbally manifest their 'disappointment' when their expectations are not met. It therefore seems acceptable to speak of a non-verbal way of thinking and expressing wishes. Likewise, children are convinced about their parents or the situations they encounter before they learn the language. Before we master speech, we already want something. Language is basically a way of describing the desires that precede it. Following this path, it can be said that animals have sufficiently developed, non-linguistic beliefs and desires. In animals, beliefs and desires do not appear on the web of other beliefs but in connection with perceptions and actions. Animals correct their beliefs and desires according to their perceptions. Based on observation of the behaviour and reactions of animals, we can conclude that the animals have a certain perception of a given state of affairs or object. These opinions are radically different from ours, but we can try to define their content by expressing it in the sentences of our speech. When attributing to a creature a belief in reality, we do not need to know how that reality presents itself to that creature, nor does that creature need to think the way we do. The concept proposed by Davidson seems to be over-intellectualized and over-exposes the linguistic aspect of beliefs. The sentences used to describe animal thoughts do not reflect these thoughts, nor do they translate them. They are, in fact, structures created by language users. People who observe sophisticated animal behaviour are inclined to imagine animal thoughts in the form of sentences and to interpret animal behaviour as an expression of these sentences.

3.2 Intentional Stance

Analysing and anticipating behaviour of living organisms, we adopt an intentional stance, treating them as entities that have their own points of view and aspirations. Looking at a living creature, we intuitively assume simply that 'there is someone there', more or less intelligent and clever (Dennett 1989). This approach comes from the fact that the paradigmatic holders of consciousness are adult people who extrapolate their way of experiencing and capturing it to other beings. This is a useful and often effective attitude, as it allows to predict the behaviour of people and other organisms. However, does it describe the processes actually taking place? Or is it rather a useful fiction? According to Dennett, this centralisation and unification of consciousness even for people is not something innate, but the result of participating in culture and using its tools, especially language.

The intentional stance can be taken not only towards living organisms but also towards machines, such as computers or inanimate beings, which behave in a regular and sophisticated manner. It works well for beings that behave in a predictable and rational way and seem to achieve their goals. Thus, treating them as rational subjects does not lead to too many mistakes in prediction (Dennett 1989). We ascribe to some subjects deliberate approach in situations when we are aware of metaphorical nature of this ascription. Sometimes we say that 'the computer cannot see the hard disk', children say that 'the doll wants to sleep', one can even claim that 'the car is tired and wants to rest'. Then it is clear that we are dealing with a metaphor. What about living organisms? Are they closer to people or machines? Is the use of intentional attitudes in the description of their behaviour metaphorical in nature, and does this result only from the intuition of common psychology, the categories of which apply to people?

Habituation and the shifting of our perspective to other non-human beings make us think of animals as if they were hidden behind a disguise of furs or feathers who know the reasons for their behaviour, who consciously plan and pursue their goals. It seems, however, that this is an unjustified anthropomorphism, which can nevertheless be a useful tool for translating and predicting the behaviour of many beings. Is it also a description of reality? It is difficult to expect animals to be aware of their own rationale of action and to make efforts in accordance with it, when even people have such difficulties. We can ask questions: When are we motivated by our own internal motives, and when are we pushed to act by reasons, reasons that we do not embrace? When are animals subject to their own inner, mental self-determination, and when are they acting under the influence of innate and evolutionary biological mechanisms that push them to act?

4 Embodied Mind

Living organisms, and this also applies to a large extent to humans, are not autonomous individuals who freely control their actions. What they are, who they are and how they act, depends on their evolutionary history and the biological context which, like them, is the result of evolutionary processes. Evolution resembles a kind of 'brainstorming', it is the coming up of ideas, some of which are more successful and others less successful. Billions of years of this science, which has tried and tested many solutions by trial and error, are in all likelihood giving optimum structures to organisms. Phylogeny or the evolutionary history of species is a kind of creation, learning, a process of gaining knowledge. It can be said that there is an accumulation of knowledge in living organisms, which has arisen in the process of evolution as a result of countless experiences. This knowledge is stored as genetic information in the structures and behavioural patterns of organisms that act and pursue their goals according to patterns and reasons of which these organisms have no idea. In a word, in all living beings, we are dealing with what can be called unconscious wisdom. The motivational rationale for the behaviour and perception of organisms is based on the whole structure of these organisms and the context of life. The consciousness, the mind of living beings is not some dimensionless centre, but is present in the body of living beings and their surroundings.

References

Agar, N.: Life's Intrinsic Value: Science, Ethics, and Nature, p. 25. Columbia University Press, New York (2001)

Bermudez, J.L.: Thinking without words: an overview for animal ethics. J. Ethics **11**(3), 319–335 (2007)

Bernstein, M.H.: On Moral Considerability: An Essay on Who Morally Matters, p. 105. Oxford University Press, New York (1998)

Block, N.: On a confusion about a function of consciousness. Behav. Brain Sci. **18**(2), 227–247 (1995)

Carruthers, P.: Invertebrate minds: a challenge for ethical theory. J. Ethics **11**(3), 275–297 (2007)

Carruthers, P.: Human and Animal Minds. The Consciousness Questions Laid to Rest, p. 177. Oxford University Press, Oxford (2019)

Chalmers, D.J.: Facing up to the problem of consciousness. J. Conscious. Stud. **2**(3), 200–219 (1995)

Davidson, D.: Subjective Intersubjective Objective, p. 98. Oxford University Press, Oxford (2001)

De Waal , F.: Primates and Philosophers: How Morality Evolved, pp. 59–67. Princeton University Press, Princeton (2006)

DeGrazia, D.: Animal Rights: A Very Short Introduction, p. 18. Oxford University Press, Oxford (2002)

Dennett, D.C.: The Intentional Stance. A Bradford Book, pp. 17–18. The MIT Press, Cambridge, London (1989)

Griffin, D.: Animal Minds, p. 260. University Of Chicago Press, Chicago (1994)

Kirk, R.: Zombies and Consciousness, p. 3. Oxford University Press, Oxford, New York (2005)

Levine, J.: Materialism and qualia: the explanatory gap. Pacific Philos. Q. **64**, 354–361 (1983)

Singer, P.: Practical Ethics, pp. 59–60. Cambridge University Press, Cambridge (2011)

Stamp Dawkins, M.: Through animal eyes: what behaviour tells us. Appl. Anim. Behav. Sci. **100**(1/2), 4–10 (2006)

Instantaneous Frequency of the EEG as a Stress Measure - A Preliminary Research

Adam Łysiak$^{(\boxtimes)}$ (iD)

Opole University of Technology, Proszkowska 76, 45-758 Opole, Poland
`a.lysiak@doktorant.po.edu.pl`

Abstract. Stress is one of the most common factors of everyday life. There is a plethora of different measures used to determine the physiological responses to stress. One of the most commonly used is the electroencephalogram (EEG). Power of various frequency bands in the EEG is often correlated with mental states of the measured subject. Particularly, decrease in the alpha band and increase in the beta range is related to the stressed, active state. In this research, a methodology to compare this common measures to the approach based on the instantaneous frequency's slope (*ifs*) was proposed.

Generally, proposed methodology is as follows. Subject, while listening to quiet rain noise, is focused on solving very easy math problems (incrementing or decrementing one-digit numbers). Suddenly, loud one-second white noise is played in the earphones of a subject. This is treated as a stressor. For the time of the study, a 14-channel EEG monitor is measuring activity of the brain.

Obtained preliminary results, based on five participants, indicated that the *ifs* can be more statistically significant than common approaches in measuring the dynamics of the stress response.

Keywords: EEG · Instantaneous frequency · Sound stressor · Stress monitoring

1 Introduction

Stress is one of the most common issues in today's society, and seems to be ever-growing [24]. It is, however, a very broad term and its different types can have both positive and negative impact on a life. Generally, chronic or long-term stress has detrimental effect on one's health and can lead to major psychological problems, like anxiety [31] or depression [14]. Short periods of stress, on the other hand, can seem motivating and have overall positive impact on one's life [20]. Furthermore, acute physiological stressors, like cold showers, were proposed as a potential treatment for depression [28]. Besides short- and long-term stress, authors of [16] proposed a division into systematic stressors, which involve immediate physiologic response and processive stressors, which need to be somehow interpreted by the higher brain structures.

S. Paszkiel (Ed.): ICBCI 2021, AISC 1362, pp. 107–118, 2021.
https://doi.org/10.1007/978-3-030-72254-8_11

There are lots of measurable physiological responses to stress in human body. They are, to some extent, more effective in quantifying stress than questionnaires [24]. One of such responses is change in the electrical activity of the brain, which can be non-invasively measured by the electroencephalogram (EEG). Signals generated by the EEG are commonly used in stress response measuring task, and will be analyzed in this study.

Most commonly, specific frequency bands are used to quantify stress response. In this research, however, another method was used - instantaneous frequency (IF) measure. It does not utilize particular frequency ranges, but includes whole spectrum of the signal in a given moment. This measure could be further used to study dynamics of a stress response or possibly quantify different methods of stress reduction.

1.1   Background

Common measure of stress defined for the EEG is the energy in the alpha frequency band (8 to 12 Hz [1]). It is usually observed to be higher in relaxed state of the brain [22]. However, at least three different sources contribute to the energy in this spectrum: visual, somatosensory and temporal cortex, associated with the alpha (α), mu (μ) and tau (τ) rhythms, respectively [22]. For this reason, interpretation of results obtained solely for this frequency band can be difficult. Nevertheless, it is used extensively to quantify stress.

In [26] authors used EEG, along with the electrooculogram (EOG) and electrocardiogram (ECG) to define physiological measure quantifying mental workload in subjects. As of EEG, the alpha waves were used to quantify the stress and proved to be statistically significant. Stress was induced by simulation of being a pilot in a landing plane. Additionally, mental arithmetic was used.

Authors of [21] studied asymmetry in alpha bands caused by examination-induced stress, leading to conclusion that left hemisphere is more involved in processing of positive emotions. Later, authors of [23] studied the effects of induced stress and binaural beats sounds on alpha asymmetry. Stress was induced by playing loud sound of traffic noise. Differences in the alpha band proved to be statistically significant for the left hemisphere, further confirming findings of [21].

In [18] authors investigated patterns in brainwaves induced by stress, measured by the ratio of difference in beta and alpha bands power. Stress was induced by mental arithmetic and color-word test. Proposed measures proved to be effective in classifying stress by Support Vector Machine classifier, obtaining accuracy up to 96%.

Authors of [2] also used alpha band power, obtained by wavelet decomposition, to differentiate between stress and control group. They obtained statistically significant results. Later, in [5] and [3], the author conducted a study on 3-level stress estimation, obtaining classification accuracy up to 94%, also utilizing alpha band power.

Other frequency bands were also being correlated with mental tasks or stress. For example, in [15] authors observed, that the power of delta and theta waves in the EEG signals were dependent on the performed task, increasing with the tasks difficulty. Theta waves were also observed to be increased in mental arithmetic task in [13].

Instantaneous frequency (IF) is another spectral analysis tool. It differs, however, from the band-pass approach seen in mentioned works. This is a time-frequency measure showing spectral contents of a signal in a given moment. At first, this concept can seem counterintuitive or, considering uncertainty principle, even self-contradictory. However, depending on the definition, this measure is not exactly instantaneous. There are many definitions and interpretations of instantaneous frequency [7]. In this study, definition of IF as the first conditional moment of the power spectrum of the EEG was used [8].

Variously defined IF measures were applied to the EEG signals. For example, in [32], authors obtained IFs of the Intrinsic Mode Functions (IMFs) generated by the Empirical Mode Decomposition (EMD). To obtain instantaneous frequencies of each IMF, they used Generalized Zero Crossing (GZC) method. Calculated IFs were used to classify infant sleep stages.

Authors of [30] used IF of the Variational Mode Decomposition to remove noise from the EEG signals, later classified into four emotional classes (happy, afraid, sad and relaxed). They used Hilbert transform to obtain the instantaneous frequency.

Generally, frequency and time-frequency analysis is ever-growing field and IF measure is just one of the many different analysis methods [9,10,19]. Moreover, besides the EEG, there is a plethora of different physiological measures used to identify stress response, for example electrooculogram (EOG) [26], electrocardiogram (ECG) [26], electrodermal activity (EDA) [27] or functional Near Infrared Spectroscopy (fNIRS) [4].

1.2   Objective of the Study

The main objective of this study was to propose a methodology, which would allow to obtained features by both frequency band and IF approaches. Then, quantifying quality of those features by statistical significance tests, compare them to each other and determine which one is more informative.

General study procedure was as follows. Participant was listening to quiet rain noise using earphones. In order to stay focused, easy math problems were given to him to be solved. Unexpectedly to the participant, loud 1-s noise was played in earphones. This stimuli was treated as an acute stressor and was induced three times in 7-min duration of the procedure. For the whole time, EEG and pulse measures were being recorded. Three-second periods prior to and following the stressor indicated that differences in the slope of the EEG's instantaneous frequency can be more statistically significant than the differences in specific frequency ranges, i.e. delta, theta, alpha, beta or gamma waves.

2 Materials and Methods

2.1 Participants

There were 5 participants in this study: two females (22 and 25 years old) and three males (22, 25 and 25 years old). There are lots of factors affecting response to the applied stressor. To somehow minimize the effects of those factors, all participants were well-rested and did not consume any drugs in the day of the study.

All participants were informed about the general idea of the study procedure prior to the actual study. Written consent was obtained from every participant.

2.2 Study Environment and Procedure

The study was conducted in aired room with natural lighting. Firstly, measure devices (EEG and heart rate monitor) were put on participant. Then, he put on the earphones (use of headphones was made impossible by the EEG monitor) and confirmed being in comfortable position. Participant was sitting in a chair in front of a laptop, on which he was solving math problems. They were defined to be as easy as possible, just enough to keep participant focused and prevent dosing off: incrementing, decrementing or adding nothing to one-digit numbers (for example: "$2 + 1 = ?$", "$9 - 1 = ?$" or "$3 + 0 = ?$"). To ensure minimal hand movement, answers to these problems were to be entered using numerical keyboard. The study was conducted in the Matlab environment, using custom script.

Exact study procedure was as follows. Firstly, the participant was asked to enter his name. After that, quiet sound of rain began to play. It was chosen for two reasons. Firstly, auditory noise is known to enhance focus and improve cognitive performance [29]. More importantly, listening to the noise ensured that all possible auditory stimuli unrelated to this study were unheard. Then, 10-s countdown ensured that the participant was in a comfortable position and ready to solve mentioned math puzzles. After this countdown, math problems were being given (in a one-by-one manner) to the participant. After about 3 min of being focused on solving problems, loud 1-s white noise was played in the subject's earphones. It was repeated after about 90 s, and then after another 90 s. Figure 1 shows exact sound played.

2.3 Measurement Devices

The electroencephalograph used in this study was Emotiv EPOC+. It has 14 channels (AF3, AF4, F3, F4, F7, F8, FC5, FC6, T7, T8, P7, P8, O1, O2 referenced to P3 and P4 electrodes) fixed in positions according to 10-10 standard [11]. It has sampling frequency of 128 Hz.

To measure the pulse, Beurer PO-80 finger pulse oximeter was used. The photoplethysmographic curve was sampled at 60 Hz with pulse being calculated every 1 s.

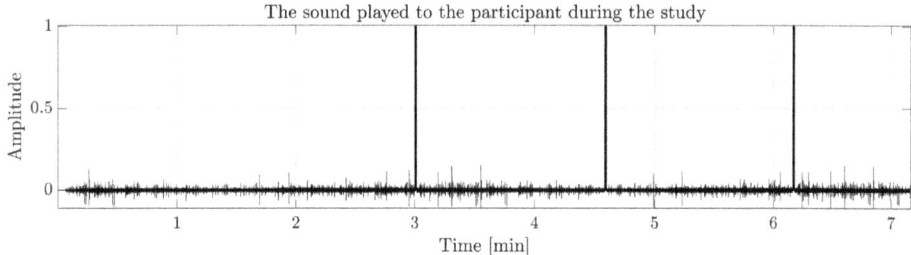

Fig. 1. The sound played to the participant during the study procedure.

2.4 EEG Processing and Defined Features

Signal Preprocessing. Signals collected by the EEG are often contaminated by unwanted components or artifacts generated by the both external (like power lines frequency or imperfect adhesion of the EEG electrodes to the scalp) and internal (like face muscle activity or heart rate) factors. To clean the raw EEG signals, the PREP pipeline was used [6]. Generally, it consist of following stages. Firstly, components generated by the power lines are removed without using typical notch-filter methods, as those are known to distort frequencies in a range greater than desired [25]. Then, robust reference of the signal is estimated. Finally, distorted channels are interpolated using this reference. Figure 2 shows exemplary signal from F7 electrode before and after preprocessing.

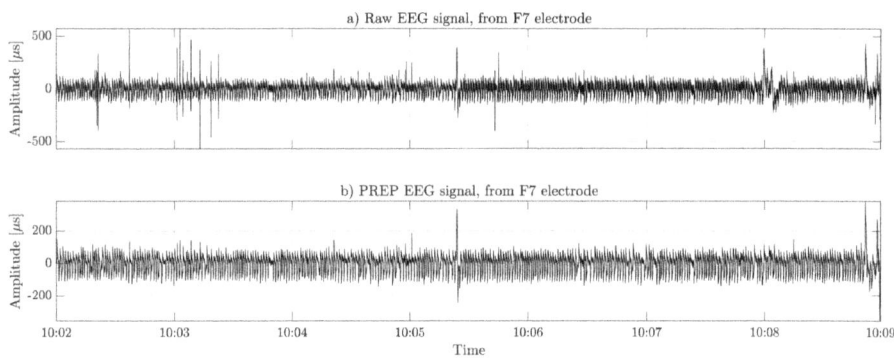

Fig. 2. Exemplary EEG signal from F7 electrode: (*a*) before and (*b*) after PREP pipeline preprocessing.

The EEG signals were preprocessed in the Matlab environment, using EEGLAB toolbox [12].

Brain Waves Frequency Bands. Exact frequency ranges were defined according to [1] in following way:

- delta (δ) waves: 0.5–4 Hz,
- theta (θ) waves: 4–8 Hz,
- alpha (α) waves: 8–12 Hz,
- beta (β) waves: 12–35 Hz,
- gamma (γ) waves: >35 Hz.

Energy of particular frequency ranges was obtained by summing squared values of the Discrete Fourier Transform (DFT) in those ranges.

Instantaneous Frequency. As mentioned before, there are many definitions and interpretations of the instantaneous frequency (*IF*) [7]. In this research, definition of the *IF* as the first conditional moment of the power spectrum of the EEG was used [8]. It was chosen because of its insusceptibility to signal noise (especially compared to *IF* definition based on the Hilbert transform) and low computational power required to calculate it. Exact definition was following:

$$
IF(t) = \frac{\sum_{f=0}^{f_{MAX}} f\, P(t, f)}{\sum_{f=0}^{f_{MAX}} P(t, f)},
\tag{1}
$$

where $P(t, f)$ is the power spectrum and f_{MAX} is its maximal frequency.

Instantaneous frequency of the preprocessed signal showed in Fig. 2 was shown in Fig. 3.

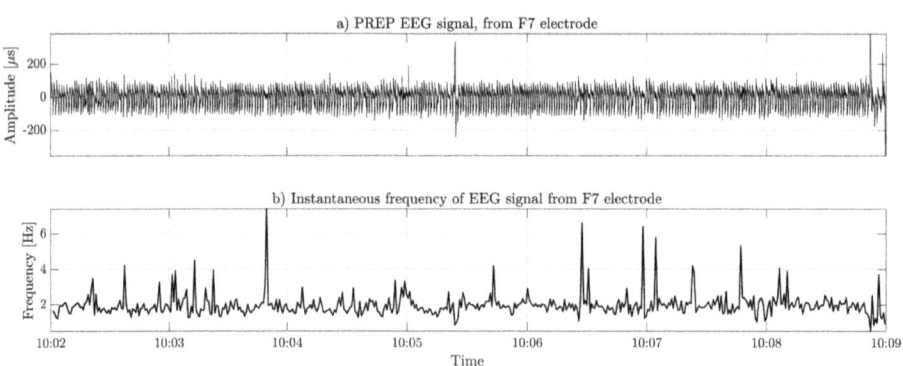

Fig. 3. Exemplary preprocessed EEG signal from F7 electrode (*a*) with corresponding instantaneous frequency (*b*).

Defined Features. Overall, 7 features were defined to differentiate between 3-s period prior to and following the stressor: energy in delta, theta, alpha, beta and gamma bands, as well as the mean value and the slope of the *IF*. Visualization of the *IF* features was given in Fig. 4. Those 7 features were calculated for both 3-s stages: prior to and following the stressor.

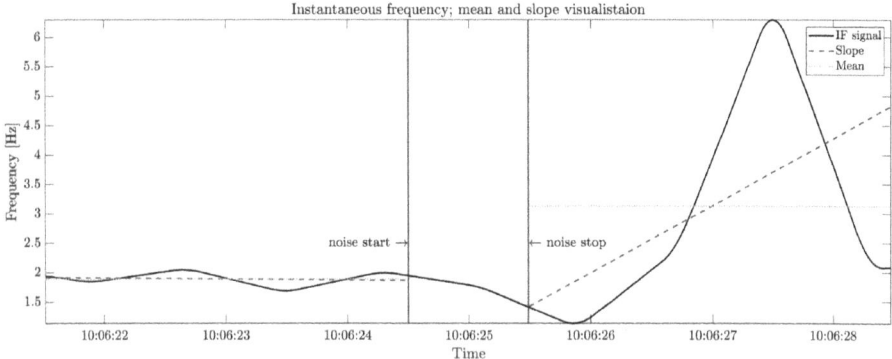

Fig. 4. Exemplary image of 3-s period prior to and following the stressor. Note, that this is a fragment of signal showed in Fig. 3 *(b)*.

Additionally, differences between those periods were calculated for the average values of the pulse and the average numbers of math problems solved. Obtained feature sets were tested for normality. Differences were used in those two features, since it was assumed that variance of pulse and math-problem solving speed would be grater between participants than between pre- and post-stressor stages of one participant.

2.5   Statistical Tests

To determine the statistical significance of the features' values, Kolmogorov-Smirnov (KS) test [17] was used. It was chosen, since being based on a maximal difference between compared cumulative distribution functions, it does not assume any specific distribution of data.

Additionally, *t*-test was used to determine normality of differences between pre- and post-stressor periods of pulse and math-problems features.

3 Results

Pulse and math problems solving speed differences proved to be normally distributed, with mean equal to zero (t-test p-value > 0.05 for both). Results (p-values) of Kolmogorov-Smirnov tests were presented in Table 1.

Boxplots of statistically significant features ($p < 0.05$) were presented in Fig. 5. Note, that despite being significantly different, distributions of features' values plotted in Fig. 5 are widely overlapped.

Table 1. Results (p-values) of K-S tests. Columns E_δ, E_θ, E_α, E_β and E_γ contain p-values for energy in corresponding frequency bands, while *ifm* and *ifs* are columns containing p-values of instantaneous frequency's mean value and slope, respectively.

Electrode	E_δ	E_θ	E_α	E_β	E_γ	*ifm*	*ifs*
AF3	0.017	0.017	0.017	0.136	0.890	0.589	0.051
F7	0.136	0.136	0.308	0.589	0.890	0.890	0.051
F3	0.136	0.136	0.136	0.890	0.136	0.890	0.017
FC5	0.589	0.890	0.890	0.890	0.890	0.890	0.017
T7	0.308	0.589	0.589	0.308	0.589	0.589	0.136
P7	0.136	0.308	0.308	0.890	0.308	0.890	0.005
O1	0.998	0.998	0.890	0.136	0.589	0.589	0.051
O2	0.890	0.890	0.589	0.890	0.308	0.136	0.017
P8	0.998	0.998	0.890	0.136	0.589	0.589	0.017
T8	0.136	0.308	0.308	0.890	0.308	0.589	0.589
FC6	0.308	0.589	0.589	0.308	0.589	0.308	0.308
F4	0.589	0.890	0.890	0.890	0.890	0.890	0.136
F8	0.136	0.136	0.136	0.890	0.136	0.890	0.051
AF4	0.136	0.136	0.308	0.589	0.890	0.589	0.136
$p < 0.05$	1	1	1	0	0	0	5

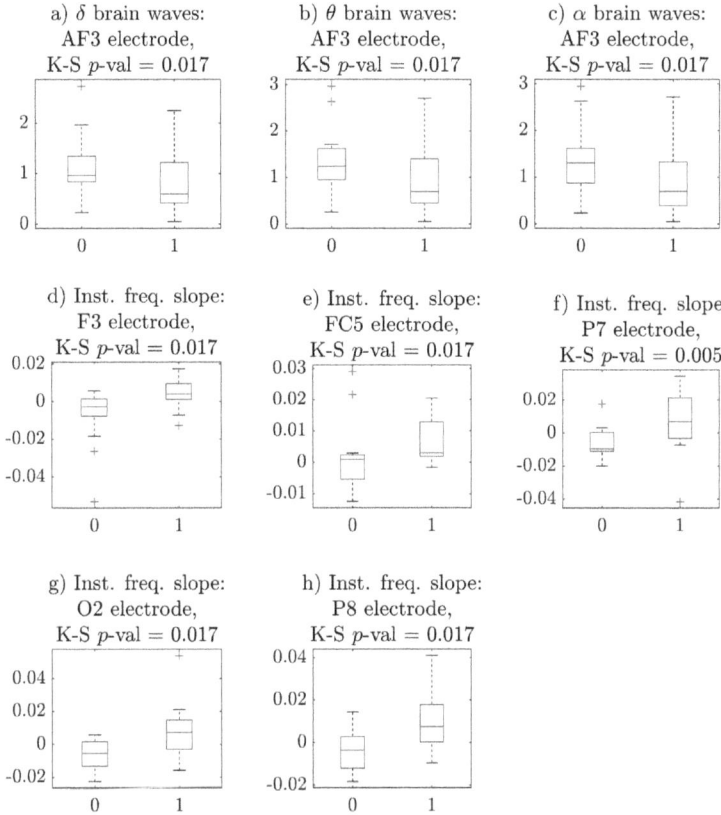

Fig. 5. Boxplots of measures which proved to be statistically significant according to Kolmogorov-Smirnov test (see Table 1). First three subplots (*a*, *b* and *c*) were obtained for AF3 electrode for delta, theta and alpha rhythms, respectively. The rest of the subplots (*d*, *e*, *f*, *g* and *h*) correspond to the slope of the instantaneous frequency for F3, FC5, P7, O2 and P8 electrodes, respectively.

4   Discussion and Conclusions

The main goal of this study was to propose a methodology allowing comparison between two approaches to measuring stress using the EEG. First, the frequency band approach, associated with strict division into different spectral ranges (delta, theta, alpha, beta and gamma range). And second, the instantaneous frequency approach, in which information about all bands are bound.

Obtained results are to be took as preliminary, and proposed methodology should be applied to a much larger sample (only five participants took part in this research). However, findings obtained by proposed methodology point out that the slope of the instantaneous frequency seem more informative than energy in theta, delta, alpha, beta or gamma waves.

Differences in numbers of solved math problems and mean pulse proved to be normally distributed around zero, meaning that proposed methodology did not allow to differentiate between those measures for pre- and post-stressor periods.

In future studies, besides larger sample, different methods of comparison between two mentioned approaches could be used. For example, obtained features could be used to train some classification algorithms, and their final accuracy could be utilized as a measure of quality. Alternatively, overlap between feature's distributions could be calculated determining how informative each feature is. Additionally, periods of different duration could be investigated.

This research constitutes a proposition of a methodology which could allow comparison between different stress measures utilizing the EEG. Additionally, obtained results indicate, that the slope of the instantaneous frequency can be used to measure stress' dynamics more informatively than common, frequency-band approaches.

Acknowledgments. Author would like to express gratitude for all the participants which took part in this study.

References

1. Abhang, P.A., Gawali, B.W., Mehrotra, S.C.: Technological basics of EEG recording and operation of apparatus. In: Introduction to EEG- and Speech-Based Emotion Recognition, pp. 19–50. Elsevier/AP, Academic Press is an imprint of Elsevier. https://doi.org/10.1016/B978-0-12-804490-2.00002-6
2. Al-Shargie, F.M., Tang, T.B., Badruddin, N., Kiguchi, M.: Mental stress quantification using EEG signals. In: Ibrahim, F., Usman, J., Mohktar, M.S., Ahmad, M.Y. (eds.) International Conference for Innovation in Biomedical Engineering and Life Sciences, IFMBE Proceedings, vol. 56, pp. 15–19. Springer, Singapore. https://doi.org/10.1007/978-981-10-0266-3_4
3. Al-Shargie, F.: Multilevel Assessment of Mental Stress Using SVM with ECOC: An EEG Approach. https://doi.org/10.31224/osf.io/7v9ks
4. Al-Shargie, F., Kiguchi, M., Badruddin, N., Dass, S.C., Hani, A.F.M., Tang, T.B.: Mental stress assessment using simultaneous measurement of EEG and fNIRS. Biomed. Opt. Express **7**(10), 3882–3898 (2016). https://doi.org/10.1364/BOE.7.003882
5. Al-Shargie, F., Tang, T.B., Badruddin, N., Kiguchi, M.: Towards multilevel mental stress assessment using SVM with ECOC: an EEG approach. Med. Biol. Eng. Comput. **56**(1), 125–136 (2018). https://doi.org/10.1007/s11517-017-1733-8
6. Bigdely-Shamlo, N., Mullen, T., Kothe, C., Su, K.M., Robbins, K.A.: The PREP pipeline: standardized preprocessing for large-scale EEG analysis. Front. Neuroinform. **9**, 16 (2015). https://doi.org/10.3389/fninf.2015.00016
7. Boashash, B.: Estimating and interpreting the instantaneous frequency of a signal. I. Fundamentals. Proc. IEEE **80**(4), 520–538 (1992). https://doi.org/10.1109/5.135376
8. Boashash, B.: Estimating and interpreting the instantaneous frequency of a signal. II. Algorithms and applications. Proc. IEEE **80**(4), 540–568 (1992). https://doi.org/10.1109/5.135378

9. Borkowski, J., Kania, D., Mroczka, J.: Comparison of sine-wave frequency estimation methods in respect of speed and accuracy for a few observed cycles distorted by noise and harmonics. Metrol. Measur. Syst. **25**, 283–302 (2018). https://doi.org/10.24425/119567

10. Borkowski, J., Kania, D.: Interpolated-DFT-based fast and accurate amplitude and phase estimation for the control of power. http://arxiv.org/abs/1601.00453

11. Chatrian, G.E., Lettich, E., Nelson, P.L.: Ten percent electrode system for topographic studies of spontaneous and evoked EEG activities. Am. J. EEG Technol. **25**(2), 83–92 (1985). https://doi.org/10.1080/00029238.1985.11080163

12. Delorme, A., Makeig, S.: EEGLAB: an open source toolbox for analysis of single-trial EEG dynamics including independent component analysis. J. Neurosci. Methods **134**(1), 9–21 (2004). https://doi.org/10.1016/j.jneumeth.2003.10.009

13. Gärtner, M., Grimm, S., Bajbouj, M.: Frontal midline theta oscillations during mental arithmetic: effects of stress. Front. Behav. Neurosci. **9**, 96 (2015). https://doi.org/10.3389/fnbeh.2015.00096

14. Hammen, C., Kim, E.Y., Eberhart, N.K., Brennan, P.A.: Chronic and acute stress and the prediction of major depression in women. Depress. Anxiety **26**(8), 718–723 (2009). https://doi.org/10.1002/da.20571

15. Harmony, T., Fernández, T., Silva, J., Bernal, J., Díaz-Comas, L., Reyes, A., Marosi, E., Rodríguez, M., Rodríguez, M.: EEG delta activity: an indicator of attention to internal processing during performance of mental tasks. Int. J. Psychophysiol. **24**(1–2), 161–171 (1996). https://doi.org/10.1016/S0167-8760(96)00053-0

16. Herman, J.P., Cullinan, W.E.: Neurocircuitry of stress: central control of the hypothalamo-pituitary-adrenocortical axis. Trends Neurosci. **20**(2), 78–84 (1997). https://doi.org/10.1016/S0166-2236(96)10069-2

17. Massey Jr., F.J.: The Kolmogorov-Smirnov test for goodness of fit. J. Am. Stat. Assoc. **46**(253), 68–78 (1951). https://doi.org/10.1080/01621459.1951.10500769

18. Jun, G., Smitha, K.G.: EEG based stress level identification. In: 2016 IEEE International Conference on Systems, Man, and Cybernetics (SMC), pp. 003270–003274. IEEE. https://doi.org/10.1109/SMC.2016.7844738

19. Kania, D., Borkowski, J.: Estimation methods of multifrequency signals with noise and harmonics for PV systems with a DSP processor. In: 2017 40th International Conference on Telecommunications and Signal Processing (TSP), pp. 524–527. IEEE. https://doi.org/10.1109/TSP.2017.8076042

20. Le Fevre, M., Matheny, J., Kolt, G.S.: Eustress, distress, and interpretation in occupational stress. J. Manag. Psychol. **18**(7), 726–744 (2003). https://doi.org/10.1108/02683940310502412

21. Lewis, R.S., Weekes, N.Y., Wang, T.H.: The effect of a naturalistic stressor on frontal EEG asymmetry, stress, and health. Biol. Psychol. **75**(3), 239–247 (2007). https://doi.org/10.1016/j.biopsycho.2007.03.004

22. Niedermeyer, E., Schomer, D.L., da Silva, F.H.L. (eds.): Niedermeyer's Electroencephalography: Basic Principles, Clinical Applications, and Related Fields, 6th edn. Wolters Kluwer, Lippincott Williams & Wilkins (2011)

23. Norhazman, H., Zaini, N.M., Taib, M.N., Omar, H.A., Jailani, R., Lias, S., Mazalan, L., Sani, M.M.: Behaviour of EEG Alpha Asymmetry when stress is induced and binaural beat is applied. In: 2012 International Symposium on Computer Applications and Industrial Electronics (ISCAIE), pp. 297–301. IEEE (2012). https://doi.org/10.1109/ISCAIE.2012.6482116

24. Paszkiel, S., Dobrakowski, P., Łysiak, A.: The impact of different sounds on stress level in the context of EEG, cardiac measures and subjective stress level: a pilot study. Brain Sci. **10**(10), 728 (2020). https://doi.org/10.3390/brainsci10100728

25. Rousselet, G.A.: Does filtering preclude us from studying ERP time-courses? Front. Psychol. **3**, 131 (2012). https://doi.org/10.3389/fpsyg.2012.00131

26. Ryu, K., Myung, R.: Evaluation of mental workload with a combined measure based on physiological indices during a dual task of tracking and mental arithmetic. Int. J. Ind. Ergon. **35**(11), 991–1009 (2005). https://doi.org/10.1016/j.ergon.2005.04.005

27. Setz, C., Arnrich, B., Schumm, J., La Marca, R., Troster, G., Ehlert, U.: Discriminating stress from cognitive load using a wearable EDA device. IEEE Trans. Inf Technol. Biomed. **14**(2), 410–417 (2010). https://doi.org/10.1109/TITB.2009.2036164

28. Shevchuk, N.A.: Adapted cold shower as a potential treatment for depression. Med. Hypotheses **70**(5), 995–1001 (2008). https://doi.org/10.1016/j.mehy.2007.04.052

29. Söderlund, G., Sikström, S., Smart, A.: Listen to the noise: noise is beneficial for cognitive performance in ADHD. J. Child Psychol. Psychiatry **48**(8), 840–847 (2007). https://doi.org/10.1111/j.1469-7610.2007.01749.x

30. Taran, S., Bajaj, V.: Emotion recognition from single-channel EEG signals using a two-stage correlation and instantaneous frequency-based filtering method. Comput. Methods Prog. Biomed. **173**, 157–165 (2019). https://doi.org/10.1016/j.cmpb.2019.03.015

31. Vyas, A., Chattarji, S.: Modulation of different states of anxiety-like behavior by chronic stress. Behav. Neurosci. **118**(6), 1450–1454 (2004). https://doi.org/10.1037/0735-7044.118.6.1450

32. Čić, M., Šoda, J., Bonković, M.: Automatic classification of infant sleep based on instantaneous frequencies in a single-channel EEG signal. Comput. Biol. Med. **43**(12), 2110–2117 (2013). https://doi.org/10.1016/j.compbiomed.2013.10.002

Measuring Stress Response via the EEG - A Review

Adam Łysiak(✉)

Opole University of Technology, Proszkowska 76, 45-758 Opole, Poland
a.lysiak@doktorant.po.edu.pl

Abstract. One of the most significant factors influencing a person's well-being is stress. It can be measured via some standardized questionnaires, however, they have some major drawbacks. Firstly, one needs time to fill out a questionnaire and secondly, one needs to interpret the questions. Since physiological measures can be obtained without active conscious involvement, they are generally much more reliable and faster than the questionnaires. One of the most commonly used physiological measures is the electroencephalograph (EEG), whose high temporal and spacial resolution allows real-time emotion recognition. This paper constitutes a summary of different EEG signal's features used to measure and quantify human response to a stressor.

Keywords: EEG stress measures · Emotion recognition · Electroencephalography stress · Stress measuring · Stress neurofeedback

1 Introduction

Stress is one of the most important factors affecting the well-being of a person. Despite being intuitively understandable, stress, as a term, is very broad. It can be divided in various ways, for example into:

- acute and chronic [23], differentiating between short- and long-lasting;
- eustress and distress [25], differentiating between states which can make positive, motivating impact on ones life, and negative, leading to maladaptive behaviors;
- systematic and processive [14], differentiating between stressors which involve immediate physiologic response, and those which need some interpretation by the higher brain structures.

Those divisions can somewhat simplify the issue and help to categorize different stimuli. However, one can imagine that those are just the extremes of some stress spectra, and should not be taken as a strict distinctions.

Everyone has some individual stress-coping mechanisms, which can further influence one's life in a positive (creative) or negative (destructive) way [6]. To maximize positive and minimize negative stress-coping effects, a lot of systems

S. Paszkiel (Ed.): ICBCI 2021, AISC 1362, pp. 119–128, 2021.
https://doi.org/10.1007/978-3-030-72254-8_12

are made. For example, mindfulness meditation is a popular stress-coping practice. However, according to [9], its effects on stress reduction are rather poorly proven. Generally, chronic stress reduction is a long-term process (for example, meditation is usually recommended for at least couple weeks [9]) and therefore its dynamics are difficult to measure. Another difficulty is measuring stress response itself. On one hand, stress is understood as being subjective. Therefore, various questionnaires were developed, such as Perceived Stress Scale (PSS) [7] or Relative Stress Scale [40] (RSS). On the other hand, questionnaires have significant disadvantages. Filling them out takes time and requires interpretation, which can distort acute stress response. Furthermore, they seem to be less informative than physiological measures [29].

There is a plethora of physiological measures used to quantify stress response [34]. One of the most commonly used ones is the electroencephalogram (EEG). It is a non-invasive method of measuring electrical activity of the brain. Its high temporal resolution and relatively precise electrode location allow real-time emotion recognition [8,37]. It was also widely used in various mental disorders resulting from acute stressors, such as post traumatic stress disorder (see [35] for review). However, measuring stress response immediately after, or even during the stressor, is quite problematic. Firstly, exposing subject to a stressor is quite disputable from the ethical point of view. Besides that, rather few people would voluntarily wear the EEG cap during some expected stressful situation. To overcome those issues, various controlled stress protocols were defined and used across studies. One example is Mental Arithmetic, during which the participant is rushed and given negative feedback concerning his mathematical skills. Another one is Stroop Color-Word Test, in which participant have to choose between color, in which given word is spelled and color it denotes (e.g. word "green" spelled with red font). During those protocols, the physiological measures can be registered.

After obtaining the EEG signals, they need to be somehow analyzed to obtain meaningful features describing the stress response. There are various approaches in the literature to achieve this, and this review constitutes an attempt to summarize them.

2   EEG-Based Stress Measures

Because of the oscillatory character of the brain activity measured by the EEG, the most common stress recognition approach is the frequency analysis, especially power fluctuations in alpha (8–12 Hz) and beta (12–35 Hz) bands. Another common approach is filtering the EEG signal in the given frequency range, either using band-pass filters or wavelets, and then analyzing it. A Hilbert-Huang Transform is yet another way of the EEG signal analysis, in which signal is decomposed into modes and instantaneous frequency of each mode is determined. Some approaches are based on the assumption that the EEG spectra from the different hemispheres of the brain response differently to a stressor. Therefore, correlations or ratios between features obtained from different hemispheres are measured. In different analyzes, self-similarity of signals is determined, utilizing fractal dimension or autocorrelation estimates. Temporal or statistical analyzes are also commonly used. In this section, those various measures will be briefly described and exemplary researches using them will be mentioned.

2.1   Features in Time Domain

There is a plethora of features defined for signals in the time domain, mostly based on statistical or complexity measures.

Statistical Analysis

Authors of [19] and [20] used a lot of various temporal features of the EEG to quantify stress response. They calculated: mean and median amplitude values, standard deviation and kurtosis, peak value, peak location, peak to peak difference, peak to root mean square (RMS) ratio, RMS level, cumulative minimum and maximum values, and total number of zero crossings. They also used sliding window on the signal, calculating smallest element and median within each window. In [41], authors also used statistical analysis applied to signal represented in temporal domain. Additionally to some typical measures (like mean value or standard deviation), they analyzed the first and second differences of the signals. These measures were also used in [17]. In [10] authors used cluster-based permutation approach to the time-frequency data obtained for all used electrodes, generating three-dimensional (electrode-time-frequency) maps.

In [16] author defined different statistical features, i.e.: activity, which, being the variance of a signal in time, can be interpreted as the surface of the power spectrum; mobility, which can indicate the mean frequency; and complexity, which represents the change in frequency. Those measures are called Hjorth's parameters (after the author) and were used in [42] to measure stress response in the EEG.

Complexity and Self-similarity Measures

Another approach of signal analysis in the temporal domain is self-similarity or complexity determination.

Fractal Dimension

It is a complexity measure of the signal defined in the temporal domain. One of the most popular methods of obtaining it is a Higuchi's method [15], which was widely used to recognize stress response via the EEG [17, 22, 41] (although [41] used modified version, so-called Generalized Higuchi's Fractal Dimension Spectrum).

Autocorrelation

Autocorrelation measures the Pearson's correlation statistic between the signal and its own, delayed copy. It is, therefore, a function of delay (or lag). Partial autocorrelation is a measure for which, instead of the Pearson's correlation, the partial correlation is being determined. It additionally removes confounding variables, making the function more robust. It was applied to the EEG to measure stress response in [42]. Also, authors of [41] used six coefficients of an autoregressive model as EEG signal features to describe stress response.

Permutation Entropy

Another measure of complexity of a time series is Permutation Entropy [5]. It is based on comparison of neighboring values' ordinals. Its modified versions (Delayed Permutation Entropy and Permutation Min-Entropy) were used in [26] to distinguish EEG signals obtained in the calm state from the ones of distressed state.

2.2   Features in Frequency Domain

Despite being quite fruitful, signal analysis in time domain is usually less descriptive than frequency domain. Usually, neural oscillations are divided into five distinct frequency bands, namely: delta (δ: 0.5–4 Hz), theta (θ: 4–8 Hz), alpha (α: 8–12 Hz), beta (β: 12–35 Hz) and gamma (γ: >35 Hz) bands [1]. Specific ranges, however, slightly differ from study to study. Frequency analysis of the EEG is undoubtedly very informative. However, limiting analysis to the spectral characteristics can sometimes prevent reaching correct conclusions. Some oscillations registered by the EEG can have similar frequency contents, yet originating from different sources [28]. For example, delta rhythms can be generated by the thalamus or the cortex. Further, frequency ranges 10 Hz can be attributed to the α waves, generated by the visual cortex, μ waves, generated by the somatosensory cortex or τ waves, generated by the temporal cortex [28]. Nevertheless, frequency band analysis constitutes the most common and informative approach to the measurement of stress response.

Usually, delta waves are correlated with sleep and anesthesia states [28]. However, some studies suggests that delta activity can indicate increased attention [13]. Yet, according to different studies [38,42] this band is rather uninformative in terms of stress recognition. Theta waves are usually correlated with memory formation, spatial navigation [28], deep relaxation and focus [1]. Fluctuations in this range were also used to differentiate stress in binary [4,10,17,20,21,24,30,42] and multiple-class classification [4,17,19,21,36,38]. Alpha and beta waves are bands most commonly correlated with stress (see [2,11,12,17,19–21,24,29–33,36,38,39,42]), sometimes with relation to each other. Gamma rhythms are usually associated with concentration [1]. They are considered to enable synchronization of various cortical areas, allowing information processing in the cortex [28]. Neural oscillations in this range were also used to recognize stress state in humans [27,30,32,36].

Summary of researches using the frequency band approach to quantify stress response was given in Table 1. Specific features based on this approach were described in following subsections.

Table 1. Summary of studies in which frequency analysis of the EEG was utilized to measure stress. In the *Stressor* column, *MA* stands for mental arithmetic, *IQ* for IQ test, *MF* for mental fatigue, *BR* for bike race, *SCT* for Stroop color-word test, *MEM* for memory, *GM* for game, *VID* for video, *CW* for working on construction site, and *PSS* for perceived stress scale. Note, that in researches using the *PSS*, stress was just measured, not induced during the study procedure. *UC* in *Participants* columns means that particular information was unobtainable or unclear. Frequency bands with subscripts were additionally divided in a given study (e.g. β_1 and β_2 in [32] means that the beta band was divided into two sub-bands).

Ref.	Year	Participants			Stressor	Classes	Freq. bands
		Count	Sex	Age			
[11]	2010	13	5f/8m	23–39	PSS	–	α, β
[38]	2012	180	UC	20–40	IQ	3	θ, α, β
[24]	2013	13	13m	UC	MF	2	θ, α, β
[10]	2015	31	31m	20–50	MA	2	θ
[17]	2015	9	UC	21–28	SCT	4 3 2	θ, α, β
[12]	2015	86	39f/47m	21–40	PSS	2	α, β
[30]	2015	28	UC	UC	PSS	2	δ, θ, α_1, α_2, β_1, β_2, γ_1, γ_2
[21]	2016	10	1f/9m	20–35	SCT MA SCT, MA	2 2 3	θ, α, β
[42]	2016	20	UC	18–23	BR	2	θ, α, β
[3]	2016	22	22m	22–30	MA	2	α, β
[27]	2016	6	UC	26.3 ± 6.4	MA	2	θ, α, β, γ
[33]	2017	9	3f/6m	21 ± 1.7	MEM	2	α, β
[31]	2017	28	19f/9m	22–33	PSS	2	β_1, β_2
[36]	2017	42	11f/31m	19–25	MA	4	δ, θ α_1, α_2 β_1, β_2, β_3 γ_1, γ_2, γ_3
[39]	2018	5	3f/2m	23–25	GM, IQ	3	α, β
[32]	2018	28	10f/18m	21–34	PSS	2	β_1, β_2, γ_1
[19]	2018	7	7m	26–50	CW	3	δ, θ, α, β
[4]	2019	28	15f/13m	18–40	PT	2 3	θ
[20]	2019	32	16f/16m	19–37	VID	2	δ, θ, α, β
		7	7m	26–50	CW	2	
[29]	2020	9	9f	22	MA	2	α

Power Spectrum Features
The most intuitive EEG signal features are defined based on the Power Spectral Density (PSD), which shows distribution of power over the frequency. There is a number of methods to estimate it (for example Welch's method or squared magnitude of the Discrete Fourier Transform). Various features are defined based on the PSD, and they will be described in the rest of this subsection.

Spectral Power
It is usually defined as a sum [4,11,13,17,20,27,29–32,36,41] or an average [3,19,21,27,33] value in a given frequency range. In some studies [36], authors used relative power, additionally dividing sum of given range by the total sum (sum of all frequencies). Different studies used ratios of selected bands (like α/β in [19]).

Central Measures
Feature locating mass' center of the spectrum (called *spectral centroid*) was obtained for the specific frequency bands in [38,39]. It can be interpreted as a weighted mean frequency, where weights are frequencies' amplitudes. Authors of [4] used, besides the mean value, standard deviation of the frequencies in each sub-band. In the [19,20] slightly different measure was used, i.e. median frequency power. It can be interpreted as a frequency, for which power below it is equal to the power above it.

Spectral Entropy
Authors of [38] used spectral entropy, which is the Shannon's Entropy, but defined for each frequency sub-band. It can be interpreted as a measure of a signal's complexity.

Filter Banks and Wavelet Analysis
Another approach of analyzing specific frequency contents of a signal is to filter it in a given frequency range [32] or decompose it using wavelet decomposition [2,3,10,42]. Then, from each obtained sub-band, statistical or power measures can be calculated.

Asymmetry and Connectivity Features
Because stress responses in two hemispheres differ, many asymmetry measures were defined to measure it. Authors of [11] used ratios in alpha and beta frequency bands ($\frac{\alpha_R - \alpha_L}{\alpha_R + \alpha_L}$ and $\frac{\beta_R - \beta_L}{\beta_R + \beta_L}$) to measure stress response. In [36], authors obtained similarly defined asymmetry: $\frac{M-N}{M+N}$, where M and N are instantaneous amplitudes of different frequency bands.

In [4] authors used three different quantities to measure asymmetry as a stress response. In time domain, they calculated correlation between asymmetric channels (e.g. $AF7$ and $AF8$ electrodes), and in frequency domain they used DASM and RASM features, defined as the difference and the ratio, respectively, between the absolute power of asymmetric channels.

Two different measures, called Arousal and Valence were used in [19] and [20]. Those features capture differences in spectral asymmetry in the frontal part of the brain and are defined as follows:

$$\text{Arousal} = \frac{\alpha(AF3 + AF4 + F3 + F4)}{\beta(AF3 + AF4 + F3 + F4)},$$

$$\text{Valence} = \frac{\alpha(F4)}{\beta(F4)} - \frac{\alpha(F3)}{\beta(F3)},$$

where α and β are powers in alpha and beta bands, and $AF3$, $AF4$, $F3$ and $F4$ are the electrodes.

Different measures, also used to quantify the connectivity between various brain regions, were utilized in [36]. As the EEG features, authors calculated measures of coherence (interpreted as a correlation between spectra, also used in [22]) and phase lag (interpreted as a synchronization of brain regions).

Hilbert-Huang Transform
Hilbert-Huang Transform (HHT) is an algorithm, which decomposes the signal into Intrinsic Mode Functions (IMF) using Empirical Mode Decomposition (EMD) and then, using the Hilbert Transform (HT), determines Instantaneous Frequency of each IFM [18]. It was widely used in the analysis of biomedical signals, including the EEG. For example, authors of [33] have use it to measure stress response, where root mean square and maximum amplitude of selected IFMs, as well as mean (and weighted mean) values of their instantaneous frequency, were utilized as signal's features.

3   Conclusions

A plethora of the EEG signal's features were defined and used in the literature to measure physiological response to stress. This publication constitutes a summary of those measures. In most of the mentioned studies, a number of participants was quite small, undermining obtained conclusions to some extent. In the future research, a series of measurements could be conducted on a larger group of subjects. Then, described features could be compared in terms of informativeness. Most descriptive ones could constitute a feature vector, which further could be used to quantify effectiveness of some stress-relieving systems.

References

1. Abhang, P.A., Gawali, B.W., Mehrotra, S.C.: Technological basics of EEG recording and operation of apparatus. In: Introduction to EEG- and Speech-Based Emotion Recognition, pp. 19–50. Elsevier/AP, Academic Press is an imprint of Elsevier. https://doi.org/10.1016/B978-0-12-804490-2.00002-6
2. Al-Shargie, F., Kiguchi, M., Badruddin, N., Dass, S.C., Hani, A.F.M., Tang, T.B.: Mental stress assessment using simultaneous measurement of EEG and fNIRS. **7**(10), 3882–3898. https://doi.org/10.1364/BOE.7.003882

3. Al-shargie, F., Tang, T.B., Badruddin, N., Dass, S.C., Kiguchi, M.: Mental stress assessment based on feature level fusion of fNIRS and EEG signals. In: 2016 6th International Conference on Intelligent and Advanced Systems (ICIAS), pp. 1–5. IEEE. https://doi.org/10.1109/ICIAS.2016.7824060

4. Arsalan, A., Majid, M., Butt, A.R., Anwar, S.M.: Classification of perceived mental stress using a commercially available EEG headband. **23**(6), 2257–2264. https://doi.org/10.1109/JBHI.2019.2926407

5. Bandt, C., Pompe, B.: Permutation entropy: a natural complexity measure for time series. https://doi.org/10.1103/PHYSREVLETT.88.174102

6. Bernstein, E.E., McNally, R.J.: Exercise as a buffer against difficulties with emotion regulation: a pathway to emotional wellbeing. **109**, 29–36. https://doi.org/10.1016/j.brat.2018.07.010

7. Cohen, S., Kamarck, T., Mermelstein, R., et al.: Perceived stress scale. **10**, 1–2 (1994)

8. Dzedzickis, A., Kaklauskas, A., Bucinskas, V.: Human emotion recognition: review of sensors and methods. **20**(3), 592. https://doi.org/10.3390/s20030592

9. Goyal, M., Singh, S., Sibinga, E.M.S., Gould, N.F., Rowland-Seymour, A., Sharma, R., Berger, Z., Sleicher, D., Maron, D.D., Shihab, H.M., Ranasinghe, P.D., Linn, S., Saha, S., Bass, E.B., Haythornthwaite, J.A.: Meditation programs for psychological stress and well-being: a systematic review and meta-analysis. **174**(3), 357–368. https://doi.org/10.1001/jamainternmed.2013.13018

10. Gärtner, M., Grimm, S., Bajbouj, M.: Frontal midline theta oscillations during mental arithmetic: effects of stress. **9**. https://doi.org/10.3389/fnbeh.2015.00096

11. Hamid, N.H.A., Sulaiman, N., Aris, S.A.M., Murat, Z.H., Taib, M.N.: Evaluation of human stress using EEG Power Spectrum. In: 2010 6th International Colloquium on Signal Processing & Its Applications, pp. 1–4. IEEE. https://doi.org/10.1109/CSPA.2010.5545282

12. Hamid, N.H.A., Sulaiman, N., Murat, Z.H., Taib, M.N.: Brainwaves stress pattern based on perceived stress scale test. In: 2015 IEEE 6th Control and System Graduate Research Colloquium (ICSGRC), pp. 135–140. IEEE. https://doi.org/10.1109/ICSGRC.2015.7412480

13. Harmony, T., Fernández, T., Silva, J., Bernal, J., Díaz-Comas, L., Reyes, A., Marosi, E., Rodríguez, M., Rodríguez, M.: EEG delta activity: an indicator of attention to internal processing during performance of mental tasks. **24**(1–2), 161–171. https://doi.org/10.1016/S0167-8760(96)00053-0

14. Herman, J.P., Cullinan, W.E.: Neurocircuitry of stress: central control of the hypothalamo–pituitary–adrenocortical axis. **20**(2), 78–84. https://doi.org/10.1016/S0166-2236(96)10069-2

15. Higuchi, T.: Approach to an irregular time series on the basis of the fractal theory. **31**(2), 277–283. https://doi.org/10.1016/0167-2789(88)90081-4

16. Hjorth, B.: EEG analysis based on time domain properties. **29**(3), 306–310. https://doi.org/10.1016/0013-4694(70)90143-4

17. Hou, X., Liu, Y., Sourina, O., Tan, Y.R.E., Wang, L., Mueller-Wittig, W.: EEG based stress monitoring. In: 2015 IEEE International Conference on Systems, Man, and Cybernetics, pp. 3110–3115. https://doi.org/10.1109/SMC.2015.540

18. Huang, N.E., Shen, Z., Long, S.R., Wu, M.C., Shih, H.H., Zheng, Q., Yen, N.C., Tung, C.C., Liu, H.H.: The empirical mode decomposition and the Hilbert spectrum for nonlinear and non-stationary time series analysis. **454**, 903–995 (1971). https://doi.org/10.1098/rspa.1998.0193

19. Jebelli, H., Hwang, S., Lee, S.: EEG-based workers' stress recognition at construction sites. **93**, 315–324. https://doi.org/10.1016/j.autcon.2018.05.027

20. Jebelli, H., Mahdi Khalili, M., Lee, S.: A continuously updated, computationally efficient stress recognition framework using electroencephalogram (EEG) by applying online multitask learning algorithms (OMTL). **23**(5), 1928–1939. https://doi.org/10.1109/JBHI.2018.2870963

21. Jun, G., Smitha, K.G.: EEG based stress level identification. In: 2016 IEEE International Conference on Systems, Man, and Cybernetics (SMC), pp. 003270–003274. IEEE. https://doi.org/10.1109/SMC.2016.7844738

22. Khosrowabadi, R., Quek, C., Ang, K.K., Tung, S.W., Heijnen, M.: A Brain-Computer Interface for classifying EEG correlates of chronic mental stress. In: The 2011 International Joint Conference on Neural Networks, pp. 757–762. https://doi.org/10.1109/IJCNN.2011.6033297

23. Koudouovoh-Tripp, P., Hüfner, K., Egeter, J., Kandler, C., Giesinger, J.M., Sopper, S., Humpel, C., Sperner-Unterweger, B.: Stress enhances proinflammatory platelet activity: the impact of acute and chronic mental stress. https://doi.org/10.1007/s11481-020-09945-4

24. Laurent, F., Valderrama, M., Besserve, M., Guillard, M., Lachaux, J.P., Martinerie, J., Florence, G.: Multimodal information improves the rapid detection of mental fatigue **8**(4), 400–408. https://doi.org/10.1016/j.bspc.2013.01.007

25. Le Fevre, M., Matheny, J., Kolt, G.S.: Eustress, distress, and interpretation in occupational stress. **18**(7), 726–744. https://doi.org/10.1108/02683940310502412

26. Martínez-Rodrigo, A., García-Martínez, B., Zunino, L., Alcaraz, R., Fernández-Caballero, A.: Multi-lag analysis of symbolic entropies on EEG recordings for distress recognition. **13**. https://doi.org/10.3389/fninf.2019.00040

27. Minguillon, J., Lopez-Gordo, M.A., Pelayo, F.: stress assessment by prefrontal relative gamma. **10**. https://doi.org/10.3389/fncom.2016.00101

28. Niedermeyer, E., Schomer, D.L., Lopes da Silva, F.H. (eds.): Niedermeyer's Electroencephalography: Basic Principles, Clinical Applications, and Related Fields, 6th edn. Wolters Kluwer, Lippincott Williams & Wilkins (2017)

29. Paszkiel, S., Dobrakowski, P., Łysiak, A.: The impact of different sounds on stress level in the context of EEG, cardiac measures and subjective stress level: a pilot study. **10**(10), 728. https://doi.org/10.3390/brainsci10100728

30. Saeed, S.M.U., Anwar, S.M., Majid, M., Bhatti, A.M.: Psychological stress measurement using low cost single channel EEG headset. In: 2015 IEEE International Symposium on Signal Processing and Information Technology (ISSPIT), pp. 581–585. https://doi.org/10.1109/ISSPIT.2015.7394404

31. Saeed, S.M.U., Anwar, S.M., Majid, M.: Quantification of human stress using commercially available single channel EEG headset. **E100.D**(9), 2241–2244. https://doi.org/10.1587/transinf.2016EDL8248

32. Saeed, S.M.U., Anwar, S.M., Majid, M., Awais, M., Alnowami, M.: Selection of neural oscillatory features for human stress classification with single channel EEG headset. https://doi.org/10.1155/2018/1049257

33. Secerbegovic, A., Ibric, S., Nisic, J., Suljanovic, N., Mujcic, A.: Mental workload vs. stress differentiation using single-channel EEG. In: Badnjevic, A. (ed.) CMBEBIH 2017, IFMBE Proceedings, vol. 62, pp. 511–515. Springer, Singapore. https://doi.org/10.1007/978-981-10-4166-2_78

34. Sharma, N., Gedeon, T.: Objective measures, sensors and computational techniques for stress recognition and classification: a survey. **108**(3), 1287–1301. https://doi.org/10.1016/j.cmpb.2012.07.003

35. Steingrimsson, S., Bilonic, G., Ekelund, A.C., Larson, T., Stadig, I., Svensson, M., Vukovic, I.S., Wartenberg, C., Wrede, O., Bernhardsson, S.: Electroencephalography-based neurofeedback as treatment for post-traumatic stress disorder: a systematic review and meta-analysis. **63**(1) (2020). https://doi.org/10.1192/j.eurpsy.2019.7
36. Subhani, A.R., Mumtaz, W., Saad, M.N.B.M., Kamel, N., Malik, A.S.: Machine learning framework for the detection of mental stress at multiple levels. **5**, 13545–13556. https://doi.org/10.1109/ACCESS.2017.2723622
37. Suhaimi, N.S., Mountstephens, J., Teo, J.: EEG-based emotion recognition: a state-of-the-art review of current trends and opportunities. https://doi.org/10.1155/2020/8875426
38. Sulaiman, N., Taib, M.N., Lias, S., Murat, Z.H., Mustafa, M., Aris, S.A.M., Rashid, N.A.: Electroencephalogram-based stress index. **2**(3), 327–335. https://doi.org/10.1166/jmihi.2012.1106
39. Sulaiman, N., Ying, B.S., Mustafa, M., Jadin, M.S.: Offline labview-based EEG signals analysis for human stress monitoring. In: 2018 9th IEEE Control and System Graduate Research Colloquium (ICSGRC), pp. 126–131. IEEE. https://doi.org/10.1109/ICSGRC.2018.8657606
40. Ulstein, I., Wyller, T.B., Engedal, K.: High score on the Relative Stress Scale, a marker of possible psychiatric disorder in family carers of patients with dementia. **22**(3), 195–202. https://doi.org/10.1002/gps.1660
41. Wang, Q., Sourina, O.: Real-time mental arithmetic task recognition from EEG signals. **21**(2), 225–232. https://doi.org/10.1109/TNSRE.2012.2236576
42. Zheng, Y., Wong, T.C.H., Leung, B.H.K., Poon, C.C.Y.: Unobtrusive and multimodal wearable sensing to quantify anxiety. **16**(10), 3689–3696. https://doi.org/10.1109/JSEN.2016.2539383

Using EEG Based Brain-Computer Interface to Control Actions in Applications – The Way to Provide New Possibilities for Disabled People

Mateusz Adamczyk$^{(\boxtimes)}$ and Szczepan Paszkiel

Faculty of Electrical Engineering, Automatic Control and Informatics,
Opole University of Technology, Prószkowska 76, 45-758 Opole, Poland
mat.adamczyk@student.po.edu.pl

Abstract. Nowadays, many researches are focusing on finding new technologies which can lead to the revolution in all branches of science. A lot of institutes focus on the dormant potential of the human's brain. The signals from the scalp through electroencephalography (EEG) are the key to obtain many new possibilities for example controlling actions in applications with human's brain. An EEG based brain-computer interface can be used to enable the disabled people doing things equal to normal people, but by use of their brain potential.

Keywords: Electroencephalography (EEG) · Brain-computer interface (BCI) · Motor imagery (MI) · Applications

1 Introduction

After the year 2000, Brain-computer interfaces (BCI) became one of the leading branch of scientific research, because its potential is huge. This technology is a chance for millions of disabled people, especially for the ones that have physical disabilities. Our current technology gave us BCI devices like: motorized wheelchair, Neural Impulse Actuator (a special device developed in 2008 by OCZ Technology for use in video games which relies primarily on EEG), Emotiv EPOC (device that allows users to play video games by using the brain), LIFESUITs exoskeletons for paraplegics and quadriplegics, etc.

A BCI system has an input, output and a signal processing algorithm that maps the inputs to the output. The major strategies that are considered for the input of a BCI system are: P300 wave of event related potentials (ERP), steady state visual evoked potential (SSVEP), slow cortical potentials and motor imagery (MI). The EEG is usually measured non-invasively, with electrodes mounted on the human scalp using conductive gel. One of the most successful BCI strategies relies on subject's ability to learn to alter the mu and central-beta components of the EEG at will [1]. The last sentence is the greatest prove that BCI will be very good solution for everyone, because with its ability to learn, every person would adjust the device to his own needs. This means that for example someone who lost his hand due to an accident will be enabled to configure his BCI

© The Author(s), under exclusive license to Springer Nature Switzerland AG 2021
S. Paszkiel (Ed.): ICBCI 2021, AISC 1362, pp. 129–137, 2021.
https://doi.org/10.1007/978-3-030-72254-8_13

prosthetic arm to his convenience and close to his lost hand's accuracy. BCI technology uses our brain's different states – for example, if we want to move something to the left, we generate thoughts which tell the interface that we want to move the object this way (Fig. 2). Of course it is much easier to sign to this kind of movement the contraction of our muscles, for example the face muscles, because the reaction of the brain is much stronger when we do actions with muscles than only think about something (Fig. 1).

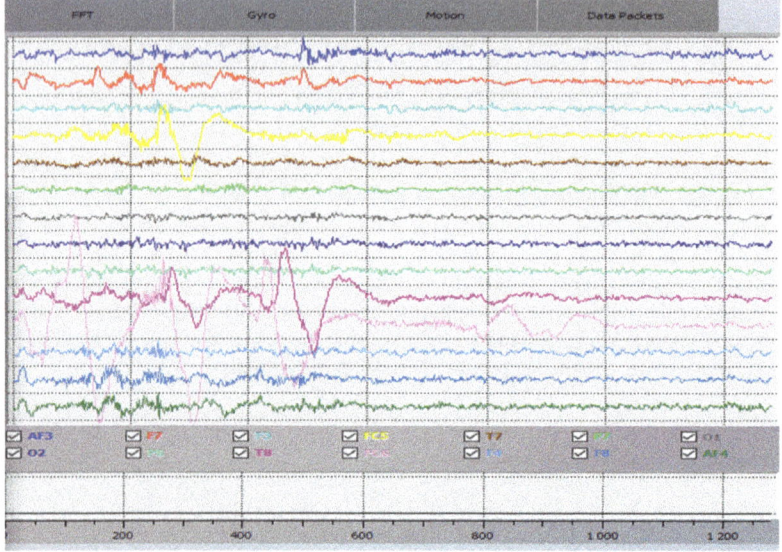

Fig. 1. Brain activity while using muscles contraction.

The ideal example to show the different level of difficulty between calibrating the interface with our focused thoughts only and our gestures only is the Emotiv EPOC + Neuroheadset device. It consists of 16 electrodes, including 14 EEG channels and two-axis gyroscope. Emotiv EPOC + Neuroheadset uses a noninvasive type brainwave monitoring system where EEG is recording the data of electric activity in an interval of 20–40 min or even less from the scalp of the brain. Noninvasive technique comes with an advantage of relieving the subject from the difficulties of operation as the subject can easily measure the neural activity through simple wearable items. EEG actually monitors the voltage fluctuations resulting from ionic current flows within the neurons of the brain and it occurs 1.5 s before the movement takes place [2].

As we see in the figures – the brain reacts stronger when we use the gestures or muscle contraction than focused thoughts only, because using muscles demands more brain power than thinking about something. So it would be easier for a disabled person to calibrate for example his prosthetic arm with gestures usage, because brain's activity is higher. The research was done during 4 h of scientific Laboratory Neuroscience using Emotiv EPOC + NeuroHeadset device.

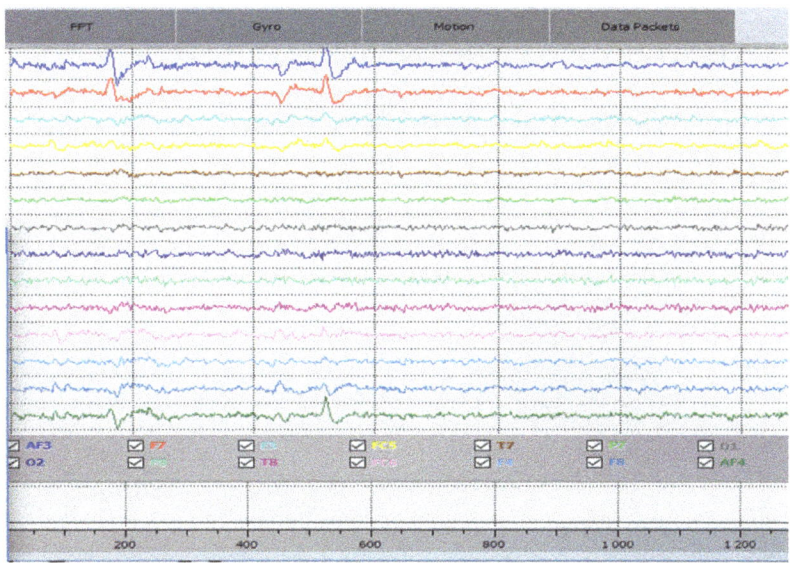

Fig. 2. Brain activity while using focused thoughts only.

2   Motor Imagery – A Chance for People with Muscle Problems

But when we talk about people who are paralyzed and are not able to move or even contract their muscles, Motor Imagery (MI) has to be a leading topic. Motor imagery is a cognitive-perceptive process and is the mental performance of movements not accompanied by any kind of muscular activity. Visual and kinesthetic forms of motor imagery are discriminated. In the former, a person produces a visual image of their own movement, seeing it as though it was a third person's - subjects generate the kinesthetic sensation of movement [3]. The accuracy of movement and the connection between the brain and the machine won't be of course in 100% the same as in the case of a normal healthy limb. The subject will be obliged to practice a lot to adapt its mind to a new leg, arm, hand, etc. The most important thing is to make our cerebrum remember the signals it is sending while we are practicing and learning for example how to move a robotic limb to the left or right side. During the scientific experiments in the laboratory, it was found that it demands plenty of time to adapt and calibrate our brain with the application that is responsible for the connection.

The application used for a research is a part of Emotiv EPOC + NeuroHeadset device and it consist of an orange cube and the panel of actions, which we can learn to control by our mind. We can move the cube to the left/right, drop/lift it (Fig. 3), pull or push it or rotate it. Every action can be learned by two steps: the first step is to clear our mind and don't think about anything – the warm up. The second step is to focus and think about what kind of movement the cube should do. The hardest think is to gain enough focus to reach expected result.

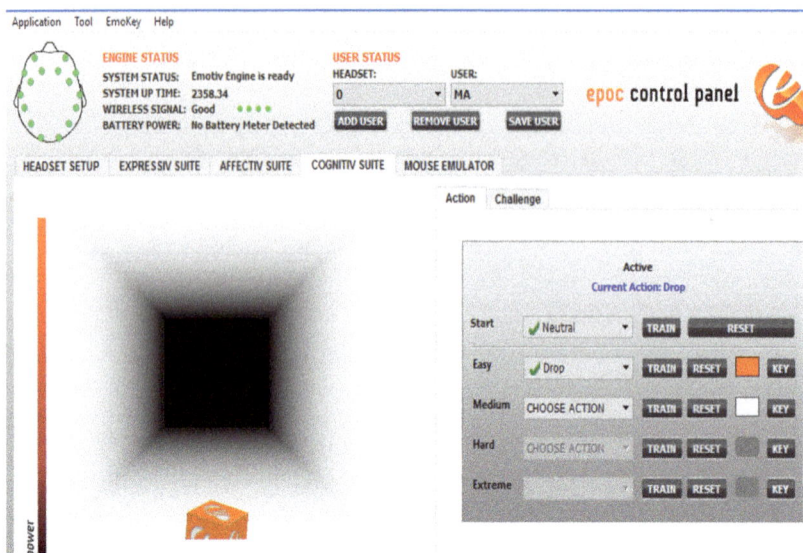

Fig. 3. The behavior of the cube while performing the Drop action with using brain.

This application can imitate the process of controlling for example Motor Imagery based Brain Computer Interface controlled wheelchair. The important thing is to imagine the movement process in our mind. But how accurate the MI with BCI can be?

3   Motor Imagery and BCI Accuracy

The current work presents a comparative evaluation of the MI-based BCI control accuracy between stroke patients and healthy subjects - five patients who had a stroke that affected the motor system participated in the experiment made in the Rehabilitation Hospital of Iasi, and were trained across 10–24 sessions lasting about 1 h each with the RecoveriX system (consists of a computer, a FES device and a biosignal amplifier with active EEG electrodes). The participants' EEG data were classified while they imagined left or right hand movements, and real-time feedback was provided on a monitor. If the correct imagination was detected, the FES (Functional Electrical Stimulation) was also activated to move the left or right hand. The grand average mean accuracy was 87.4% for all patients and sessions. All patients were able to achieve at least one session with a maximum accuracy above 96%. Both (accuracy and the maximum accuracy) were surprisingly high and above results seen with healthy controls in prior studies. Importantly, the study showed that stroke patients can control a MI-BCI system with high accuracy relative to healthy persons. This may occur because these patients are highly motivated to participate in a study to improve their motor functions. Participants often reported early in the training of motor improvements and this caused additional motivation. However, it also reflects the efficacy of combining motor imagination, seeing continuous bar feedback, and real hand movement that also activates the tactile and proprioceptive systems. Results also suggested that motor function could improve even if classification accuracy did not, and suggest other new questions to explore in future work [4]. This shows that with specific technology of stimulation, disabled people can reach high accuracy using MI-BCI technology. So this technology can get them close to the healthy ones and provide better conditions of living.

4 Multitasking in BCI Technology

The next important question is: how hard is to learn multitasking with BCI based device. Will it be a problem for an average person to reach this? Application from Emotiv EPOC + Neuroheadset was used again, but this time the main goal was to calibrate the machine to remember more than one action. We know that multitasking is possible, when we do more than 1 action in the same time. So first we have to teach the machine more than 1 action in the same session (Fig. 4). The hardest thing to do was to achieve more states of the brain – one state for one action. Human brain has to send different signals to make the cube move to the left than to the right (Fig. 5). It demands a lot of time to do it on a minimum level, but it's not impossible. The process of using Emotiv EPOC + Neuroheadset device is simple – firstly, we have to use a special conductive gel on all 16 electrodes. Secondly, we have to connect a special PenDrive which enables connection between PC and Emotiv Headsets. Then we will see on the interface of the application, if all 16 electrodes are connected – the circles should be green. After that all steps, we can start using the device.

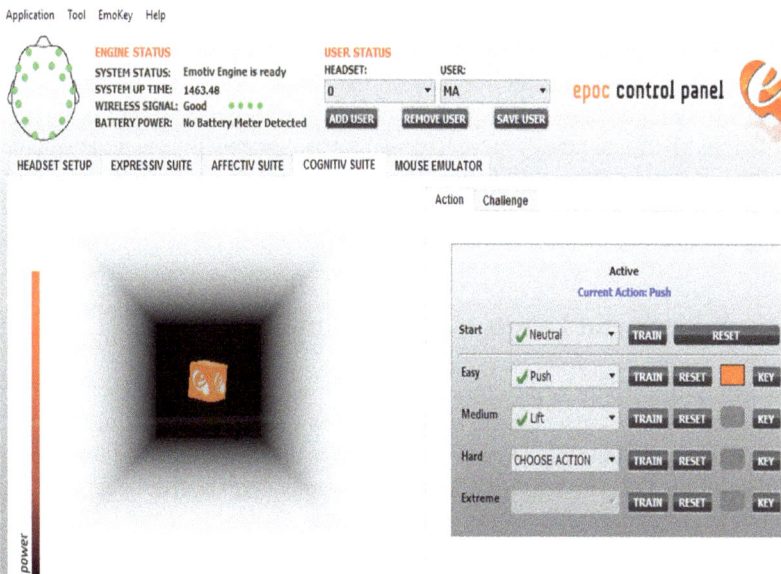

Fig. 4. Teaching machine to remember Lift and Push action in the same session.

As we see in the figures, that session was really overwhelming for the brain, but it is very presumable that more practice would adapt it to the conditions. The BCI technology

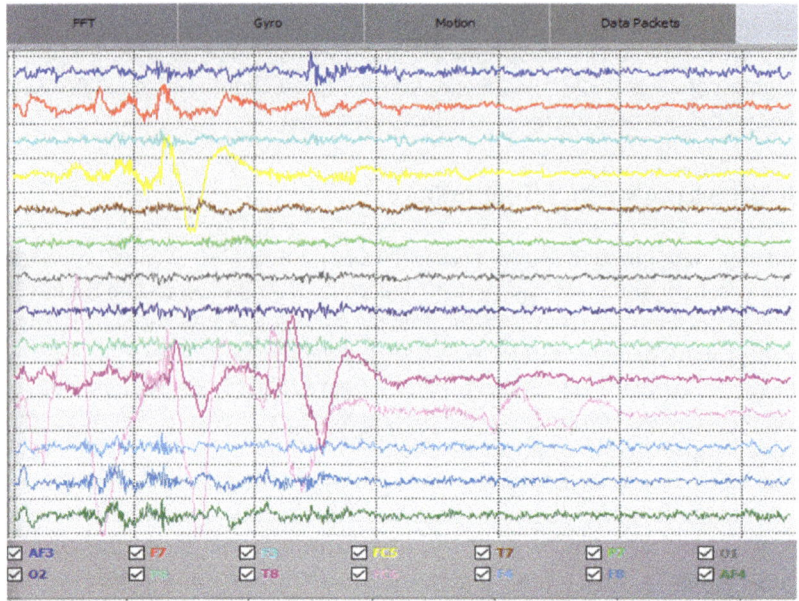

Fig. 5. Brain activity in the session, where 2 actions where remembered by the machine.

is trying to make the device more flexible for multitasking. There are many researches which tend to construct a BCI that can be used without any specific subject-dependent calibration to make that technology more efficient [5]. But the most important thing is to mention that the user must be focused on the single task or action paradigm being performed, so quickly adapting cognitive processes, which enable multitasking, aren't utilized to make BCI technology more effective for use for now [6]. As we know, human's inventions are getting better and better, so in the future we can expect this topic to be continued. For now, we know that something like for example prosthetic BCI arm can easily satisfy basic needs and it is for a chance for a disabled person to live a normal life.

5   Convenience and Limits of BCI Technology

Another thing to consider is the convenience of using such technology. We know, that such complicated devices, which use BCI systems can also have their limits. As we know today, there exist systems which consist of modules that can overcome for example the problem of wiring [7]. The next limit for BCI is a person with neurological and cognitive disabilities [8]. In this case the communication between the brain and BCI can be more complicated. Furthermore, there's a very important limit of BCI – the hardware. Many devices now demand electrodes that need conductive gel and aren't comfortable enough, so the time of using is also limited [9]. The reliability of the BCI technology is also a problem, because machines aren't as dependable as limbs in muscle-based actions. Without this improvement, BCI prosthetic devices remain limited. Also the look of that machines won't be natural, because the skin won't be the same, the feeling won't be the same and of course the behaviour will be different – a prosthetic limb won't bleed for example. For some people, especially for that kind who lost their limbs due to an accident, it would be uncomfortable to adapt because of the machine's limits.

6   P300 Speller – Description and Appliance

As an example of a BCI device which helps people with their disability I want to describe BCI with the P300 Speller. It helps people who struggle with ALS (amyotrophic lateral sclerosis) – a progressive neurodegenerative disease, which leads to a destruction of the cells in the area of the spinal cord. P300 Speller is a virtual keyboard connected with the BCI system [10]. It uses OpenViBE platform and runs on Windows system. The keyboard displays on a monitor and has 43 symbols including punctuation marks and a backspace key. Characters flash in order to induce a P300 response for the attended character. P300 is a brain wave which is elicited in a decision making process. An interesting parameter of the P300 wave is its amplitude, because it can give us a lot of information about someone, for example when we are involved in a driving a car – evoked P300 amplitudes have been found to be larger in poor navigators than good ones [11]. Characters flash randomly in a letter matrix and P300 Speller types them using a P300 component [12]. Thanks to this, people with ALS will have a chance for functioning normally, because the main problem of an ALS disease is the communication ability loss [13]. The change of the character size, motion and sharpness on the P300 Speller interface has a big impact on the stimulus used on our brain in this device [14].

7   The Luke Arm – A Bionic Arm from Star Wars, Similar to a True Limb in a Real Word

As we know, a machine won't substitute a true natural limb, but it can be similar to it, so it could even enable disabled people to feel a stimulus from the world around them. That kind of things could be possible thanks to The LUKE Arm – a bionic arm. The arm itself has been developed by DEKA Research & Development Corporation.

It consisted of a metal motors and a part of skin made from silicon. The system which is used by this arm has been developed by the University of Utah team. This system allows to tap into nerves and send signals to the arm to move. The Array consists of 100 microelectrodes and wires that are implanted into the amputee's nerves – the forearm nerves, and connected to a computer. The Array reads signals from the still-remaining arm nerves and transfers it to the computer, which translates them to digital signals that tell the arm to move. Of course the machine has to learn how to "feel" the objects, if we want to pick something up with this device. Everything is to let the machine know, how much pressure should it use to pick something or hold something. The LUKE Arm isn't available yet, because there's many things to develop. For example, the main problem is that the device isn't portable – it has to be wired to a computer outside the body, but the developers believe it will be ready to give it for a home testing for 3 subjects in 2020 or 2021 [16, 17].

8 Summary

The current technology is still being developed. Especially the main focus of a significant bunch of research is around the BCI technology. Human brain is still not examined in 100% [21]. There is a potential in it, which is unknown for us in this moment. The one thing is certain – this organ is very powerful. This means that it can "adapt" to the circumstances that is to say it can control things, if only we calibrate it with a special BCI technology. So namely, you will need only a cerebrum to control all your body movements, which is a huge opportunity for a disabled people, who lost their limbs or were born without. That also is guaranteed by Motor Imagery, which doesn't demand any muscular activity.

Of course, when someone has the neurological problems and due to them isn't possible for him to overwhelm his brain that much, there's a solution, which was also described in this chapter – the muscle contraction combined with BCI. It was proved with experiments in the laboratory presented on the screens, that our brain reacts strong for impulses generated by the muscle contraction.

So generally the BCI technology is developed on this level, that it can fulfil basic human needs and even (if practiced) apply simple multitasking.

It is important to add, that people with such diseases and problems can get close to the level of a healthy ones thanks to the still developing MI-BCI systems.

Due to its still existing limits, like for example the wiring problem or the hardware problem, BCI is a light at the end of the tunnel for overcoming disability – one of the biggest problems of the human race. The current concepts or prototypes of the devices like The LUKE Arm, which is going to be released in 2020/2021 or the existing machines like P300 speller (which is currently in use) show that in the future, the disability will be another problem to forget about.

But the main problem in the future and probably the only problem hard to solve will be the cost of such technology. As we can conclude, such devices will be very expensive. For example The LUKE Arm is being provided to cost about 150 000–200 000$.

References

1. Irimia, D.C., Ortner, R., Krausz, G., Guger, C., Poboroniuc, M.S.: BCI application in robotics control. IFAC **45**(6), 1869–1874 (2012)
2. Paszkiel, S., Szpulak, P.: Methods of acquisition, archiving and biomedical data analysis of brain functioning. In: Hunek, W.P., Paszkiel, S. (eds.) Biomedical Engineering and Neuroscience. Book Series: Advances in Intelligent Systems and Computing, vol. 720, pp. 158–171 (2018). https://doi.org/10.1007/978-3-319-75025-5_15
3. Mokienko, O.A., Chernikova, L.A., Frolov, A.A., Bobrov, P.D.: Motor imagery and its practical application. Neurosci. Behav. Physiol. **44**(5), 483–489 (2014)
4. Irimia, D.C., Ortner, R., Poboroniuc, M.S., Ignat, B.E.: High classification accuracy of a motor imagery based brain-computer interface for stroke rehabilitation training. Front. Robot. AI **5**, 130 (2018)
5. Paszkiel, S., Sikora, M.: The use of brain-computer interface to control unmanned aerial vehicle. In: Szewczyk, R., Zielinski, C., Kaliczynska, M. (eds.) Automation 2019: Progress in Automation, Robotics And Measurement Techniques. Advances in Intelligent Systems and Computing, vol. 920, pp. 583–598 (2020). https://doi.org/10.1007/978-3-030-13273-6_54

6. Paszkiel, S., Hunek, W.P., Shylenko, A.: Project and simulation of a portable device for measuring bioelectrical signals from the brain for states consciousness verification with visualization on LEDs. In: Szewczyk, R., Zielinski, C., Kaliczynska, M. (eds.) Challenges in Automation, Robotics and Measurement Techniques. Advances in Intelligent Systems and Computing, vol. 440, pp. 25–35 (2016). https://doi.org/10.1007/978-3-319-29357-8_3

7. Paszkiel, S.: The use of facial expressions identified from the level of the EEG signal for controlling a mobile vehicle based on a state machine. In: Szewczyk, R., Zielinski, C., Kaliczynska, M. (eds.) Automation 2020: Towards Industry of the FutureBook Series. Advances in Intelligent Systems and Computing, vol. 1140, pp. 227–238 (2020). https://doi.org/10.1007/978-3-030-40971-5_21

8. Paszkiel, S.: Data acquisition methods for human brain activity. In: Analysis and Classification of EEG Signals for Brain-Computer Interfaces. Studies in Computational Intelligence, vol. 852, pp. 3–9 (2020). https://doi.org/10.1007/978-3-030-30581-9_2

9. Paszkiel, S.: Using the raspberry PI2 module and the brain-computer technology for controlling a mobile vehicle. In: Szewczyk, R., Zielinski, C., Kaliczynska, M. (eds.) Automation 2019: Progress in Automation, Robotics and Measurement Techniques. Advances in Intelligent Systems and Computing, vol. 920, pp. 356–366 (2020). https://doi.org/10.1007/978-3-030-13273-6_34

10. Guy, V., Soriani, M.H., Bruno, M., et al.: Brain computer interface with the P300 speller: usability for disabled people with amyotrophic lateral sclerosis. Ann. Phys. Rehabil. Med. (2017)

11. Li, F., Tao, Q., Peng, W., Zhang, T., et al.: Inter-subject P300 variability relates to the efficiency of brain networks reconfigured from resting – to task-state: evidence from a simultaneous event-related EEG-fMRI study. NeuroImage **205**, 116285 (2019)

12. Won, K., Kwon, M., Jang, S., et al.: P300 speller performance predictor based on RSVP multi-feature. Front. Hum. Neurosci. **13**, 261 (2019)

13. Paszkiel, S.: Characteristics of question of blind source separation using moore-penrose pseudoinversion for reconstruction of EEG signal. In: Szewczyk, R., Zielinski, C., Kaliczynska, M. (eds.) Automation 2017: Innovations in Automation, Robotics and Measurement Techniques. Advances in Intelligent Systems and Computing, vol. 550, pp. 393–400 (2017). https://doi.org/10.1007/978-3-319-54042-9_36

14. Speier, W., Deshpande, A., Cui, L., Chandravadia, N., Roberts, D., Pouratian, N.: A comparison of stimulus types in online classification of the P300 speller using language models. PLoS ONE **12**(4), (2017)

15. Researchgate. https://www.researchgate.net/figure/Screenshot-of-6-x-6-P300-Speller-and-experimental-environment_fig1_220366680. Accessed 10 Nov 2018

16. George, J.A., Kluger, D.T., Davis, T.S., Wendelken, S.M., Okorokova, E.V., et al.: Biomimetic sensory feedback through peripheral nerve stimulation improves dexterous use of a bionic hand. Sci. Robot. (2019)

17. University of Utah: Motorized prosthetic arm can sense touch, move with your thoughts (2019)

18. Mobius bionics page. http://www.mobiusbionics.com/luke-arm/. Accessed 11 Nov 2018

19. Fatherly page. https://www.fatherly.com/news/robotic-arm-inspired-by-luke-skywalker-gives-amputees-the-ability-to-touch-and-feel-again/. Accessed 11 Nov 2018

20. Mobius bionics page. http://www.mobiusbionics.com/wpcontent/uploads/2019/09/Mobius-Bionics-LUKE-Product-Spec-Sheet.pdf. Accessed 11 Nov 2018

21. Paszkiel, S., Dobrakowski, P., Lysiak, A.: The impact of different sounds on stress level in the context of EEG, cardiac measures and subjective stress level: a pilot study. Brain Sci. **10**(10), 728 (2020). https://doi.org/10.3390/brainsci10100728

COVID-19 Disease During Pregnancy - Presentation of Two Cases and Literature Review

Dariusz Kowalczyk[1]([✉]), Szymon Piątkowski[2], Agata Wysocka[2], Patrycja Trentkiewicz[2], Justyna Kordek[2], and Zuzanna Pokorna[2]

[1] Department of Anatomy, Faculty of Medicine, University of Opole, Opole, Poland
kontakt@dariuszkowalczyk.pl
[2] Students Scientific Association of Gynecology and Obstetrics, University of Opole, Opole, Poland

Abstract. COVID-19 disease is currently an extremely difficult problem for healthcare systems in Poland and in the world. The growing number of infections makes one suspect that the incidence of COVID-19 cases will also increase in the population of pregnant women. The article presents case reports of two patients with positive results of the SARS-CoV-2 test and a literature review on COVID-19 during pregnancy.

Keywords: COVID-19 · Pregnancy · SARS-CoV-2 · Treatment

1 Introduction

In December 2019, the first cases of COVID-19 disease appeared in China. The notice about the first confirmed case of severe acute respiratory syndrome caused by new type of coronavirus, called SARS-CoV-2, was announced on the first December 2019 in Wuhan, China [1]. In the following months, cases of the disease were confirmed on all continents, and the WHO, due to the increase in cases and deaths, announced a pandemic on March 11, 2020 [2].

SARS-CoV-2 virus infection in pregnant women is related to many potential complications that may affect the course of pregnancy and childbirth. The period of pregnancy is associated with the weakening of the cell-type immune response, which may increase the risk of developing the disease and the severe course of viral infections [3]. However, no differences were found between the mortality of pregnant and non-pregnant women. On the other hand, there are more frequent hospitalizations in the intensive care unit and an increased number of mechanical ventilation among pregnant women [4]. SARS-CoV-2 infections occur via droplets (mainly) and through contact with the surface on which it has settled [5]. The most common symptoms of COVID-19 disease in pregnant women include: cough, fever, headache, dyspnoea, nausea and vomiting, abdominal pain, runny nose, and loss of taste and/or smell [4]. Due to the non-specific nature of symptoms, laboratory and imaging diagnostics have crucial importance.

© The Author(s), under exclusive license to Springer Nature Switzerland AG 2021
S. Paszkiel (Ed.): ICBCI 2021, AISC 1362, pp. 138–146, 2021.
https://doi.org/10.1007/978-3-030-72254-8_14

2   Two Case Reports of Patients with SARS-CoV-2 Infection, Treated in the Clinical Department of Gynecology and Obstetrics in Nysa

2.1   Patient 1, 36 Years, Gravida 2, Para 2

The patient in the 7th week of pregnancy was admitted to the department with a delayed miscarriage. So far, she has not been seriously ill. She gave birth naturally once, the child was born alive and healthy. During the admission, she completed an epidemiological interview card in which she did not indicate any risk factors for SARS-CoV-2 infection and reported no clinical symptoms of infection (fever, cough, dyspnoea). After admission, an ultrasound examination confirmed the miscarriage in the 7th week of pregnancy. Pharmacological induction of miscarriage with misoprostol has been proposed. Testing for the presence of SARS-CoV-2 virus was also ordered. The patient was given 2 tablets of misoprostol vaginally, after which the miscarriage was completed. During this time, a positive test result for SARS-CoV-2 was obtained. The patient was discharged home in good condition, with a recommendation to go to isolation. After the miscarriage, the patient did not develop clinical signs of COVID-19 disease.

The staff taking care of the patient were equipped with surgical masks and gloves. Unfortunately, in connection with this case, the staff of the department became infected. However, there was no need for hospitalization. As it turned out, the patient misled the staff on purpose because she had contact with someone suffering from SARS-CoV2 before admission.

2.2   Patient 2, 30 Years, Gravida 6, Para 3

The patient, gravida 6, para 3, was admitted to the department on March 1st, in 29th week of pregnancy with symptoms of dyspnoea, tachycardia and threatened preterm labour. She was hospitalized before March 10th, 2020 (at that time tests for SARS-CoV-2 infection were not yet commonly performed in Poland). After the implementation of a typical treatment, which inhibits contractile uterine activity. Internal medicine consultation raised the suspicion of pulmonary embolism. On physical examination dyspnoea and chronic fatigue were observed. No changes in an auscultation of lungs and electrocardiography were detected.

In the chest X-ray, an ongoing inflammatory process was suspected. The picture showed both lung fields with peribronchial streaky condensations, representative for viral pneumonia (Fig. 1). The treatment included 600 mg of lutein intravaginally, magnesium in an intravenous infusion and low-molecular-weight heparin in a therapeutic dose. The patient received a full course of steroid therapy to stimulate the maturation of the fetal lungs. In the cervical swab Escherichia coli was cultured, whereas Staphylococcus aureus in the pharyngeal swab. Blood oxygen saturation was then at level of 94%. During the consecutive days of hospitalization, the general health of the patient vitally deteriorated. Major fatigue occurred – the patient stopped getting up from bed, dyspnoea, coughing and decrease of saturation to 81%. Regular contractions of uterus appeared in the CTG record, testifying the start of a preterm labour. Saturation was stable at the level of 80%. During the contractile uterine activity fetal heart rate was dropping below 90 beats per minute. Considering the deteriorating general health condition of the patient with respiratory failure and the risk of fetal asphyxia, the patient was qualified to an urgent cesarean section. An alive, preterm male neonate was born, weighing 1460 g, rated on 10 points in Apgar scale. After the cesarean section, the patient was transferred to the intensive care unit, where she required treatment with 100% oxygen for 24 h. Computed tomography angiography (CTA) of lungs was performed, where minor embolic changes in peripheral segments of pulmonary vessels and ground-glass opacity were verified (Fig. 2). The patient was transferred back to the obstetric ward after 24 h. In the maternity ward administration of unfractionated heparin in continuous intravenous infusion, an antibiotic therapy and a passive oxygen therapy were being continued. An intensive treatment was maintained for 5 days after the cesarean section. Afterwards the dose of heparin was gradually reduced and an oxygen therapy was being applied on request. In the 9th day postpartum, the patient was discharged from the hospital in good general health. The newborn, in good general health, was being abided in the neonatal unit due to features of preterm birth.

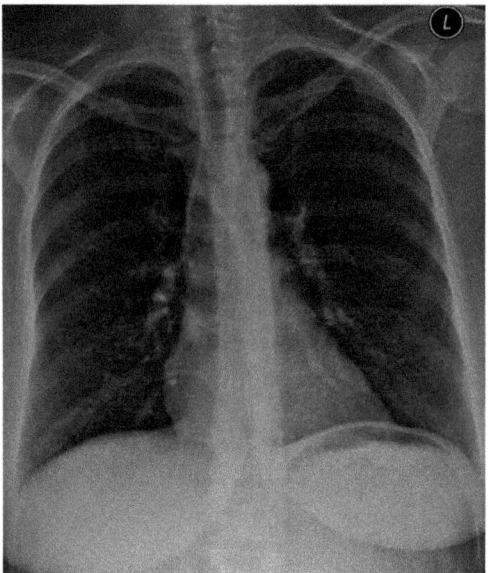

Fig. 1. Chest X-ray in anteroposterior view. Peribronchial streaky condensations, representative for viral pneumonia, are visible in both lungs.

At that time, the staff wasn't fully secured and in some physicians and midwifes periodically appeared dyspnoea, without elevated temperature, but this was not yet associated with SARS-CoV-2 infection.

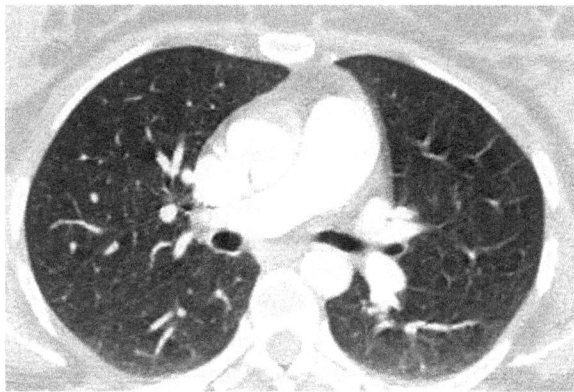

Fig. 2. Computed tomography angiography (CTA). The picture reveals ground-glass opacities in both lungs.

3 Characteristics of the Etiological Factor

SARS-CoV-2 belongs to the coronaviruses; this is a group of enveloped RNA viruses including MERS-CoV and SARS-CoV. SARS-CoV-2 is a zoonotic virus with a characteristically high virulence. Originally appeared at a seafood market in Wuhan, China. It is said that bats were probably its host [6]. It is indicated that environmental and cultural factors played an important role in the spread of the virus. Low temperatures and humidity during the winter 2019 may have contributed to the survival of the virus, and the consumption of wild animals facilitated its transmission to the human body [7]. The spike protein seems to be crucial in the course of infection. It is a membrane protein that binds to a specific cell receptor through its S1 subunit, and then participates in the fusion of the virus and the host's cell membrane through the S2 subunit [6]. From the clinical point of view, it is also important that the SARS-CoV-2 virus binds to the angiotensin II converting receptor (ACE2), which is strongly expressed especially on type II pneumocytes and intestinal epithelial cells [8].The virus is transmitted between people through respiratory droplets or by contact with the surface on which droplets from the respiratory system of an infected person have deposited [9]. There is currently no evidence of vertical transmission, so we can assume that, as with other coronaviruses, it will not be reported [10]. Therefore, it should be suspected that cases of transmission to a child concern the postpartum period.

4 The Course of the Disease in Pregnant Women

There are contradictory reports in the literature regarding complications and the disease's course in pregnant women; however, there is much information that may prove useful in everyday clinical practice. The differences in the course of the disease between pregnant and non-pregnant women include reduced frequency of headache, muscle pain, fever and diarrhea in pregnant women. Chronic lung disease, diabetes, and cardiovascular disease are reported as comorbidities in pregnant women with COVID-19. They are also more common than in non-pregnant women. It is worth noticing that pregnant women's hospitalization ratio is also significantly higher than in non-pregnant women. They require admission to the intensive care unit and mechanical ventilation more often [4]. Among the complications of pregnancy, premature rupture of the membranes, coagulopathies and liver dysfunction are reported. There is also an increased number of cesarean sections and premature deliveries [11].

Currently, COVID-19 disease in pregnant women is not an indication for termination of pregnancy [12]. Mild and stable cases should be monitored, and the pregnancy continued, and in critical situations, when there is a high risk to the life of the fetus and mother, termination of pregnancy should be considered [11]. A multidisciplinary team should manage cases of pregnancy in women with COVID-19. It is also necessary to monitor hospitalized pregnant women's mental state because the risk of anxiety and depression increases due to isolation.

It is worth considering the use of psychological care through telephone contact or other indirect methods [11]. Recommendations for management of labor in women with SARS-CoV-2 infection are presented in the table (Table 1). Both SARS-CoV-2 infection and COVID-19 are not an obligatory indication for cesarean section, but in practice, it is often preferred due to technical difficulties in protecting medical stuff and the newborn from infection during natural delivery.

Table 1. Recommendations for management of labor in women with SARS-CoV-2 infection.

Recommendations for management of labor in women with SARS-CoV-2 infection
The woman in labor should wear personal protective equipment (mask with a filter and protective gloves) throughout the hospital
After leaving the room where the patient is staying, remove all protective clothing and secure it correspondingly. Before re-entering the room, put on a new set of protective clothing
The newborn should be washed and placed in an incubator as soon as possible after birth
The mother should be separated from the newborn for 14 days after labor
24 h after labor, the newborn should be tested for SARS-CoV-2
Family births should be abandoned
Health professionals in the delivery room should be kept to the minimum necessary
Unless the SARS-CoV-2 test is negative, skin-to-skin contact of the newborn with the mother should be avoided
Health professionals should be equipped with and use personal protective equipment throughout the hospital
Only the necessary equipment should be found in the labor ward
A patient in labor without anesthesia should wear a mask with a filter
Breastfeeding is not recommended
The mother should use a breast pump to maintain lactation
Healthcare professionals should be alerted to their own potential symptoms of COVID-19

5   Diagnostic and Therapeutic Possibilities

Taken into consideration non-specific symptoms of SARS-CoV-2, infection is diagnosed by laboratory tests and computed tomography (CT). Nasopharyngeal and oropharyngeal swabs are generally used for antigen testing or RT-PCR. These methods allow the detection of genetic material of the virus. Secretions from the lower respiratory tract also can be used for testing. Laboratory tests carried out on pregnant women during COVID-19 disease shows frequently lymphocytopenia, elevated C- reactive protein, increase of alanine aminotransferase (ALT) and aspartate aminotransferase (AST) [11]. Computed tomography plays a significant role in the diagnosis of COVID-19 disease. Due to CT we can diagnose COVID-19 pneumonia at different stages of the disease [11]. Ground-glass opacity with is the most common finding, observed by radiologists. The higher sensitivity of computed tomography scan comparing to RT-PCR is visibly noted [13].

Vaccines have appeared recently on the market, which gives an optimistic approach to look forward to way out of the crisis related to the health care system. However, until a satisfactory level of social immunity would be achieved, new incidents of the disease will continue to appear. For COVID-19 disease, treatment options are limited and based on symptom relief. Treating pregnant women is especially difficult due to the limitation of the use of certain drugs during pregnancy. Hypoxia during pregnancy is a risk of threat to both the mother and the fetus. Oxygen therapy is frequently used to avoid decreased saturation and dangerous complications of oxygen deficiency. The therapy uses oxygen mask or nasal cannulas. This method is however ineffective in severe cases when a large part of the lungs has been damaged. In difficult and complex cases, it is worth to consider the use of mechanical ventilation or high-frequency ventilation [14]. It must be emphasized that there are no unambiguous evidence that would prove effectiveness and safety of implementing extracorporeal membrane oxygenation (ECMO/ECLS). Thus, in cases when not only assisted ventilation is impossible but also other procedures of increasing blood saturation failed, it is worth considering carrying out ECMO [15].The use of steroids for pregnant women with COVID-19 can postpone the removal of the virus from the organism. Steroids should be administered at risk of preterm labor in order to promote lung maturation [11]. There is an increased risk of thrombosis both in pregnancy and in the course of COVID-19 [16]. It is vital to implement anticoagulant therapy and monitor blood clotting parameters throughout the entire pregnancy and a course of disease. The use of antiretroviral drugs should be also considered. In China Lopinavir/Ritonavir treatment is routinely used with nebulized α-interferon inhalation [18].

Another therapeutic strategy may include convalescent plasma, however, there is no clear evidence for the safety and efficacy of this treatment in pregnant patients. However, there are isolated cases of a very good response to such treatment [19].

6   The Role of Diagnostic Imaging in the Described Cases

In the described cases diagnostic imaging played a significant role. The most common method of chest and lung imaging is chest X-ray. The X-ray tube emits radiation that the patient is exposed to. Some of the radiation is retained on the human organs, and the rest passes through and is recorded. An image is created in the form of a negative [20].

X-ray is a method that has been used for a long time and is still very popular, mainly due to its simplicity, accessibility and a relatively low dose of radiation that the patient receives. The classic examination of the chest, which was taken in patient No. 2, allowed the diagnosis and treatment. The phenomenon of attenuating X-rays in contact with the patient's body parts is also used by computed tomography. The patient is moved into the area where the X-ray tube emits X-rays. It circulates around the patient's body, creating images from many angles. Then, the computer processes the obtained photos and presents the image in the form chosen by the radiologist assistant [20]. The CT angiography method used in patient No. 2 enables precise visualization of pulmonary arteries and the differential diagnosis between pulmonary and vascular diseases.

An extremely important imaging technique in a daily practice of obstetricians and gynecologists is ultrasound. Ultrasound uses sound waves to create an image. The probe emits sound waves that reflect off the tissues and organs and then return to the emitter. Sound waves are converted into electrical impulses, digitally processed and presented as an image on a monitor screen [20]. In the first case, ultrasound was used to confirm the miscarriage, and in the second, to assess the fetal well-being.

7 Summary

In urgent cases, it is worth considering diagnostics using computed tomography, because the time between the examination and treatment application is shortened. The principle of limited confidence in new patients should be applied and tests for SARS-CoV-2 infection should be performed regardless of the interview. In all cases of confirmation or suspicion of SARS-CoV-2 infection, isolation of the pregnant woman should be strictly applied. At the same time, the patient's condition and the fetal well-being must be monitored.

References

1. Ludwig, S., Zarbock, A.: Coronaviruses and SARS-CoV-2: a brief overview. Anesth. Analg. **131**(1), 93–96 (2020). https://doi.org/10.1213/ANE.0000000000004845. PMID: 32243297; PMCID: PMC7173023
2. Pollard, C.A., Morran, M.P., Nestor-Kalinoski, A.L.: The COVID-19 pandemic: a global health crisis. Physiol. Genomics **52**(11), 549–557 (2020). https://doi.org/10.1152/physiolgenomics.00089.2020. Epub 29 Sept 2020. PMID: 32991251; PMCID: PMC7686876
3. Zhang, L., Dong, L., Ming, L., Wei, M., Li, J., Hu, R., Yang, J.: Severe acute respiratory syndrome coronavirus 2(SARS-CoV-2) infection during late pregnancy: a report of 18 patients from Wuhan, China. BMC Pregnancy Childbirth **20**(1), 394 (2020). https://doi.org/10.1186/s12884-020-03026-3. PMID: 32641013; PMCID: PMC7341473
4. Ellington, S., Strid, P., Tong, V.T., Woodworth, K., Galang, R.R., Zambrano, L.D., Nahabedian, J., Anderson, K., Gilboa, S.M.: Characteristics of women of reproductive age with laboratory-confirmed SARS-CoV-2 infection by pregnancy status - United States, 22 January–7 June 2020. MMWR Morb. Mortal. Wkly Rep. **69**(25), 769–775 (2020). https://doi.org/10.15585/mmwr.mm6925a1. PMID: 32584795; PMCID: PMC7316319

5. Fathizadeh, H., Maroufi, P., Momen-Heravi, M., Dao, S., Köse, Ş., Ganbarov, K., Pagliano, P., Esposito, S., Kafil, H.S.: Protection and disinfection policies against SARS-CoV-2 (COVID-19). Infez Med. **28**(2), 185–191 (2020). PMID: 32275260

6. Liu, J., Xie, W., Wang, Y., Xiong, Y., Chen, S., Han, J., Wu, Q.: A comparative overview of COVID-19, MERS and SARS: review article. Int. J. Surg. **81**, 1–8 (2020). https://doi.org/10.1016/j.ijsu.2020.07.032. Epub 26 July 2020. PMID: 32730205; PMCID: PMC7382925

7. Sun, Z., Thilakavathy, K., Kumar, S.S., He, G., Liu, S.V.: Potential factors influencing repeated SARS outbreaks in China. Int. J. Environ. Res. Public Health **17**(5), 1633 (2020). https://doi.org/10.3390/ijerph17051633. PMID: 32138266; PMCID: PMC7084229

8. Shang, J., Ye, G., Shi, K., Wan, Y., Luo, C., Aihara, H., Geng, Q., Auerbach, A., Li, F.: Structural basis of receptor recognition by SARS-CoV-2. Nature. **581**(7807), 221–224 (2020). https://doi.org/10.1038/s41586-020-2179-y. Epub 30 Mar 2020. PMID: 32225175; PMCID: PMC7328981

9. Ortiz, E.I., Herrera, E., De La Torre, A.: Coronavirus (COVID 19) infection in pregnancy. Colomb. Med. (Cali) **51**(2), (2020). https://doi.org/10.25100/cm.v51i2.4271. PMID: 33012886; PMCID: PMC7518733

10. Schwartz, D.A., Dhaliwal, A.: Infections in pregnancy with COVID-19 and other respiratory RNA virus diseases are rarely, if ever, transmitted to the fetus: experiences with coronaviruses, HPIV, hMPV RSV, and influenza. Arch. Pathol. Lab Med. (2020). https://doi.org/10.5858/arpa.2020-0211-sa. PMID: 32338533

11. Marim, F., Karadogan, D., Eyuboglu, T.S., Emiralioglu, N., Gurkan, C.G., Toreyin, Z.N., Akyil, F.T., Yuksel, A., Arikan, H., Serifoglu, I., Gursoy, T.R., Sandal, A., Akgun, M.: Lessons learned so far from the pandemic: a review on pregnants and neonates with COVID-19. Eurasian J. Med. **52**(2), 202–210 (2020). https://doi.org/10.5152/eurasianjmed.2020.20118. PMID: 32612432; PMCID: PMC7311134

12. Qi, H., Luo, X., Zheng, Y., Zhang, H., Li, J., Zou, L., Feng, L., Chen, D., Shi, Y., Tong, C., Baker, P.N.: Safe delivery for pregnancies affected by COVID-19. BJOG **127**(8), 927–929 (2020). https://doi.org/10.1111/1471-0528.16231. Epub 28 Apr 2020. PMID: 32219995

13. Gong, X., Song, L., Li, H., Li, L., Jin, W., Yu, K., Zhang, X., Li, H., Ke, H., Lu, Z.: CT characteristics and diagnostic value of COVID-19 in pregnancy. PLoS ONE **15**(7), (2020). https://doi.org/10.1371/journal.pone.0235134. PMID: 32614854; PMCID: PMC7331988

14. Ai, T., Yang, Z., Hou, H., Zhan, C., Chen, C., Lv, W., Tao, Q., Sun, Z., Xia, L.: Correlation of chest CT and RT-PCR testing for coronavirus disease 2019 (COVID-19) in China: a report of 1014 cases. Radiology. **296**(2), E32–E40 (2020). https://doi.org/10.1148/radiol.2020200642. Epub 26 Feb 2020. PMID: 32101510; PMCID: PMC7233399

15. Jiang, B., Wei, H.: Oxygen therapy strategies and techniques to treat hypoxia in COVID-19 patients. Eur. Rev. Med. Pharmacol. Sci. **24**(19), 10239–10246 (2020). https://doi.org/10.26355/eurrev_202010_23248. PMID: 33090435

16. Zeng, Y., Cai, Z., Xianyu, Y., Yang, B.X., Song, T., Yan, Q.: Prognosis when using extracorporeal membrane oxygenation (ECMO) for critically ill COVID-19 patients in China: a retrospective case series. Crit. Care **24**(1), 148 (2020). https://doi.org/10.1186/s13054-020-2840-8. PMID: 32293518; PMCID: PMC7156900

17. Thachil, J., Tang, N., Gando, S., Falanga, A., Cattaneo, M., Levi, M., Clark, C., Iba, T.: ISTH interim guidance on recognition and management of coagulopathy in COVID-19. J. Thromb. Haemost. **18**(5), 1023–1026 (2020). https://doi.org/10.1111/jth.14810. Epub 27 Apr 2020. PMID: 32338827

18. Liang, H., Acharya, G.: corona virus disease (COVID-19) in pregnancy: what clinical recommendations to follow? Acta Obstet. Gynecol. Scand. **99**(4), 439–442 (2020). https://doi.org/10.1111/aogs.13836. Epub 5 Mar 2020. PMID: 32141062
19. Jafari, R., Jonaidi-Jafari, N., Dehghanpoor, F., Saburi, A.: Convalescent plasma therapy in a pregnant COVID-19 patient with a dramatic clinical and imaging response: a case report. World J. Radiol. **12**(7), 137–141 (2020). https://doi.org/10.4329/wjr.v12.i7.137. PMID: 32850016; PMCID: PMC7422528
20. Pruszyński, B., Cieszanowski, A.: Radiologia. Diagnostyka obrazowa. RTG, TK, USG i MR. (B. Pruszyński & A. Cieszanowski) s. 7–61 (2014)

Identification and Valuation of Invisible Hospital Assets in the Light of Their Reporting Presentation on the Example of a Psychiatric Hospital

Joachim Fołtys[1], Marzena Filus-Strojek[2], Marek Ksol[1], and Wiktoria Frącz[3]([envelope])

[1] SP ZOZ Health Care Center State Hospital for Mental Diseases in Rybnik, Rybnik, Poland
[2] University of Economics in Katowice, Katowice, Poland
[3] Medical University of Łódź, Łódź, Poland
`wiktoria.fracz@stud.umed.lodz.pl`

Abstract. Healthcare entities, including hospitals, operate, among others, using their material resources. Assets of healthcare entities may have various forms: tangible, intangible, financial and in the form of settlements. The resources disclosed in the financial statement are called assets. Accounting law, in particular the Accounting Act (the Act) and International Accounting Standards/International Financial Reporting Standards (IAS/IFRS) precisely define assets, indicating the conditions that an asset must meet in order to be included in assets and disclosed in balance sheet. In terms of financial reporting, it is important that some of the resources owned and used in healthcare entities are not disclosed in their financial statements (Evans et al. 2015; Mućko 2007; Szewieczek 2011). Healthcare entities, including hospitals, are those entities that generally possess assets of significant value.

Keywords: Assets · Financial statement · Hospital's assets · Intangible assets · Psychiatric hospitals

1 Introduction

The assets of hospitals include, among others, tangible fixed assets classified as fixed assets, such as buildings, medical equipment, diagnostic equipment, technical equipment, which due to their high unitary and aggregate value are generally the main component of assets recognised in the balance sheet. There are also intangible assets in hospitals. Szewieczek (2011) indicates the basic analytical structure of intangible assets of medicinal entities as rights to computer programs (licenses) and other intangible assets, which include property rights and know-how (K-H).

Research conducted in Poland shows that the solutions adopted in hospital accounting policy most often focus on the use of instruments not related to fixed assets (e.g. regarding cost and income accounting over time) and to a much lesser extent on fixed assets and intangible assets (Kludacz-Alessandri 2016, p. 90).

S. Paszkiel (Ed.): ICBCI 2021, AISC 1362, pp. 147–157, 2021.
https://doi.org/10.1007/978-3-030-72254-8_15

Authors of many publications indicate that intangible assets play an increasingly important role in the global economy and are increasingly ahead of fixed assets in this process (Zeghal and Maaloul 2011; Wyatt et al. 2001). Hospitals in the USA recognise the following as the most common intangible assets: trade names, certificate of need, state licensure, medicare certification, patents (Newsad et al. 2014). Research conducted by Rider et al. (2019) points to the growing importance of intangible assets in medical entities and focuses on ways of identifying them.

Psychiatric hospitals are hospitals with a high specific needs in terms of their structure and the value of assets presented in the financial statement. Generally, this group of hospitals has fewer diagnostic devices and medical apparatus than others. At the same time, due to the specific nature of medical services, these hospitals create and use various types of non-material property components, for example, therapeutic procedures aligned with individual types of mental disorders, individual patients (their environmental characteristics), the severity level of the disease entity and others, which are highly specialised procedures that position a given hospital on the map of medical services in Poland and in the world.

At the same time, this type of intangible assets of the hospital is not disclosed in its financial reporting, because it does not meet the conditions for including them in intangible assets, as specified in the accounting law. The literature refers to them as invisible or hidden assets (Bąk 2011). For example, the balance sheet drawn up on 31.12.2018 of the Psychiatric Hospital in Starogard Gdański reported the value of the presented intangible assets of PLN 0.

The balance sheet of the Psychiatric Hospital in Rybnik on 30.06.2018 reported less than 97 thousand PLN for the balance sheet total of PLN 38 879 thousand PLN, which was mainly computer software.

As a result, one of the most important resources due to the specific nature of psychiatric services (diagnostics, treatment and rehabilitation), which are strategic resources, is not reflected in the assessment of the material and financial situation of this type of hospital – Psychiatric hospitals. The indicated situation may translate into the assessment of hospital management by the founding body (owner), and significantly affects the efficiency of asset management, and in extreme cases its dysfunctionality.

On the other hand, in terms of the accounting system of a psychiatric hospital, the question should be raised whether the information gap regarding the assets presented in the financial statement in this case does not violate the primary accounting principle, which is the principle of true and fair view? Problems related to understanding and applying this principle in practice have been indicated in the literature for years (e.g. Flint 1982; Alexander and Archer 2003; Penmam Erb and Pelger 2015). Solutions included in the accounting law regarding the disclosure of intangible assets are complex and regarded as debatable (Kotyla 2009, Gos and Hońko 2011).

This paper addresses a range of issues related to the disclosure (shaping) and valuation of hidden (intangible) assets of Hospitals and their reporting presentation.

In the opinion of the authors of this study, reporting disclosure of hidden intangible assets typical of a psychiatric hospital will increase its value and may prove to be crucial in terms of assessing its actual financial and material situation. This approach can be a starting point for changes in the process of managing assets of Psychiatric Hospitals and will develop some premises for the asset management of Hospitals with different characteristics of health services.

The paper aims at assessing the possibility of demonstrating intangible assets of a psychiatric hospital in its financial statement and at indicating a proposal for their valuation. The paper also contains the authors' recommendations on how to present information on this type of resources as part of financial and non-financial reporting.

The structure of the paper is adapted to the implementation of the above aim.

2 Characteristics of Hidden (Unnoticed) Assets of a Psychiatric Hospital

A literature search showed that psychiatric hospitals operating in Poland practically do not identify intangible resources. The consequences of this approach for hospital reporting will be presented later in the article, while the next paragraphs will focus on an attempt to identify and systematize these resources.

A psychiatric hospital, like any institution, has a number of intangible goods (computer programs, accounting, management skills, etc.) that can be generally described as administrative. Due to their non-specificity, they do not need to be discussed in greater detail here. It seems more interesting to try to delve into more specific values.

The work of a psychiatrist is distinct from somatic medicine, and a psychiatric hospital must combine general medical and psychiatric elements. Due to the long duration of psychiatric hospitalizations and frequent coexistence of somatic diseases, general medical measures cannot be avoided. Already this - cursory - view allows us to list at least two categories of intangible assets: general medical and psychiatric procedures and skills. In both of these categories, subcategories, or rather functions, must be distinguished: diagnostic, therapeutic, rehabilitation and organizational.

Moreover, each psychiatric center - especially large, old hospitals - over the years developed their own local methods of work, which determine the specificity of the center. In Poland, there are examples of centers with the greatest experience in treating personality disorders, using electroconvulsive therapy, art therapy, etc. Some psychiatric centers (hospitals and clinics) have also developed unique scientific or educational traditions. It is no secret that doctors starting their professional careers recognize the profiles of training centers, trying to match their choices to the planned directions of development. The same category of intangible assets should include the management style of such a specific structure as a psychiatric hospital, the team structure developed over the years, and the mechanisms of its functioning. In the era of standardization and constant reform of psychiatry, there is a temptation to depreciate local traditions, but it is they that determine the unique value of the center, in accordance with the old rule "if two do the same - it does not turn out to be the same".

An analogy is drawn here with old, tradition-based companies, in which know-how is passed on from generation to generation. European Union documents define know-how as a set of practical information resulting from experience and research which is classified, i.e. not generally known or easily accessible; significant, i.e. important and useful from the point of view of the production of products, and identified, i.e. described in a sufficiently comprehensible manner [25]. This definition cannot be used literally to describe this category of resources, because describing the mechanisms governing the team in a way that allows them to be copied is very difficult or even impossible, and the place of confidentiality is taken by a high level of complexity and subtlety of the behaviors and relationships developed. Of course, these traditions - as the highest-ranked intangible assets of a hospital - need to be protected against devastation in the form of, for example, breaking the continuity of the team, excessive unification of procedures or the centralized imposition of unwanted reforms.

Based on the above considerations, an attempt can be made to classify intangible assets of hospitals. Such a classification should be at least biaxial, taking into account, on the one hand, the level of specificity of an asset, and on the other, its function. Table 1 presents a proposal for such a classification with examples of intangible assets. Creating an exhaustive list of invisible hospital assets would break the framework of this article, so the presented proposal is only a starting point for their identification, and items placed in various places should be treated as examples. Naming and describing intangible assets can be difficult at first, but the difficulty must be overcome if we do not want to give in to the notion that years - and sometimes decades - of skills and traditions are meaningless and worthless.

3   Presentation of Intangible Assets on the Face of the Balance Sheet of the Hospital on the Example of a Psychiatric Hospital

In the context of the examples of therapies in a psychiatric hospital presented in the previous part of the paper, it is worth asking the question whether individual, e.g. therapeutic procedures developed in-house, can be recognised in the balance sheet as components of intangible assets? The answer to this question is connected with the analysis of the possibility of capitalising the value of these therapies as intangible assets (IA) of a psychiatric hospital.

The creation of hospital assets is closely linked to its organisational and legal form as well as the manner of establishment (Hass-Symotiuk 2013). The property of most hospitals operating as Independent Public Health Care Entities (SPZOZ) can be acquired through the free use of resources (real estate) separated from State Treasury assets, municipal property or can be earned on their own, which particularly concerns movable property (Klich 2008, p.119). The hospital can also become the owner of assets as a result of a donation and receiving appropriate funds. Hospitals operating as commercial law companies have more freedom in shaping their assets in relation to SPZOZ although the differences in the presentation of intangible assets in the financial statement are minor.

Klich (2008) draws attention to dysfunctions in the area of management of SPZOZ, which relate to two spheres: regulatory and functional. Within the regulatory sphere, the possibility of enforcing property rights is of particular importance. In the case of the functional sphere, it points to the behaviours of the healthcare system actors.

Solutions regarding intangible assets are crucial in terms of the stated purpose of the paper. Niklewicz-Pijaczyńska (2017) emphasises different approaches to intangible resources, one for accounting purposes and another for the entity's management process.

Table 1. Proposal for the classification of invisible hospital assets.

Function / The level of specificity	Organization	Diagnostic	Therapy	Rehabilitation	Different, e.g. educational, scientific
Common to different institutions					
Common to hospitals		Diagnostic treatments. General medical procedures	Techniques of somatic treatment		
Common to psychiatric hospitals		The ability to adequately select and use psychological diagnostic tools		Occupational therapy techniques	
Common to psychiatric clinics					Experience in teaching students Experience in resident education. Skills in planning and carrying out research work
Common to departments at general hospitals			Therapy of mental disorders in somatic diseases		
Common to large hospitals				Organization of rehabilitation laboratories	
Site specific	An efficient therapeutic community. Good organization of the work of residents	Experience in the diagnosis of personality disorders	Skills and experience in the treatment of eating disorders. Experience in electroconvulsive therapy	The tradition of art therapy	Experience in genetic/psychosomatic research

This applies to both their identification and valuation. Such a distinction also applies to healthcare entities – including hospitals.

The Accounting Act defines assets as: assets controlled by the entity with a reliably determined value that have arisen as a result of past events and which will bring economic benefits to the entity. Intangible assets are defined as those assets that have been acquired and constitute property rights suitable for economic use, with an expected economic useful life of more than a year. They must also be intended for use by the entity. According to Act, IA include:

- copyrights, related rights, licenses and concessions,
- rights to inventions, patents, trademarks, utility models and decorative designs,
- know-how (K-H),
- acquired goodwill,
- costs of completed development works (completed with a positive result).

Similarly, IAS 38 'Intangible assets' defines intangible assets and indicates additional conditions for including them in assets:

a) they can be separated from the enterprise's assets and sold, transferred, licensed, rented or exchanged,
b) arises from contractual or other legal rights, regardless of whether they are transferable or separable from the enterprise.

Under the accounting law, the method of acquiring intangible assets is particularly crucial. The basic forms of obtaining them include:

a) acquisition for consideration,
b) internal generation (creation),
c) received free of charge,
d) non-cash contribution to a company.

At the same time, the internal generation concerns only research and development works, properly documented.

The 'invisible' IA include: intellectual capital of employees, employee knowledge, developed brand, organisational culture as well as the reputation and image of the entity. However, formalised therapies in the form of developed medical procedures associated with specific diseases and disease entities go beyond these types of assets.

Bearing in mind the stated purpose of the paper, the issue of the possibility of recognising intangible assets – e.g. therapeutic procedures – generated internally as balance sheet assets is particularly crucial. According to the definition of IA, this possibility may optionally apply to two items only: development costs and K-H. Capitalisation of development costs is limited by numerous conditions. In this light, the process of acquiring intangible assets under development works is divided into two stages: research and appropriate development work. It is important that the costs incurred in the first stage, such as the search for alternative materials in production, equipment, processes, systems, their design, assessment and final selection should be recognised as operating costs, thus reducing the financial result of the period. Only the costs incurred in the second stage, such as design, manufacture, testing of prototypes and experimental models before their pre-production, design of tools, instruments, moulds, dies can be disclosed in the balance sheet as intangible assets.

According to the Act and IAS, these costs must meet additional conditions consisting in confirming, among others, the possibility of completing the intangible asset so that it is suitable for use or sale, the ability to use or sell the intangible asset, and generating future economic benefits by the intangible asset, availability of technical, financial and other means used to ensure the completion of development works, reliability in determining the expenditure incurred.

The nature of formalised, e.g. therapeutic procedures, differs from the costs of technical, technological and organisational development works. In practice, the above-mentioned conditions for recognising these procedures as development costs can be very difficult to meet even at the stage of their creation. In this case, the recognition of therapies already created and used in a given hospital as a development cost for several years is excluded.

Recognising these components as K-H seems to be slightly more beneficial. Examples of K-H could be: the concept of reducing production costs, technology, recipe, research method (Gos and Hońko 2011). In tax regulations (the Personal Income Tax Act, Article 29) K-H is defined as knowledge or experience in the field of economic activity. Pursuant to the judgment of the Supreme Administrative Court of 31 July 2003 (registered as Case No. III 1661/01), it is a set of confidential information that is relevant and identified in an appropriate form. These features have been interpreted as follows:

– the term 'confidential' means that the subject of the contract is not publicly available and known,
– the term 'confidential' means that the information is important and original,
– he term 'identified' means that K-H is described or recorded in such a way that it can be checked whether it meets the criteria of confidentiality and materiality.

In the case of specialist medical procedures in the form of e.g. therapeutic procedures, all three features seem to be met, with the third of these being particularly important, which comes down to detailed, also formal documentation. Golat (2013) points out that K-H is one of the intangible assets that can be divided into two groups:

a) patentable,
b) unpatentable and thus not having the formal possibility of their protection under specific, separate exclusive rights.

Developing therapies can be included in the second group of assets and at the same time it is difficult to apply the provisions of the Industrial Property Law in this case.

According to Michalczuk (2009) K-H is a capitalisation resource only in the case of its acquisition. However, if it is internally generated, it is the 'hidden' value of the property. Gos and Hońko (2011) took a different view, allowing for a situation in which the capitalisation of this resource is possible. They subject this possibility not to the form of K-H acquisition, but to its purpose. According to the researchers, the path of verification of the criteria specified for assets as such and then for IA may be used.

According to the authors of this publication, therapeutic procedures developed and used in a psychiatric hospital, as a selected example in the multitude of procedures used by the Hospital, meet the conditions for recognition as assets (intangible assets) indicated in the definition of the Act (and also IAS 38) cited in the first part of this section. They are controlled by an entity which is a psychiatric hospital, they are the effect of past events and thanks to them future benefits will be achieved in the form of settlements with the National Health Fund, and perhaps some of these services can be provided on commercial terms.

A crucial condition is also a reliably determined value, but according to Gos and Hońka (2011) it is not necessarily determined as the cost of production or purchase price. A valuation based on estimates can also be reliable. The authors of this publication have included in the paper a proposal for the valuation of therapy based on the example of the selected procedure, by the replacement cost method. Considering additional criteria set out in IAS 38, it can also be stated that the therapeutic procedures indicated meet these conditions because:

– they are of non-monetary nature,
– they are characterised by a lack of physical form,
– it is possible to identify them, i.e. to separate them from the property for sale or to indicate contractual or legal rights to that asset.

Compliance with all of the above-mentioned conditions, according to the authors, should provide the opportunity to recognise the therapeutic procedures described above in the balance sheet. It is also necessary, though not sufficient, to prepare a separate module for the presentation of 'invisible' (hidden) assets in the psychiatric hospital management report. This solution is indicated in the National Accounting Standard No. 9 Management Report. Pursuant to the Act, the management report should contain significant information about the property and financial standing, including an assessment of the results obtained. Certainly, an important element for the future position of the psychiatric hospital on the market of medical services, as well as its reputation, is the possibility of recognising as a property of self-developed specialised medical procedures, including unique therapeutic procedures. The list of medical procedures together with their estimated value should be supplementary to balance sheet information regarding hospital assets. From a management point of view, they testify to the entity's development in the context of improving the process of treating hospital patients. In the case of a psychiatric hospital, when developing medical procedures, including therapeutic procedures, close cooperation between team members consisting of psychologists, therapists and nurses under the guidance of a psychiatrist is required. The cooperation of the hospital medical team with selected scientific entities, e.g. Medical Universities, Medical Institutes (Psychiatric Rehabilitation, Neurology, Psychiatry, Psychology) seems necessary in this approach. The approach presented in the paper can be the basis for developing a very effective incentive system for specific groups of hospital staff, including medical staff first of all.

It should be noted that there is an ongoing discussion among academics and practitioners of accounting about the method of valuation, accounting records and reporting presentation of intangible assets (Niemczyk 2013, Chojnacka and Wiśniewska 2015). One of the proposed courses of action is the establishment of a new accounting branch.

4 Summary

A significant information gap is emerging regarding real, used assets generating economic benefits, e.g. through settlements with the National Health Fund, as well as potential medical services of a commercial nature. The concept of truth in accounting is quite elusive (Flint 1982; Erb and Pelger 2015; Maruszewska et al. 2015). Accounting law solutions for recognising intangible assets in the balance sheet are not adapted to the modern, dynamically changing, knowledge-based world economy The currently functioning financial accounting system presented by the accounting law is dominated by traditional factors of economic processes, especially manufacturing (Rodov and Leliaert 2002; Niemczyk 2013; Chojnacka and Wiśniewska 2015). It does not sufficiently take into account the importance of intangible assets and the wide range of opportunities for their acquisition by an enterprise. However, even if the current regulations were adopted as valid as to their basic assumptions, they do not sufficiently legibly and clearly (and even do not fall under any exclusion cases) refer to specific, special cases that go beyond the basic set out framework of activity.

Psychiatric hospitals are an example of entities that are primarily based on the knowledge of doctors, psychologists, therapists, and not on infrastructure. It is the intangible resources of individual hospitals that determine their reputation and, above all, the effectiveness of the patient treatment process. In our opinion, not displaying such significant and at the same time very high property value in the psychiatric hospital financial statement leads to distortions of the true and fair view, and this principle should be the basis of the financial statement. It is also worth emphasising that IFRS Conceptual Framework indicates as the main feature of financial information its decision-making utility, which manifests itself in a faithful presentation, having such attributes as, e.g. materiality, no errors, completeness. The omission in the balance sheet of key and valuable resources for the functioning of the psychiatric hospital may undermine its compliance with the above-mentioned qualitative features of the financial statement. This entails the consequences of an erroneous assessment of the financial standing of such an entity. Disclosure of information about such resources in the management report is not sufficient, as they are not included in the financial analysis that assesses the entity's financial standing.

These consequences may relate to settlements and corporate governance as well as the effectiveness of asset management. They may also be associated with the possibility of obtaining foreign capital. The importance of intangible resources in the activities of various entities, including hospitals, will grow in the coming years, and this will translate into their value. In addition, the identified intangible resources can form the basis for building a very effective incentive system for technical staff in the medical field, through participation in commercialisation (sales) of other defined and valued medical procedures – therapeutic procedures in the case of a psychiatric hospital – to other medical entities, surgeries, clinics, retirement homes. In addition, the authors would like to draw attention to the growing importance of the virtual environment (both as VR and as AR) which today is a very current and efficient source for identifying intangible assets. In case of Hospitals we are talking about Tele - medicine, and in case of psychiatric hospitals, we are basically talking about a new area of psychiatry, we are talking about cyber psychiatry. Hence, the process of increasing the significance of intangible resources will particularly affect psychiatric hospitals.

In our opinion, the accounting law should define intangible assets, including know-how, in a more flexible and transparent way, combining them with the possibility of disclosing them in the financial statement. The interpretation of the applicable provisions adopted by the authors of the paper is also a contribution to the discussion on changes in the regulatory sphere regarding intangible assets.

References

Alexander, D., Archer, S.: On economic reality, representational faithfulness and the "true and fair override". Account. Bus. Res. **33**(1), 3–17 (2003)

Barth, M.E., Clinch, G.: Revalued finanacial, tangible and intangible assets: associations with share prices and non-market-based value estimates. J. Acount. Res. **36**, 199–233 (1998)

Bontis, N.: Assessing knowledge assets: a review of the models used to measure intellectal capital. Int. J. Manage. Rev. **3**(1), 41–60 (2002)

Bąk, M.: Aktywa niewidzialne przedsiębiorstwa – istota i znaczenie, „Zeszyty Teoretyczne Rachunkowości" **62** (118), 41–55 (2011)

Chojnacka, E., Wiśniewska, J.: Wybrane zagadnienia pomiaru, wyceny, ewidencji i ujęcia spra-wozdawczości kapitału intelektualnego. Zeszyty Teoretyczne Rachunkowości, tom **83**(139), 35–64 (2015)

Gos, W., Hońko, S.: Know-how jako składnik aktywów. Prace Naukowe Uniwersytetu Eko-nomicznego we Wrocławiu **190**, 82–96 (2011)

Erb, C., Pelger, Ch.: Twisting words? A study of the construction and re construction of reliability in financial reporting standard-setting. Account. Organ. Soc. **40**, 13–40 (2015)

Evans, J.M., Brown, A., Baker, G.R.: Intellectual capital in the healthcare sector: a systematic review and critique of the literature. BMC Health Serv. Res. **15**(1), 556 (2015)

Flint, D.: A True and Fair View in Company Accounts. Gee & Co., United Kingdom (1982)

Kotyla, C.: Ograniczenia bilansowe możliwości ujawniania zasobów niematerialnych, Zeszyty Naukowe Uniwersytetu Szczecińskiego. Finanse, Rynki Finansowe, Ubezpieczenia **17**, 95–101 (2009)

Mućko, P.: Problemy uznawania wartości niematerialnych w rachunkowości. w: Zeszyty Naukowe Uniwersytetu Szczecińskiego nr 832 Finanse, Rynki Finansowe, Ubezpieczenia **71**(2014), 121–134 (2007)

Golat, R.: Prawne aspekty know-how, Raport SEKOCENBUD.PL, raportsekocenbud.pl/artykul/pokaz/news/prawne-aspekty-know-how/ (2013)

Hass-Symotiuk, M.: Przekształcenie samodzielnych publicznych zakładów opieki zdrowotnej w spółki kapitałowe a efektywność gosodarowania zasobami opieki zdrowotnej, Zeszyty Naukowe Uniwersytetu Szczecińskiego nr757. Finanse, Rynki, Ubezpieczenia **58**, 57–67 (2013)

Klich, J.: Wokół dysfunkcji zarzadzania samodzielnymi publicznymi zakładami opieki zdrowot-nej. Zarządzanie PubliczneNr **3**(5), 119–133 (2008)

Kludacz-Alessandri, M.: Stosowanie narzędzi polityki rachunkowości w polskich szpitalach. Zeszyty Teoretyczne Rachunkowości, tom **89**(145), 77–94 (2016)

Kotyla, C.: Ogranizenia bilansowe możliwści ujawniania zasobów niematerialnych, Zeszyty Naukowe Uniwersytetu Szczecińskiego. Finanse, Rynki, Ubezpieczenia **17**, 95–101 (2009)

Krajowy Standard Rachunkowości nr 9 Sprawozdanie z działalności (2014)

Maruszewska, E.W., Szewieczek, A., Strojek-Filus, M.: Truth, faithfullness and reliability vague-ness In the accounting theory as a challenge for accounting teachers. Prob. Educ. 21st Century **68**, 36–51 (2015)

Michalczuk, G.: Możliwości identyfikowania i pomiaru know-how w rachunkowości, Zeszyty Naukowe Uniwersytetu Szczecińskiego. Finanse, Rynki, Ubezpieczenia **18**, 301–308 (2009)

International Accounting Standards Board – IASB. Intangible assets. International Accounting Standards No. 38, London, UK (2004)

Niklewicz-Pijaczyńska, M.: Wycena wartości niematerialnych i prawnych przedsiębiorstwa na przykładzie praw własności przemysłowej. Zarządzanie i Finanse, **15**(2), 99–108 (2017)

Niemczyk, L.: Rachunkowość finansowa aktywów kompetencyjnych i kapitału intelektualnego. Nowy dział rachunkowości. Pacioli Institute, Rzeszów (2013)

Newsad, N.A., Matuga, T.A., Mello, T.J.: HealthCare Appraisers, White Paper, Intangible Assets i Healthcare, Healthcare Appraisers, Inc. (2014)

Rodov, J., Leliaert, P.: FiMIAM: financial method of intangible assets measurement. J. Intellect. Capital **4**(2), 181–190 (2002)

Commission Regulation (EU) No. 316/2014 of 21 March 2014 on the application of Art. 101 paragraph. 3 of the Treaty on the Functioning of the European Union to the category of technology transfer agreements, https://eur-lex.europa.eu/legal-content/PL/TXT/PDF/?uri=CELEX: 32014R0316&from=PL. Accessed 15 Sept 2020

Rider, E.A., Coneau, M., Truog, R.D., Boyer, K., Meyer, E.C.: Identifying intangible assets in interprofessional healthcare organizations: feasibility of an asset inventory. J. Interprofessinal Care **33**(5), 583–586 (2019)

Szewieczek, A.: Elementy rachunkowości w podmiocie leczniczym, Wydawnictwo Uniwersytetu Ekonomicznego w Katwicach (2011)

Ustawa z dnia 29 września 1994 r. o rachunkowości, ujednolicony tekst Dz. U. 2019, poz. 351

Ustawa z dnia 26 lipca 1991 r. o podatku dochodowym od osób fizycznych, tekst ujednolicony Dz. U. 2019, poz. 1387

Wyrok NSA z 31.07.2003 r. III SA 1661/01

Wyatt, A., Matolcsy, Z., Strokes, D.: Capitalization of intangibles – a review of current practice and the regulatory framework. Aust. Account. Rev. **11**(2), 22–38 (2001)

Zeghal, D., Maaloul, A.: The accounting treatment of intangibles – a critical review of the literature. Account. Forum **35**, 262–274 (2011)

Penman, S.H.: Accounting for intangible assets: there is also an income statement. Abacus **45**(3), 358–371 (2009)

Zarzecki, D.: Metody wyceny wartości niematerialnych i prawnych. http://e-rachun-kowosc.pl/artykul.php?view=404&part=2

https://bip.kocborowo.pl/struktura-wlasnosciowa-i-majatkowa-szpitala.html. Accessed 20 Apr 2020

https://bip.psychiatria.com/sites/default/files/zalaczbudzmaj/majatek_stan_30.06.2018.pdf. Accessed 20 Apr 2020

Logical Sentential Calculi Inspired by the Chrysippean Sentential Calculus

Robert Sochacki[✉]

Opole University, Copernicus Square 11a, 45-040 Opole, Poland
rsochacki@uni.opole.pl

Abstract. The aim of the present paper is to consider an approach, different from that presented by J. Łukasiewicz, concerning the interpretation of the so-called stoic undemonstrables, which were given by Chrysippus. Stoic undemonstrables have been interpreted in two different ways: using the notion of "negation of a sentence" (Łukasiewicz) and using the notion of "a sentence inconsistent with a given one" (Mates). According to the Stoics, two sentences are inconsistent if one of them is negation of the other. The Mates' interpretation generates five different inference rules. Based on one of these rules we can consider (with other undemonstrables) four different stoic propositional calculi. Taking into account the interpretation of Łukasiewicz, we obtain Stoic logic as a fragment of classical sentential logic, but with the interpretation of Mates, Stoic logic is the whole of classical sentential logic.

Keywords: Stoic undemonstrables · Inference rules · A sentence inconsistent with a given one · Deductive reasoning

1 Introduction

Greek ancient world is most often associated with richly developed mythology associated with religious beliefs. However many societies believe that in ancient Greece there was an outburst of extraordinary intellectual forces, which, with their breadth and range, paved the way for future civilizations. It is enough to mention the efforts of Greek intellectuals to systematize and expand knowledge. The works of Euclid (4th-3rd century BC) are known to form a compact, axiomatic system of geometry. This system was up-to-date and slightly improved at the beginning of the last century by D. Hilbert. Even after the Second World War, school children learned about geometry as an axiomatic Euclidean system [11]. A great succes was also the construction by Aristotle (275-180 BC) of the axiomatic syllogistic system and the construction by Chrysippus from Soloi (around 281/277-208/204 BC) of a (nonaxiomatic) compact system of propositional calculus, called Stoic logic. Aristotle's syllogistic (calculus of names) has almost dominated (to the detriment of the classical calculus of sentences and a narrower functional calculus) logical problems considered in higher schools on non-mathematical fields of study.

In any logic, the object of study is reasoning. This reasoning activity can be done well (correctly) and it can be done badly (incorrectly). Logic is the discipline that aims

S. Paszkiel (Ed.): ICBCI 2021, AISC 1362, pp. 158–165, 2021.
https://doi.org/10.1007/978-3-030-72254-8_16

to distinguish good reasoning from bad one. For over two thousand years, both Stoic logic and Aristotle's syllogistics constituted a solid basis for the so-called deductive reasonings. In a deductively valid reasoning, the truth of the premises guarantees the truth of the conclusion. In other words, to say that a reasoning is deductively valid is to say that it is impossible for all of its premises to be true and its conclusion to be false. The period between the ancient logic and the beginning of the nineteenth century is generally regarded as barren by historians of logic. Logic revived in the mid-nineteenth century, at the beginning of a revolutionary period, when the subject developed into a rigorous and formal discipline which took as its exemplar the exact method of proof used in mathematics, hearkening back to the Greek tradition. Progress in mathematical logic in the first few decades of the twentieth century had a significant impact on analytic philosophy and philosophical logic, particularly from the 1950s onwards, in subjects such as modal logic, temporal logic, deontic logic, and relevance logic [10]. It turned out that the emergence of these logics was necessary because the reasoning based on classical logic did not encompass the richness of natural language. It is worth emphasizing that deductive reasoning is used in every natural science, in every social science, and in all applied sciences. In all of these sciences, the kind of deductive reasoning characteristic of mathematics has an important role, but deductive reasoning can also be found entirely outside of mathematical contexts. We also find attempts at such reasoning throughout the law and in theology, economics, and everyday life.

The development of many new technologies resulted in the inflow of a large volume of information that had to be properly processed. People began to use reasonings that was not deductive: inductive reasoning, abductive reasoning, statistical reasoning, causal reasoning and others [8, 28]. The achievements of different logics (in particular the so-called fuzzy logic [29]) began to be used in such areas of knowledge as psychology, computer science and philosophy of mind. Some of the main questions for the philosophy of mind are metaphysical questions about the nature of minds and mental states. Philosophers of mind, along with cognitive psychologists, information scientists, and neuro-scientists have begun to work out detailed explanations of how our physical brains realize and carry out the functions of many mental states [23].

2   Some Remarks on the Philosophy of Stoics

Two trends prevailed in Greek philosophy - an idealist trend from Plato (about 427-347 BC), a student of Socrates, founder of the Plato Academy and a materialistic trend from Zenon from Kition (around 312/311-262 BC) and Chrysippus. Platonism assumed the existence of an immortal soul and the existence of a creator who used the mechanism of building the world. In turn, stoicism, opposed to Plato's philosophy, was based on rationalism, the assumption of the materiality of the world and the non-existence of supernatural forces. The philosophical position that only the existing, self-existent being is matter previously occupied by the atomist Democritus (about 460-370 BC).

Stoicism assumed the cognition of the world and the reliability of human knowledge. In ethics, the thesis was put forward that human life should be wholly compatible with reason and nature, as far as possible free of affects and emotions. In the well-known saying "stoic calmness" there is a deep content – self-control, endurance, stubbornness, calmness, freeing oneself from emotions. The rational attitude of the Stoics provoked them to creatively learn about the world and accept the judgments based mainly on the way of reasoning. Hence, the main emphasis was placed on the development of logic concerning the judgments (sentences), in contrast to Aristotle's concept of building a logical system concerning names. In opposition to empiricism and irrationalism, the Stoics saw in understanding the main source of knowledge and the main criterion of truth. They assumed the opportunity to know the world independently of experience and they rejected knowledge that can not be verified (intuition, revelation, hypothesis). Even in ethical and aesthetic spheres, they appealed to the intellect. The Stoic school was active for several centuries (4th century BC – 3rd century). It was founded in Athens by Zenon, who at the beginning was her lecturer and manager. The name of the school comes from the cloister in Athens called Stoa Pojkile (the Freak Portico), where lectures were held. In the history of the Stoic school there are three periods of its activity. The first period of its activity, the period of the so-called old Stoic school, lasts from the end of the fourth century and throughout the third century BC. The second period of activity of the Stoic school, the so-called the period of the average Stoic school (mediostoicism), included the 2nd and the 1st century BC. The last, third period (neostoicism), falls on the first two centuries of the Christian era. This is the period of so-called the Roman Stoic school or the new Stoic school. It should be emphasized, however, that while the views of successive representatives of the Stoic school in the sphere of physics and ethics underwent more or less significant modifications, in the sphere of logic they were established at the beginning by Chrysippus and were not subject to any changes. Quite extensive information about Stoic philosophy from the periods of: the old Stoic school, mediostoicism and neostoicism can be found in [25, 26]. Typically logical problems are treated there rather fragmentarily.

3   Chrysippean Undemonstrable

Speaking of ancient logic, we mean both independent logic systems: Aristotle's syllogistic and Chrysippean propositional calculus. In this work, I will discuss the stoic propositional calculus built by Chrysippus. The main elements of this calculus are "undemonstrables", "themes" and "chains reasoning". The undemonstrables as the initial rules are unprovable, the themes are meta-rules, and the chains reasoning are proofs of rules and thesis.[1] Building chains reasoning in Stoic logic was similar to proving in the modern natural deduction system of classical logic, originating from Stanisław Leśniewski and perfected by Jerzy Słupecki and Ludwik Borkowski. Many researchers have studied Stoic logic, among others: Bobzien [1, 2], Bocheński [3, 4]; Bryll [5], Burski [6]; Corcoran [7]; Frede [9]; Kostrzycka [5], Kotarbiński [13]; Krokiewicz [14]; W. Kneale and M. Kneale [15]; Łukasiewicz [17–19]; Mates [20]; Majewski [21]; Mignucci [22]; Prantl [24]; Scholz [27] and Żarnecka-Biały [30–32].

Among the people involved in studying Stoic logic, there is a certain fundamental difference in reading the content of undemonstrables. An eminent expert on ancient logic, J. Łukasiewicz, read undemonstrables (rules adopted a priori) using the notion "of

---

[1] Examples of "chains reasonings" and examples of applications of "undemonstrables" in practice can be found in [14].

negation of a given sentence" [18], and B. Mates as the concept of "a sentence inconsistent with a given sentence" [20]. This difference in the interpretation of undemonstrables causes that we get two (and even more) systems of Stoic logic. Taking into account the interpretation of J. Łukasiewicz, we obtain Stoic logic as a fragmentary system of classical propositional calculus, while in the case of B. Mates' approach, Stoic logic appears as the full classical calculus. It is worth to mention that Mates' method of reading undemonstrables is consistent with the texts of Sextus Empiric, and the way they are read by Łukasiewicz is compatible with the texts of Chrysippus. In this work, we will rely mainly on the conclusions resulting from the Mates' interpretation of undemonstrables.

4 The Mates' Interpretation of Undemonstrables

Let us remind that in the Mates' interpretation two sentences are called inconsistance when one of them is the negation of the other. Thus, a sentence inconsistent with the sentence α (not beginning with negation) is the sentence $\sim \alpha$, while sentences inconsistent with the sentence $\sim \alpha$ are sentences α and $\sim\sim \alpha$.

According to the Mates' interpretation, undemonstrables have the following form [5]:

The first (unprovable inference) gives a successor from the conditional period and its predecessor;

The second – from a conditional period and a sentence inconsistent with its successor, gives a sentence inconsistent with its predecessor;

The third – from the negation of conjunction of two sentences and from a sentence of this conjuction, gives a sentence inconsistent with the remaining one;

The fourth – from the exclusive disjunction of two sentences and from a sentence of its part, gives a sentence inconsistent with the remaining one;

The fifth – from the disjunction of two sentences and from a sentence inconsistent with the one of them, gives the remaining one;

Using the symbol $\neg$ as the symbol of inconsistancy, the above undemonstrables may be rewritten in more formalized forms:

$$\alpha \Rightarrow \beta, \alpha / \beta \tag{1}$$

$$\alpha \Rightarrow \beta, \neg\beta / \neg\alpha \tag{2}$$

$$\sim (\alpha \wedge \beta), \alpha / \neg\beta, \quad \sim (\alpha \wedge \beta), \beta / \neg\alpha \tag{3}$$

$$\alpha \perp \beta, \alpha / \neg\beta, \quad \alpha \perp \beta, \beta / \neg\alpha \tag{4}$$

$$\alpha \vee \beta, \neg\alpha / \beta, \quad \alpha \vee \beta, \neg\beta / \alpha \tag{5}$$

where the symbols: $\Rightarrow, \sim, \wedge, \vee$ are the connectives of: implication, negation, conjuction, exlusive disjunction and disjunction respectively. Scheme 2 corresponds to four transposition rules (contraposition):

$$\alpha \Rightarrow \beta, \sim \beta / \sim \alpha \tag{2a}$$

$$\alpha \Rightarrow \sim \beta, \beta/\sim \alpha \tag{2b}$$

$$\sim \alpha \Rightarrow \beta, \sim \beta/\alpha \tag{2c}$$

$$\sim \alpha \Rightarrow \sim \beta, \beta/\alpha \tag{2d}$$

5 Consequence of Mates' Interpretation

Let A be a set of axioms for the positive implicational Hilbert calculus. This calculus may be based on the following set of axioms:

$$A = \{\alpha \Rightarrow (\beta \Rightarrow \alpha), (\alpha \Rightarrow \beta) \Rightarrow [(\beta \Rightarrow \gamma) \Rightarrow (\alpha \Rightarrow \gamma)], [\alpha \Rightarrow (\alpha \Rightarrow \beta)] \Rightarrow (\alpha \Rightarrow \beta)\}$$

Rules (2a)–(2d) correspond to the following laws of the classical propositional calculus:

$$(\alpha \Rightarrow \beta) \Rightarrow (\sim \beta \Rightarrow \sim \alpha) \tag{a1}$$

$$(\alpha \Rightarrow \sim \beta) \Rightarrow (\beta \Rightarrow \sim \alpha) \tag{a2}$$

$$(\sim \alpha \Rightarrow \beta) \Rightarrow (\sim \beta \Rightarrow \alpha) \tag{a3}$$

$$(\sim \alpha \Rightarrow \sim \beta) \Rightarrow (\beta \Rightarrow \alpha) \tag{a4}$$

These laws and axioms belonging to A, generate four stoic propositional calculi ST1–ST4. Their axioms are as follows:

$$A1 = A \cup \{a1\}$$

$$A2 = A \cup \{a2\}$$

$$A3 = A \cup \{a3\}$$

$$A4 = A \cup \{a4\}$$

Let T1–T4 be the sets of thesis for the stoic propositional calculi ST1–ST4 respectively. Then the following inclusions are true [5]:

$$T \subset T1 \subset T2 \cap T3 \subset T2 \subset T_{INT} \subset T4$$

$$T3 \subset T4$$

where T is the set of thesis of the positive implicational Hilbert calculus and T_{INT} is the set of thesis of intuitionistic logic. The mutual location of the above mentioned Stoic calculi can be presented in the figure below (Fig. 1).

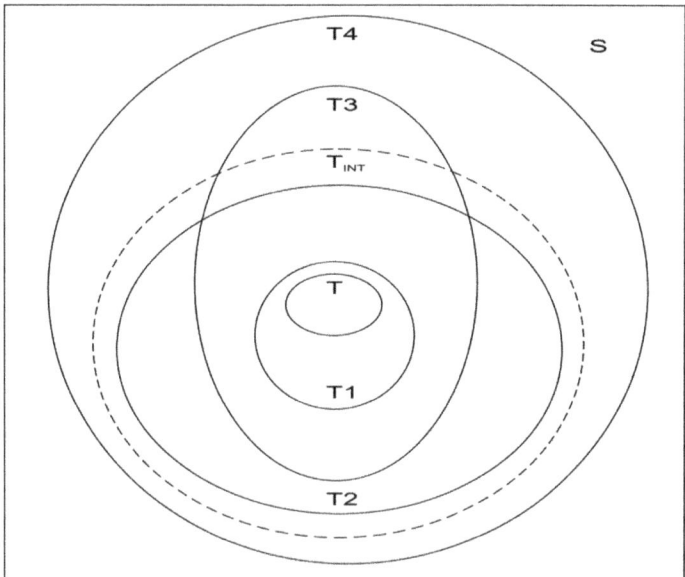

Fig. 1. Mutual relations between Stoic calculi

The ST2 calculus is the implicational-negational fragment of the Johansson's minimal sentetntial calculus and the ST4 calculus is the classical implicational-negational sentential calculus. The ST2 calculus may be obtained from the ST1 calculus by adjoining the axiom $(\alpha \Rightarrow \sim \alpha) \Rightarrow \sim \alpha$, and the ST3 system from the ST1 system by adjoining the axiom $\sim\sim \alpha \Rightarrow \alpha$. In turn, by strengthening the ST2 calculus with the axiom $\alpha \Rightarrow (\sim \alpha \Rightarrow \beta)$, one can get the intuitionistic calculus, while by strengthening with the axiom $\sim\sim \alpha \Rightarrow \alpha$, the classical sentential logic can be obtained. The intuitionistic sentential calculus intersects with the ST3 calculus. Let us recall that in the intuitionistic calculus the law of double negation $\sim\sim \alpha \Rightarrow \alpha$ is not a thesis, while in the ST3 calculus there is no the thesis of a strong right of double negation $\alpha \Rightarrow\sim\sim \alpha$. Therefore, the mentioned listed calculi are in the opposition's relation. Enriching ST1–ST4 calculi with conjunction and disjunction functors (ordinary and exclusive) and appropriate axioms, one can receive calculi with an increasingly richer set of theses.

6   Conclusion

The above considerations show that, regardless of the interpretation of undemonstrables, Chrysippus should be treated as the founder of the basics of modern classical logic. In addition to extensional functors, the modalic functors of necessity and possibilities were also considered in Stoic logic (modalities were also in the Aristotle's syllogistic). However, the above mentioned modalic functors were not mutually definable [22]. Such mutual definability can be found, for example, in Lewis S1-S5 modal systems [16]. Interesting considerations on Stoic modal logic can be found in the paper [1]. In this work the author examines relationships between the modal systems of the three ancient logicians: Chrysippus, Philo and Diodorus. Chrysippus skillfully combined Philo's and Diodorus' modal notions, with making only a minimal change to Diodorus' concept of possibility. Finally, the similarities and differences between three modal systems are the following:

a) propositions which cannot be false or cannot be true (i.e. are true or false because of their nature) meet the criteria for necessity and impossibility of all three logicians;
b) consistent propositions which are externally hindered from being false from now on are Philonian non-necessary, but necessary for Chrysippus and Diodorus (and accordingly for false propositions);
c) propositions which are always true from *now* on but not always externally hindered from being false are necessary only for Diodorus (and accordingly for false propositions);
d) propositions which change their truth-value at some future time are granted contingency by all three logicians.

According to the author, Chrysippus' concept of possibility comes fairly close to the ordinary language use of the word - at any rate closer than those, of both Philo and Diodorus.

Acknowledgements. I'm greatly indebted to the reviewer of this paper for valuable comments.

References

1. Bobzien, S.: Die stoische Modallogik. Königshausen + Neumann, Würzburg (1986)
2. Bobzien, S.: Stoic sequent logic and proof theory. Hist. Philos. Logic **40**(3), 234–265 (2019)
3. Bocheński, J.M.: Ancient Formal Logic. North Holland Publishing Company, Amsterdam (1951)
4. Bocheński, J. M.: History of Formal Logic. Notre Dame UP, Indiana (1961)
5. Bryll, G., Kostrzycka, Z.: Formal research on Stoic logic. Academic Outbuilding, Warsaw (1988). (Original: Bryll, G., Kostrzycka, Z.: Badania formalne nad logiką stoicką. Oficyna Akademicka, Warszawa (1988)).
6. Burski, A.: Dialectica Ciceronis. Łęski, M., Zamość (1604)
7. Corcoran, J.: Remarks on stoic deduction. In: Corcoran, J. (ed.). Ancient Logic and Its Modern Interpretations. D. Reidel Publishing Company, Dordrecht (1974)
8. Flack, P.A., Hadjiantonis, A.M. (eds.): Abduction and Induction. Springer, Dordrecht (2000)
9. Frede, M.: Die stoische Logik. Vandenhoeck und Ruprecht, Göttingen (1974)
10. Gabbay, D.M., Guenthner, F. (eds.): Handbook of Philosophical Logic, vol. 2. Springer, Dordrecht (2001)
11. Iwaszkiewicz, B.: Elementary geometry (part I, II). State School Publishing Institutions, Warsaw (1950). (Original: Iwaszkiewicz, B.: Geometria elementarna (cz. I, II). Państwowe Zakłady Wydawnictw Szkolnych, Warszawa (1950))
12. Iwaszkiewicz, B.: Elementary geometry (part III). State School Publishing Institutions, Warsaw (1952). (Original: Iwaszkiewicz, B.: Geometria elementarna (cz. III). Państwowe Zakłady Wydawnictw Szkolnych, Warszawa (1952))
13. Kotarbiński, T.: Lectures on the History of Logic. Polish Scientific Publishers, Warsaw (1985). (Original: Kotarbiński, T.: Wykłady z dziejów logiki. PWN, Warszawa (1985))

14. Krokiewicz, A.: On Stoic logic. Philos. Q. **17**, 173–197 (1948). (Original: Krokiewicz, A.: O logice stoików. Kwartalnik Filozoficzny 17, 173–197 (1948))
15. Kneale, W., Kneale, M.: The Development of Logic. Clarendon Press, Oxford (1962)
16. Lewis, C.I., Langford, C.H.: Symbolic Logic. Dover Publications, New York (1959)
17. Łukasiewicz, J.: On stoic logic. Philos. Rev. **30**, 278–279 (1927). (Original: Łukasiewicz, J.: O logice stoików. Przegląd Filozoficzny 30, 278 – 279 (1927))
18. Łukasiewicz, J.: From the history of propositional logic. Philos. Rev. **37**, 417–437 (1934). (Original: Łukasiewicz, J.: Z historii logiki zdań. Przegląd Filozoficzny 37, 417 – 437 (1934))
19. Łukasiewicz, J.: Aristotle's Syllogistic from the Standpoint of Modern Formal Logic. Claredon Press, Oxford (1951)
20. Mates, B.: Stoic Logic. University of California Press, Berkeley and Los Angeles (1961)
21. Majewski, M.: Bursius's characterization of formal Stoic logic. In: Voise, W., Skubała-Tokarska, Z. (eds.). From the history of Polish logic. Ossolineum, Wroclaw (1981). (Original: Majewski, M.: Charakterystyka stoickiej logiki formalnej według Bursiusa. [w:] Voise, W., Skubała-Tokarska, Z. (red.). Z historii polskiej logiki. Ossolineum, Wrocław (1981))
22. Mignucci, M.: Sur la logique modale des stoiciens. In: Brunschwig, J. (ed.) Les stoiciens et leur logique. Actes du Colloque de Chantilly. Libraire Philosophique J. Vrin, Paris (1978)
23. Nelson R.,J.: The Logic of Mind, 2nd edn. Springer, Kluwer (2012)
24. Prantl, C.: Geschichte der Logik im Abendlande. Fock, Leipzig (1927)
25. Reale, G.: History of Ancient Philosophy, vol. 3. KUL Press, Lublin (2010). (Original: Reale, G.: Historia filozofii starożytnej (t.3). Wydawnictwo KUL, Lublin (2010))
26. Reale, G.: History of Ancient Philosophy, vol. 4. KUL Press, Lublin (2012). (Original: Reale, G.: Historia filozofii starożytnej (t.4). Wydawnictwo KUL, Lublin (2012))
27. Scholz, H.: Abriss der Geschichte der Logik. Polish Scientific Publishers, Warsaw (1965). (Original: Scholz, H.: Zarys historii logiki. PWN, Warszawa (1965))
28. Stadler, F. (ed.): Inductions and Deductions in the Science. Springer, Dordrecht (2004)
29. Trillas, E.: On the Logos: A Naive View on Ordinary Reasoning and Fuzzy Logic. Springer (2017)
30. Żarnecka-Biały, E.: Stoic Logic as investigated by Jan Łukasiewicz. Rep. Philos. **2**, 27–40 (1979)
31. Żarnecka-Biały, E.: History of the Past Logic. Series "Dialogikon". Jagiellonian Uniwersity Press, Cracow (1995). (Original: Żarnecka-Biały, E.: Historia logiki dawniejszej. Seria „Dialogikon". Wydawnictwo UJ, Kraków (1995))
32. Żarnecka-Biały, E.: Noises in the History of Logic. Series "Dialogikon". Jagiellonian Uniwersity Press, Cracow (1995)

Brain-Computer Interface in Lie Detection

Julia Świec(✉) 🆔

Faculty of Electrical Engineering, Automatic Control and Informatics,
Opole University of Technology, Prószkowska 76, 45-758 Opole, Poland
julia.swiec@student.po.edu.pl

Abstract. The article describes the theoretical aspect of using Brain-Computer Interfaces in Lie-Detection. Various invented methods used to increase the result accuracies are presented. Researches on the basis of the P300 wave are described as well as Guilty Knowledge Test implementations and usage of Questioning Techniques. The results of other scientific articles are presented to show the outcome of different methods.

Keywords: P300 · Lie detection · Polygraph

1 Introduction

Since the very first creation of a polygraph in 1921, the process of detection lies drastically changed. New technologies came into force and started to outperform the initial polygraph invented by physiologist John A. Larson and a Californian policeman. The aim of the lie-detecting device was to simultaneously measure steady, uninterrupted changes in heart rate, respiration rate and blood pressure to support the detection of lies [1]. Along with the development progress, the application and results of the lie-detector have become more valuable so that in 2008 the first polygraph results were used as evidence in an Indian court [3]. A modern polygraph machine works by measuring blood pressure, breathing and skin conductance, since the creators base on the assumption that lying creates autonomic nervous system responses like fear, emotional excitement which follow by an increase of blood pressure, respiration rate, and skin resistance [4]. Because of that, a lie-detection machine consists of multiple measuring-components, each dedicated to a different measurement.

Nevertheless, researches confirm that lie detection performances cannot be fully reliable since on average only 54% of the true/false statement results are being accurately assigned [6]. A. Vrij and S. Graham noted that the accuracy of truthfully results can vary from 20% up to 70% [7]. This can be due to various factors like character traits of the examined person: shyness, extraversion, social anxiety, dysphoria, extremal stress or others [8]. Additionally, there are strategies that enable the fraud to take the lie detector test and pass, even if he was lying. Some of the strategies focus on catching breath continuously, whether they are lying or not, to not show any differences between those two states. Others describe the methods of gripping the hands hard together to mislead the hand-gripping test which is a part of polygraph [9]. Therefore, new technology alternatives for lie detection have been created, improving the results accuracy.

S. Paszkiel (Ed.): ICBCI 2021, AISC 1362, pp. 166–175, 2021.
https://doi.org/10.1007/978-3-030-72254-8_17

2 Cognitive Lie-Detection Technologies

New lie-detection technologies analyze the uncontrolled cognitive brain activities, disabling the manipulation of polygraphic results by own strategized methods and actions. Creating lies is more cognitively demanding than telling the truth, since the fraud must come up with a reliable answer and uphold it throughout the session to assure consistency [10]. Additionally, because of the unawareness of the receptions by the examiners, the fraud subconsciously controls the reaction of the investigator to assure that his story is believed in. This also requires cognitive resource. Moreover, the constant remaining themselves of the history the suspects have told, and the concealment of truth are also cognitively demanding and require more mental effort [11]. Although using brain activities to detect lies is a more modern approach, the first researches began in the 1980s when J. Peter Rosenfeld a Northern University psychologist developed one of the earliest methods to track brain waves [12]. This method used a special type of brain activity – P300. P300 wave is a component of ERPs (endogenous event-related potential) that is naturally evoked by an unusual stimulus [13]. This wave occurs in the 0.15–5 Hz frequency range and can be observed in an EEG recording machine. Usually its peak appears between 300 ms to 500 ms after the stimuli appearance [14] (Fig. 1). In other words, if the examined person recognizes a shown to him/her image, the emitted P300 wave will occur about 300 ms after the stimuli in the EEG recording- therefore the name P300. Although the origin of P300 is not yet discovered, it is probably related to the end of cognitive processing or the passing of information to consciousness [15]. Rosenfeld's idea was to use this technology to interrogate criminals by showing them pictures of a stolen object or murder weapon and track their brain waves to determinate, whether they are guilty or not.

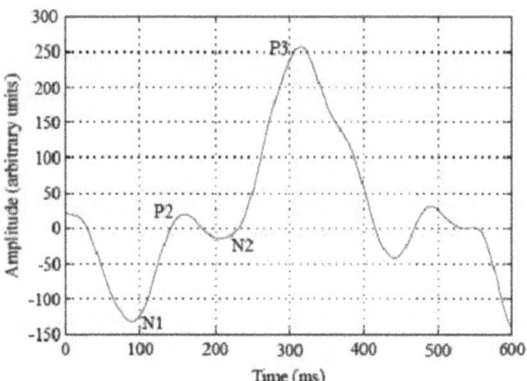

Fig. 1. A typical P300 wave.

3 Guilty Knowledge Test

A more modern approach of using P300 brain activities in lie-detection is the GTK, short for "Guilty Knowledge Test". GTK, also called the CTI (Concealed Information

Test) is a psychological research technique that bases on detection of concealed crime information that only the criminal and the investigators would know. Instead of focusing on detection of told lies, the method focuses on the capture of "guilty knowledge", by creating a research based on a series of multi-choice questions [17]. Guilty Knowledge Tests usually use the following three types of stimuli: "Probes" (P), "Irrelevant" (I) and "Target" (T). "Probe" stimuli contain classified information, which only the guilty person or the investigators would know. "Irrelevant" are stimuli that are completely irrelevant to the criminal investigation, therefore will be not known by neither the guilty nor the innocent. "Targets" may be known by all subjects for other reasons then taking part in the crime and if they appear the examined person is asked to perform a task, for example, to press a button [18]. These stimuli are created to provide a model of the examinee's reactions to irrelevant or known information about the crime. The "Probes" stimuli will induce the P300 wave if the investigated person is guilty or behaves the same as in cases of "Irrelevant" stimuli if the suspect is innocent [19]. An example of a GKT question may be "If you committed the robbery, then you know in what kind of car the robbers got away. Was the car (a) a yellow Toyota, (b) a grey Honda, (c) a red Opel, (d) a green Renault, or (e) a blue Peugeot?" (example from [20]). The main idea behind the GTK test is that the guilty suspect shows an increased psychophysiological reaction to the correct answer, since he is emotionally involved in the crime. If the suspects' reactions to the "Probes" stimuli are persistently greater than to the "Irrelevant" stimuli, the examinee must have some involvement in the crime (assuming the information was not previously leaked).

4   Questioning Techniques

Modern studies have established different questioning-techniques as well as the proof that some specific asked questions increase the cognitive load more in liars in comparison to the subjects that tell the truth. The article "Outsmarting the Liars: Toward a Cognitive Lie Detection Approach" by Aldert Vrij, Par Anders Granhag, Samantha Mann, and Sharon Leal [11] presents that these questioning methods contain: The Devil's-Advocate Approach, the SUE Approach and others.

The Devil's-Advocate Approach bases on identifying deception in the expressed subject's opinion. The subject is being asked to present arguments in a concrete topic that support the presented view, followed by a "devil's advocate" inquiry to present arguments against the view. Since the subject is more emotionally involved in the truth, he will share more information about the "in-favor" opinion than about the "devil's advocate" opinion. To a guilty subject, the "devil's advocate" view is more reliable thus he will share more details. This method has been proven to correctly classify 75% of truthful subjects and 78% of liars.

Another questioning method is the strategic use of evidence (SUE). The SUE strategy bases on detection cues about a crime, while interviewing the suspect in order to increase the chance of a correct truth/lie statement. The idea behind this method came from the assumption that liars and suspects who tell the truth have different mind-sets during the interrogation. Liars are afraid of sharing sensitive crime details, that could incriminate them, whereas the innocent are concerned about not being believed when answering questions. The SUE Method aims to recognize and manipulate by the interviewers these strategies by asking selected questions adjusted to the current situation. These questions could involve asking about an activity, that the guilty would probably hide or lie about, knowing that there is evidence of it and monitoring their reaction and response. Innocent suspects are more likely to mention that activity in opposition to liars, who subconsciously will deny it or not mention it. The results of a research carried out by Hartwig, Granhag, Stromwall, and Kronkvist [22] showed that in the case of trained about the techniques interviewers, the accuracy rate was equal to 85,4%. When examining the research with non-trained SUE examiners the accuracy was 56,1%.

5   Cognitive Load Measuring Device and Software's

To carry out brainwave-based lie-detection researches many examiners use the Emotiv EPOC+ NeuroHeadset as the test device. Emotiv EPOC+ NeuroHeadset is the most known open market Brain-Computer Interface (BCI) system in the world (Fig. 2). Its popularity has been earned not only due to its cost-effectiveness, availability on the open market but also because of a simple and efficient design which provides professional access to a range of applications. Due to a multi-channel electroencephalography (EEG system) and a wireless connection system various researches can be easily made, to achieve precise results in many fields of studies, polygraph researches among others [23].

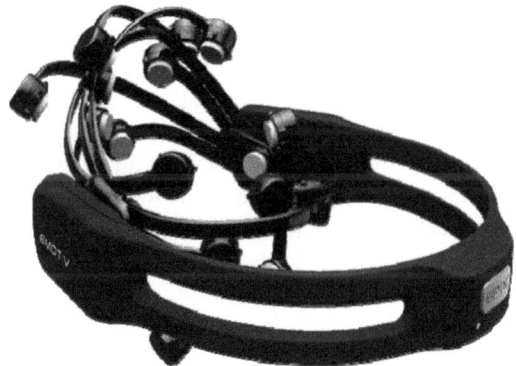

Fig. 2. Emotiv EPOC+ NeuroHeadset.

The device works via transmitting data collected from 16 electrodes attached to the scalp into a workstation equipped with a receiver signal. Those electrons include 14 EEG channels with saline based electrodes to enable a non-sticky whole brain sensing. Additionally, the 9 axis motion sensors detect any head movements to make the research results as precise as possible. The passed through signal after transmitting and processing can be used to investigate the brain activities during different tasks and to draw conclusions. A great asset of the device is its LITHIUM researchable battery and 50/60 Hz 100–250 VAC Battery Charger which provides an endurance up to 12 h.

The examiners often place the EPOC+ Neuroheadset on the subject's head in the 10–20 System with the following channel locations: AF3, AF4, F3, F4, F7, F8, FC5, FC6, P7, P8, T7, T8, O1, and O2. The International 10–20 system is a system for EEG placement of electrons on the head, which aim is to maintain standardized testing methods that would ensure that each research's results could be compared with each other. The 10–20 System places the electrons on the head in a specified positioning in distances of 10% or 20% relative to anatomical landmarks (Fig. 3) [24].

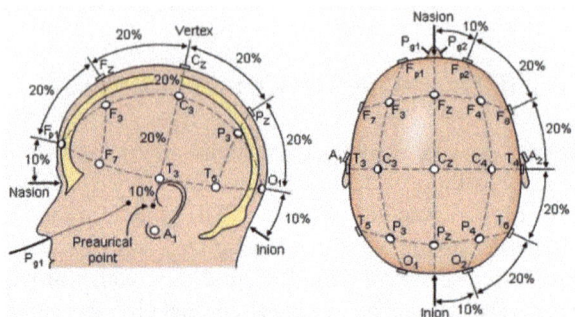

Fig. 3. Placement of electrons according to the 10–20 system.

Usually when carrying out the research the device is used at a sampling rate of 128 Hz to record EEG signals. The shortcut EEG stands for "electroencephalography" and is an electrophysiological process that records the electrical activity of the brain. It measures the changes being made in the electrical activity produced by the brain in microvolts (μV). The voltage changes are being made from an ion current that exists in brain cells and between them. The signals are mostly categorized as delta, theta, alpha, beta and gamma dependent on the signal frequencies which range from 0.1 Hz to more than 100 Hz. EEG sensors are able to record even up to several thousands of snapshots of electrical activity per second. The noted brain waves are sent to amplifiers and later to a computer or database to enable the further analysis.

EEG waveform can be monitored on the Xavier TestBench software which is provided within the NeuroHeadset package. Emotiv Xavier TestBench is an independent software that can collect data form the connected USB Transmitter device and analyze them as well as display, record and play EEG signals (Fig. 4). It can be used independently besides EPOC Control Panel to monitor the brain wave activities while performing research on the examined person [25].

The function of the EEG Tab placed in the Emotiv Xavier TestBench is to display in form of graphs real time changes in brain wave signals of 14 channels (AF3, F7, F3, FC5, T7, P7, O1, O2, P8, T8, FC6, F4, F8, AF4) and the Marker graph (Fig. 5). On the upper graph each selected brainwave signal for the corresponding sensor has a different color. The user has the ability to select and display from 1 up to all 14 channels. The lower graph displays Marker events, which change when the user wants to mark important signal slots to review. This is really helpful when later reviewing and analyzing the results. The left side contains various configuration buttons for changing the parameters of the both graphs presented on the right side.

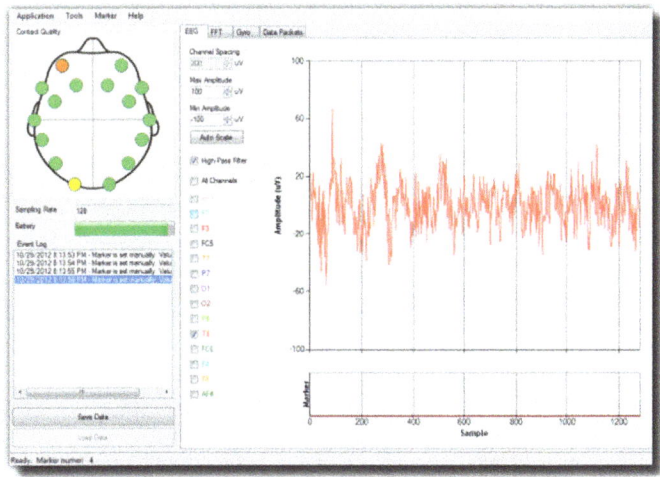

Fig. 4. Emotiv Xavier TestBench interface.

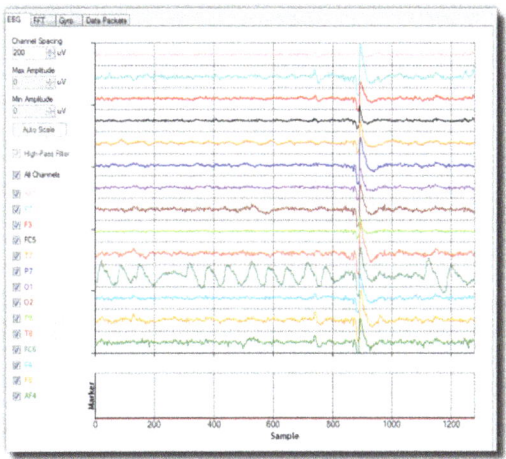

Fig. 5. EEG suite interface.

A tool for EEG signal processing and analysis is EEGLAB. EEGLAB is a MATLAB toolbox, which processes EEG, MEG and other electrophysiological information (Fig. 6) The toolbox has the ability to import different kind of files (e.g. CSV files exported from TestBench) and process the data to create a visualization of brain activities in single trails, perform time/frequency analysis (FTA) and statistics as well as perform ICA. ICA stands for Independent Component Analysis and is a blind signal separation method, which can be used to separate a set of source signals from a set of background noise or other mixed signals [26]. EEGLAB gives the user the ability to subtract ICA elements from the data. It provides various methods to visualize event-related brain activities as well as a programming environment created for accessing and manipulating EEG data. All researches and implemented methods can be shared in the EEGLAB platform with other users [27].

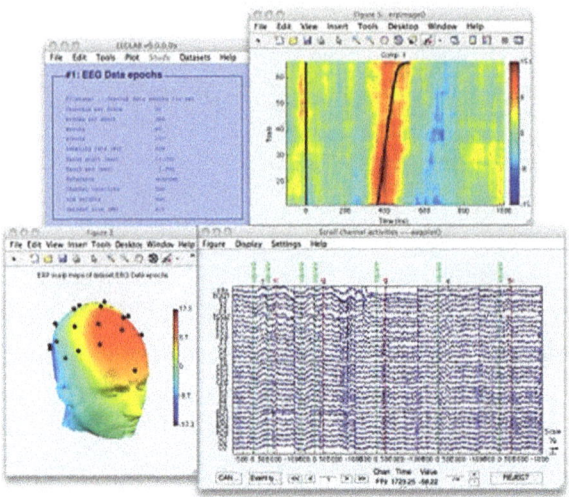

Fig. 6. MATLAB EEGLAB toolbox.

6 Lie-Detection Research Result Analysis

Various BCI and P300 brainwave-based researches carried out by different scientists had shown similar results. One of these researches described in the scientific article "Evaluation of P300 based Lie-Detection Algorithm" by Syed Kamran Haider, Malik Imran Daud, Aimin Jiang, Zubair Khan [21] was based on studying 20 subjects – 15 males and 5 females aged between 20–25 years. The created in the research scenario contained a theft of some precious items stolen from Subject1 by Subject2. After collecting some training data to determine the valid lies and valid truths, the research began. Additionally, the subjects were asked to count the number of asked questions and to present the value at the end of the research. The aim of that was to confirm that the mental activity and the focus was placed on the task and that the subject had no distracting thoughts. The used measuring device was Emotiv EPOC+ NeuroHeadset. The device was used to monitor and collect EEG brain activities. Later a MATLAB software was used to simulate the results. The overall results have shown an accuracy value in lie-detection of the presented subjects equal to 85%. This is a significant increase in accuracy compared to the basic, non-BCI based Lie-Detection technique, whose accuracy rate was equal to an average of 54%.

Another example of an carried out Lie-Detection research is the written by Vahid Abootalebi, Mohammad Hassan Moradi, Mohammad Ali Khalilzadeh article "A new approach for EEG feature extraction in P300-based lie detection" [28]. There, sixty-two subjects- 59 men and 3 women had been studied. The filtered in the 0.3–30 Hz range EEG was monitored with Ag/AgCl electrodes placed in the 10–20 international system. Brain activities were analyzed offline with the MATLAB software. Again, a theft scenario has been created: the subject received a box with a jewel and was supposed to decide whether he is guilty or not. If he chose to be guilty, he could open the box in order to memorize the "stolen" jewel details. The innocent subject could not open the box thus he contained no information about the item. During the research the subject was shown 5 jewelry pictures: 1 was a "Target" stimulus (an object shown before in the training process), one "Probe" stimuli (the stolen jewel) and three "Irrelevant" stimuli's (not previously shown to any subject). The results of this research presented an 86% accuracy rate in detection of the valid guilty/innocent subject.

7 Summary

Polygraph exams have been studied for over a hundred of years, continuously improving and increasing the accuracy rate. There is no doubt that due to a vast number of researches and test-data it developed and widespread to the point of even using the exam results in court as evidence or helping solving crimes. Throughout the time scientists have invented different techniques and methods for improving the accuracy rates, mainly focusing on the cognitive aspect of lie-detection. The results of this intentions have been widely successful increasing the accuracy rate from an average of 54% with no BCI Lie-Detection to an average of about 85% with implemented BCI technologies. Common Cognitive Lie-Detection technologies focus on the P300 wave which occurs about 300 ms after a presented stimuli or the Guilty Knowledge Test. Moreover, various Questioning Techniques have been developed to assure the best methods of asking questions during an interrogation to increase cognitive load in liars' brains. This had brought great results with accuracy rates of 75%–85%. Many BCI Lie-Detection researches share the same measuring environment: EPOC Emotiv+ NeuroHeadset as a measuring device and Emotiv Xavier TestBench or MATLAB EEGLAB ToolBox as software. These solutions are popular due to their cost-efficiency and professional access to a range of tools which can be used in various research fields, not only polygraphy. Overall a great progress has been made assuring that the polygraph exam is the most effective way of verifying truth and detecting lies.

It is also worth noting that Brain-Computer Interfaces are also used in many other applications, including: designing devices for visualization of brain activity [29], correlation with Augmented Reality technology [30], UAV control [31], identification of facial expressions in conjunction with machine learning [32].

References

1. Larson, J.A.: Lying and its detection: a study of deception and deception tests. Int. J. Ethics (1932). https://doi.org/10.1001/jama.1933.02740050060047
2. http://www.kellypolygraphe.com/polygraph-history.php. Accessed 05 Jan 2021
3. Griharadas, A.: India's novel use of brain scans in courts is debated. Neurosci. Courtroom Int. Concern, **53** (2008)
4. Rao, P.N.: Applications of BCIs. IEEE Signal Process. Mag. **27**(4) (2013). https://doi.org/10.1017/cbo9781139032803.017

5. https://people.howstuffworks.com/lie-detector.htm. Accessed 05 Jan 2021
6. Bond, C.F., De Paulo, B.M.: Accuracy of deception judgments. Pers. Soc. Psychol. Rev. **10**(3), 214–234 (2006). https://doi.org/10.1207/s15327957pspr1003
7. Vrij, A., Graham, S.: Individual differences between liars and the ability to detect lies. Expert Evid. **5**(4), 144–148 (1997). https://doi.org/10.1023/a:1008835204584
8. Meissner, C.A., Kassin, S.M.: He's guilty!: investigator bias in judgments of truth and deception. Law Hum Behav. **26**(5), 469–480 (2002). https://doi.org/10.1023/A:1020278620751
9. Marston, W.M.: Can you beat the lie detector? Polygraph **14**(4) (1935)
10. Granhag, P.A., Hartwig, M.: A new theoretical perspective on deception detection: on the psychology of instrumental mindreading. Psychol. Crime Law **14**(3), 189–200 (2008). https://doi.org/10.1080/10683160701645181
11. Vrij, A., Granhag, P.A., Mann, S., Leal, S.: Outsmarting the liars: towards a cognitive lie detection approach. Curr. Dir. Psychol. Sci. **20**(1), 23–32 (2011). https://doi.org/10.1177/0963721410391245
12. Grubin, D., Marsen, L.: Lie detection and the polygraph: a historical review. J. Forensic Psychiatry Psychol. **16**(2), 357–369 (2005). https://doi.org/10.1080/14789940412331337353
13. Lafuente, V., Gorriz Juan, M., Ramirez, J., Gonzalez, E.: P 300 brainwave extraction from EEG signals: an unsupervised approach. Expert Syst. Appl. **74**, 1–10 (2017). https://doi.org/10.1016/j.eswa.2016.12.038
14. Citi, L., Poli, R., Cinel, C.: Documenting, modeling and exploiting p300 amplitude changes due to variable target delays in donchin's speller. J. Neural Eng. **7**(5), (2010). https://doi.org/10.1088/1741-2560/7/5/056006
15. Bernat, E., Shevrin, H., Snodgrass, M.: Subliminal visual oddball stimuli evoke a P300 component. Clin. Neurophysiol. **112**(1), 159–171 (2001). https://doi.org/10.1016/s1388-2457(00)00445-4
16. https://medium.com/@AliOztas/hawkings-spelling-device-and-p300-3d7693e693d2. Accessed 6 Jan 2021
17. Ben-Shakhar, E.: The Guilty Knowledge Test (GKT) as an application of psychophysiology: future prospects and obstacles (2002)
18. Lawrence, A.: Farwell: brain fingerprinting classification concealed information test detects US Navy military medical information with P300. Front. Neurosci. **8**, 410 (2014). https://doi.org/10.3389/fnins.2014.00410
19. Farwell, L.A.: Method and apparatus for truth detection (1995)
20. Verschuere, B., Crombez, G., De Clercq, A., Koster, E.H.: Autonomic and behavioural responding to concealed information: differentiating orienting and defensive responses. Psychophysiology **41**(3), 461–466 (2004). https://doi.org/10.1111/j.1469-8986.00167.x
21. Abootalebi, V., Moradi, M.H., Khalilzadeh, M.A.: A new approach for EEG feature extraction in P300-based lie detection. Comput. Methods Programs Biomed. **94**(1), 48–57 (2009). https://doi.org/10.1016/j.cmpb.2008.10.001
22. Hartwig, M., Granhag, P.A., Stromwall, L., Kronkvist, O.: Strategic use of evidence during police interrogations: when training to detect deception works. Law Hum Behav. **30**(5), 603–619 (2006). https://doi.org/10.1007/s10979-006-9053-9
23. Paszkiel, S.: The use of facial expressions identified from the level of the EEG signal for controlling a mobile vehicle based on a state machine. In: Szewczyk, R., Zielinski, C., Kaliczynska, M. (eds.) Automation 2020: Towards Industry of the Future. Advances in Intelligent Systems and Computing, vol. 1140, pp. 227–238 (2020). https://doi.org/10.1007/978-3-030-40971-5_21
24. Paszkiel, S.: Data acquisition methods for human brain activity, analysis and classification of EEG signals for brain-computer interfaces. In: Series: Studies in Computational Intelligence, vol. 852, pp. 3–9 (2020). https://doi.org/10.1007/978-3-030-30581-9_2

25. Paszkiel, S.: Using the raspberry PI2 module and the brain-computer technology for control-ling a mobile vehicle. In: Szewczyk, R., Zielinski, C., Kaliczynska, M. (eds.) Automation 2019: Progress In Automation, Robotics And Measurement Techniques. Advances in Intelli-gent Systems and Computing, vol. 920, pp. 356–366 (2020). https://doi.org/10.1007/978-3-030-13273-6_34

26. Paszkiel, S.: Characteristics of question of blind source separation using moore-penrose pseu-doinversion for reconstruction of EEG signal. In: Szewczyk, R., Zielinski, C., Kaliczynska, M. (eds.) Automation 2017: Innovations In Automation, Robotics And Measurement Tech-niques. Advances in Intelligent Systems and Computing, vol. 550, pp. 393–400 (2017). https://doi.org/10.1007/978-3-319-54042-9_36

27. Paszkiel, S., Szpulak, P.: Methods of acquisition, archiving and biomedical data analysis of brain functioning. In: Hunek, W.P., Paszkiel, S. (ed.) Biomedical Engineering and Neu-roscience. Advances in Intelligent Systems and Computing, vol. 720, pp. 158–171 (2018). https://doi.org/10.1007/978-3-319-75025-5_15

28. Haider, S.K., Daud, M.I., Jiang, A., Khan, Z.: Evaluation of P300 based lie detection algorithm (2017). https://doi.org/10.5923/j.eee.20170703.01

29. Paszkiel, S., Hunek, W.P., Shylenko, A.: Project and simulation of a portable device for measuring bioelectrical signals from the brain for states consciousness verification with visu-alization on LEDs. In: Szewczyk, R., Zielinski, C., Kaliczynska, M. (eds.) Challenges in Automation, Robotics and Measurement Techniques. Advances in Intelligent Systems and Computing, Vvol. 440, pp. 25–35 (2016). https://doi.org/10.1007/978-3-319-29357-8_3

30. Paszkiel, S.: Augmented reality of technological environment in correlation with brain com-puter interfaces for control processes. In: Szewczyk, R., Zielinski, C., Kaliczynska, M. (eds.) Recent Advances in Automation, Robotics and Measuring Techniques. Advances in Intelli-gent Systems and Computing, vol. 267, pp. 197–203 (2014). https://doi.org/10.1007/978-3-319-05353-0_20

31. Paszkiel, S., Sikora, M.: The use of brain-computer interface to control unmanned aerial vehicle. In: Szewczyk, R.,, Zielinski, C., Kaliczynska, M. (eds.) Automation 2019: Progress in Automation, Robotics and Measurement Techniques. Advances in Intelligent Systems and Computing, vol. 920, pp. 583–598 (2020). https://doi.org/10.1007/978-3-030-13273-6_54

32. Paszkiel, S.: Using neural networks for classification of the changes in the eeg signal based on facial expressions, analysis and classification of eeg signals for brain-computer interfaces. In: Studies in Computational Intelligence, vol. 852, pp. 41–69 (2020). https://doi.org/10.1007/978-3-030-30581-9_7

Brain Activity During Competitive Games

Patryk Mróz[(⊠)] [iD]

Faculty of Electrical Engineering, Automatic Control and Informatics,
Opole University of Technology, Prószkowska 76, 45-758 Opole, Poland
patryk.mroz@student.po.edu.pl

Abstract. Based on results from EEG using Emotiv EPOC + NeuroHeadset device this paper will analyze brain activity during playing competitive games, and training. The study will be conducted on independent test subjects with different skill levels in AimLab, software environment where subjects will compete with each other. Results will show how brain is active as certain skills levels are achieved and if muscle memory have effect on it in any way.

Keywords: NeuroHeadset · Gaming · Games · EEG · Brain activity · Muscle memory · Aimlab · Emotiv · EPOC + NeuroHeadset

1 Introduction

Gaming become very popular, games industry revenue grew to $120.1 bn in 2019. Many players see in this opportunity to earn money in this filed by playing in tournaments or streaming. But becoming pro is not easy and takes a lot of time and practice. Player base in popular competitive games is growing and so it's harder to stand out in the crowd. To be good in games like counter-strike, Player Unknown's Battlegrounds, call-of-duty or Apex legends every player needs to have some skills regarding shooting and for example these are: Tracking – that refers to keeping player crosshair stuck on to target at all times, and following them as they move, Flickshots - that involve flicking player crosshairs at target then returning to a neutral position or controlling recoil of the weapons.

Brain activity is showed in 3D Brain Visualizer and its representing by channels alpha, delta, beta, gamma. AimLab is a shooting training assistant with specially designed training courses and very precise statistics and analyst build inside which can help players understand their weaknesses. Muscle memory can affect aiming signify knowing your mouse DPI and monitor resolution. Brain activity with training alone and muscle memory is not sufficient if players will just learn by trial-and-error method. That's why in this paper analytical summary will be created in the bases of these two methods to gather information's.

EEG is an process to record the electrical activity of the brain. EEG measures changes in the electrical activity of the brain produced. Voltage changes come from ionic current within and between some brain cells called neurons. An EEG test evaluates the electrical activity of the brain. EEG scans are performed by placing EEG sensors – small metal discs also called EEG electrodes – on your scalp. These electrodes pick up and record the

S. Paszkiel (Ed.): ICBCI 2021, AISC 1362, pp. 176–186, 2021.
https://doi.org/10.1007/978-3-030-72254-8_18

electrical activity in your brain. The collected EEG signals are amplified, digitized, and then sent to a computer or mobile device for storage and data processing. Analyzing EEG data is an exceptional way to study cognitive processes: It can help doctors establish a medical diagnosis, researchers understand the brain processes that underlie human behavior, and individuals to improve their productivity and wellness.

In the current state of knowledge of brain activity during competitive games there is great emphasis on the addiction aspects like that one which accrues in gambling. This is shown in article "Brain activity and desire for Internet video game play" [1]. The study in this article confirms the possibility of brain activity in the same regions when playing games. "The present findings suggest that cue-induced activation to Internet video game stimuli may be similar to that observed during cue presentation in persons with substance dependence or pathologic gambling."

With regard to muscle memory, one can refer to the article "Random Numbers and Gaming" [2] in which it is shown how in various FPS games the procedure introduces random and pseudo-random numbers. Even though these numbers are labeled, as "random" player is able to learn specific pattern that is written in any e.g., weapons. "In Counter Strike: Global Offensive spray pattern control becomes a muscle memory to a player after long periods of playing. It's a design choice that makes the gunplay between players more about instant crosshair placement with the faster player usually winning."

Not only brain activity has an impact on gaming, but also the attitude and behavior of players. This has been proven in the article "Master Maker: Understanding Gaming Skill Through Practice and Habit From Gameplay Behavior." [3] after 7 months of intensive research. "Broadly, play intensity, breaks in play, and skill change over time affect a player's skill in Halo Reach. Players with the most efficient skill gain are likely to play with moderate frequency and to avoid long breaks between matches".

2   Description of the Measurement Process

First calibration process with the device Emotiv EPOC + NeuroHeadset to gather brain activity data. Plug provided emotiv USB transceiver into USB port in any compatible computer. After a few second single LED should slowly flashing and then single bright LED when transceiver is paired with a nearby NeuroHeadset [4]. It doesn't need any external drivers, transceiver should install on computer automatically. When the system detects the device, turn on Emotiv EPOC + using switch at the rear of the headband. When properly paired a more dim LED will indicate successful data transfer. Next is to Install and open EPOC Control Panel [5]. If operating system will ask user to add this application to the local network check yes. In the firewall of operating system unblock that application. Before putting the EPOC NeuroHeadset, be sure that each of the 16 electrode sockets are fitted with a sensor unit with a moist felt pad [6]. If the pads are not wet, wet them using saline solution. Expand the EPOC NeuroHeadset and then place it head. Spread the sensors as shown on pictures provided by the developer. Sensors need to be placed very carefully and in the correct places. Front sensors should be 2–2.5 in., 50–60 mm or about three finger widths above the eyebrows [7]. Reference Sensors must have good contact with the subject to assess the Contact Quality of the remaining sensors. Inside EPOC Control Panel is EmoEngine status panel which will show four indicators: system status, system uptime, wireless signal and battery power [8]. In regard to the calibration process we are interested in system status which shows A summary of the general EmoEngine status and wireless signal that displays the quality of the connection between the EPOC + NeuroHeadset and the Emotiv wireless USB receiver connected to your machine. In user status controls select headsets that are connected and manage user profiles. Here it needs to be added user by clicking "add user", name that user and then select headset to that user. It can have many headsets connected but view only one at the time. If we want to view more than one, it needs to open a new instance of application with different user and selected different headset.

Before moving to other tabs beside "headset setup", it needs to achieve the best contact quality for best results and connection to all sensors. If the previous actions did not bring such results we should check if everything was done correctly. When connection isn't as expected, check the step by headset setup guide (Fig. 1) that is shown in this tab to make sure everything was set up correctly.

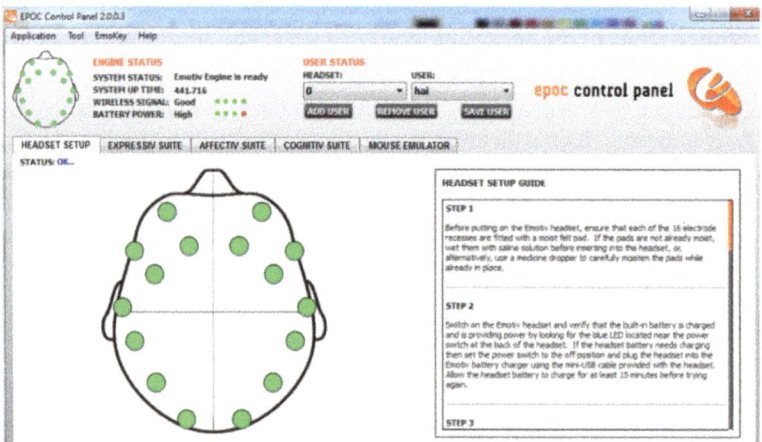

Fig. 1. EPOC control panel – headset setup suite.

The training in AimLab consists of four exercises, two out of two categories [9]. The categories are: Flicking which includes SpiderShoot and MicroShoot training and the Tracking category which includes CircleShoot and StrefeShoot. Each exercise is done three times and after each data are gathered and average is calculated. When training subjects wear Emotiv EPOC + NeuroHeadset device and all data about brain activity is gathered from Emotiv Xavier TestBench, 3D Brain Visualizer and Brain Activity Map. Each of subjects trains on their own hardware for the best results. Map is "Gray box" and weapon is "9 mm pistol" for every training.

3   Description of the Measuring Device

The structure of Emotiv EPOC + NeuroHeadset consist of: two electrode arms each containing 9 locations (7 sensors + 2 references. Two sensor locations (M1/M2). Electrodes are unremovable, so only fone pads must be applied. Opening at the rear of each sensor that allows easy rehydration. Headband can be rotated. On the rear of the head-band is mini USB port for charging, turn on/off switch and two status LED's. It has 9 axis motion sensors. Saline based electrodes.

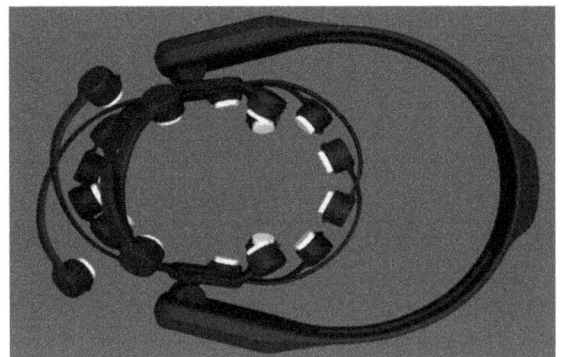

Fig. 2. Emotiv EPOC + NeuroHeadset

It comes with EEG sensors: 14 channels: AF3, F7, F3, FC5, T7, P7, O1, O2, P8, T8, FC6, F4, F8, AF4 [10]. Two references: CMS/DRL references at P3/P4; left/right mastoid process alternative. Sensor material: Saline soaked felt pads. Motion sensors: IMU part: ICM-20948, quaternions: normalized, 4D and resolution: 16 bits. Connectivity of this device: Wireless - Bluetooth® Low Energy, Bluetooth® 5.0, Bluetooth® SMART. Proprietary USB receiver: 2.4 GHz band, USB is used to change headset settings & Extender. EEG signals: Sampling method: Sequential sampling, single ADC and Sampling rate: 2048 internal down sampled to 128 SPS or 256 SPS. Resolution: LSB = 0.51 μV (14 bits mode), 0.1275 μV (16 bits mode). Bandwidth: 0.16 – 43Hz, digital notch filters at 50 Hz and 60 Hz [11, 12]. Filtering: Built in digital 5th order Sync filter, dynamic range of 8400 μV(pp) and coupling mode: AC coupled. Power: Battery - Internal Lithium Polymer battery 595mAh. Battery life: up to 12 h using USB receiver, up to 6 h using Bluetooth Low Energy. Sensor Material: Ag/AgCl + Felt + Saline. Weight of the device is 170g and dimensions as follow 9 × 15 × 15 (cm) [13].

4   Presentation of Measurement Results

First subject: Gender – Male, Age – 22, Hardware settings are set as follows, monitor resolution is 1920:1080, Field of view 90, and DPI (Dots Per linear Inch) 500. Flicking results from SpiderShot are: General score from first try 56397, accuracy 89%, time to kill 567 ms, targets 106, hits 105, misses 13, kills per second 1.75, gain 0,97, error size 7,99 and precision 90.03%. Scores and performance by movement direction in this exercise below in Fig. 3.

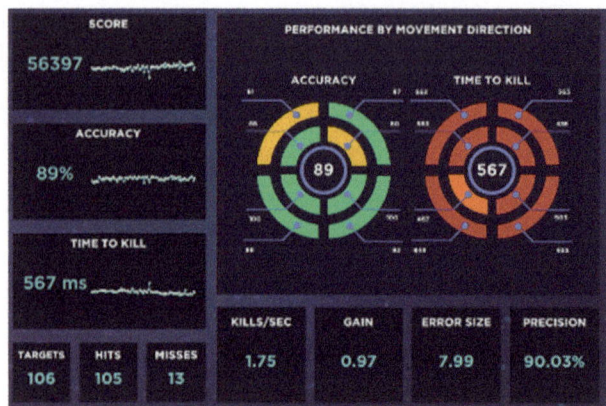

Fig. 3. Score and performance by movement direction by first test subject in Spider-Shot

General score from second try 62196, accuracy 90%, time to kill 528 ms, targets 114, hits 113, misses 12, kills per second 1.89, gain 0.99, error size 8,68 and precision 92.3%. Scores and performance by movement direction in this exercise below in Fig. 4.

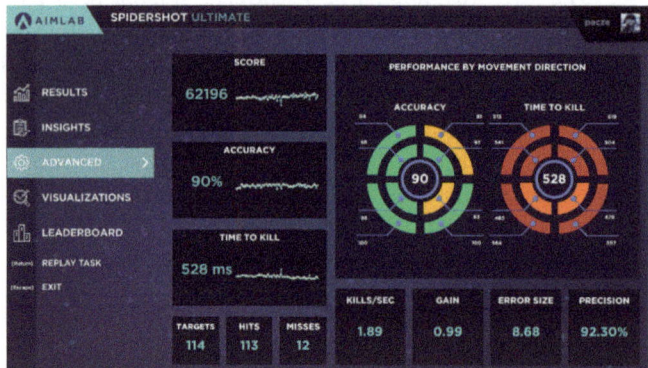

Fig. 4. Score and performance by movement direction by first test subject in SpiderShot.

General score from third try 62535, accuracy 93%, time to kill 533 ms, targets 112, hits 112, misses 8, kills per second 1.87, gain 1.00, error size 7,49 and precision 93.3%. Scores and performance by movement direction in this exercise below in Fig. 5.

Flicking results from MicroShot are: General score from first try 53794, accuracy 92%, time to kill 422 ms, targets 142, hits 142, misses 12, kills per second 2.37, gain 1,01, error size 7,08 and precision 93.01%. Scores and performance by movement direction in this exercise below in Fig. 6.

General score from second try 53191, accuracy 92%, time to kill 432 ms, targets 142, hits 141, misses 13, kills per second 2.35, gain 0.97, error size 7,56 and precision 92.72%. Scores and performance by movement direction in this exercise below in Fig. 7.

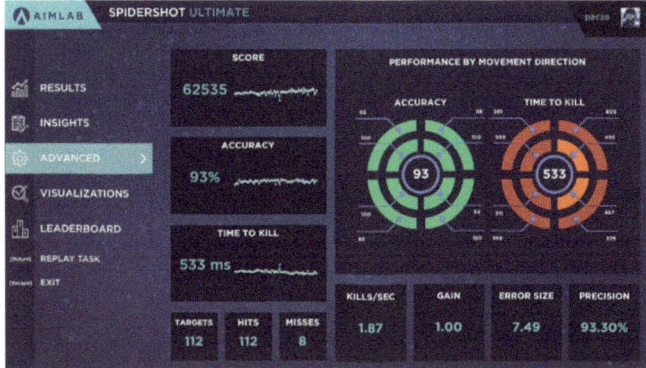

Fig. 5. Score and performance by movement direction by first test subject in SpiderShot.

Fig. 6. Score and performance by movement direction by first test subject in MicroShot

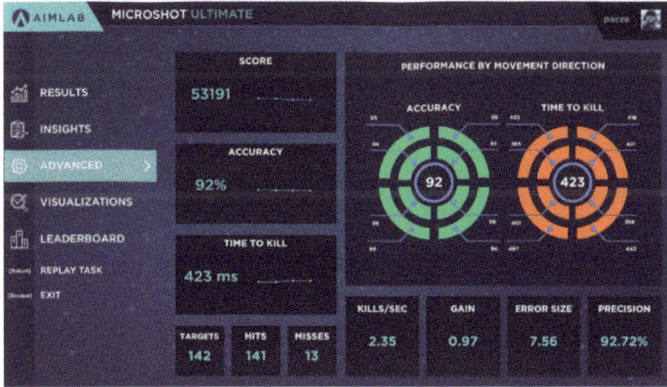

Fig. 7. Score and performance by movement direction by first test subject in MicroShot

General score from third try 59769, accuracy 95%, time to kill 389 ms, targets 154, hits 152, misses 8, kills per second 2.55, gain 0.97, error size 5,92 and precision 93.73%. Scores and performance by movement direction in this exercise below in Fig. 8.

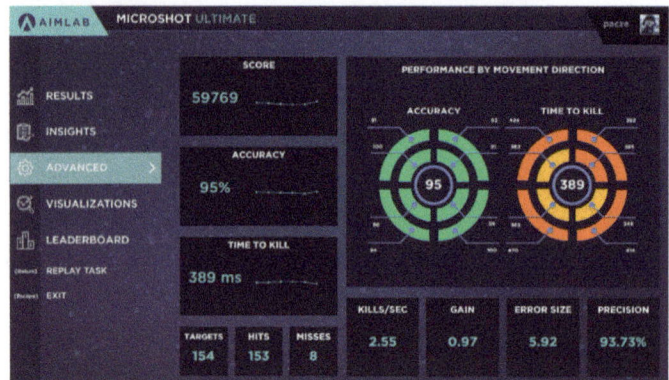

Fig. 8. Score and performance by movement direction by first test subject in MicroShot

Tracking results from CircleShot are: General score from first try 134510, accuracy 68%, time to kill 1,267 ms, kills per second 0.78, kills 47, shots per kill 4.38, hits 141, misses 64. Score and Error Bias below in Fig. 9.

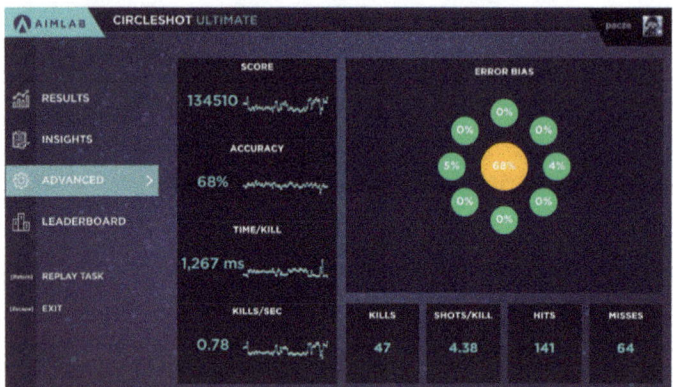

Fig. 9. Score and Error Bias by first test subject in CircleShot

General score from second try 134971, accuracy 67%, time to kill 1,208 ms, kills per second 0.78, kills 47, shots per kill 4.51, hits 143, misses 68. Score and Error Bias below in Fig. 10.

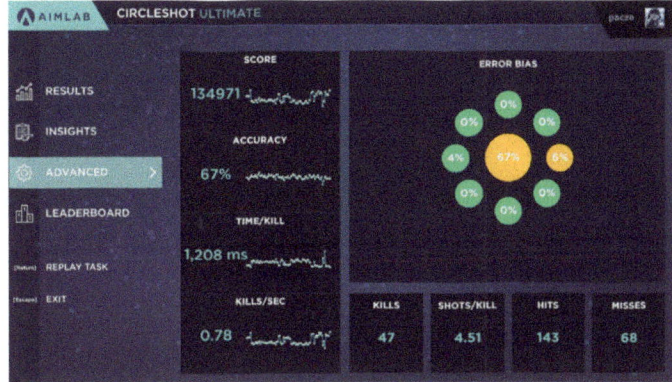

Fig. 10. Score and Error Bias by first test subject in CircleShot

General score from third try 153209, accuracy 74%, time to kill 1,152 ms, kills per second 0.85, kills 51, shots per kill 4.10, hits 154, misses 56. Score and Error Bias below in Fig. 11.

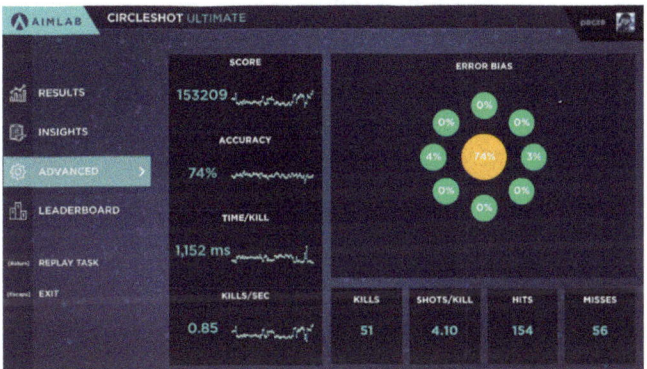

Fig. 11. Score and Error Bias by first test subject in CircleShot

Tracking results from StrafeShot are: General score from first try 129958, accuracy 64%, time to kill 1,256 ms, kills per second 0.78, kills 47, shots per kill 4.72, hits 142, misses 83. Score and Error Bias below in Fig. 12.

General score from second try 143303, accuracy 70%, time to kill 1,208 ms, kills per second 0.82, kills 49, shots per kill 4.37, hits 149, misses 65. Score and Error Bias below in Fig. 13.

General score from third try 141080, accuracy 70%, time to kill 1,217 ms, kills per second 0.80, kills 48, shots per kill 4.33, hits 146, misses 59. Score and Error Bias below in Fig. 14.

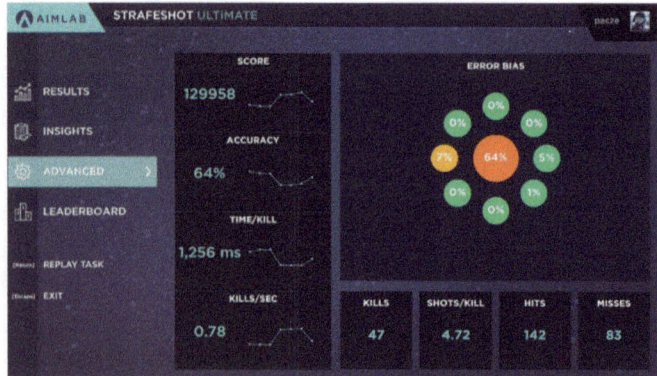

Fig. 12. Score and Error Bias by first test subject in StrafeShot

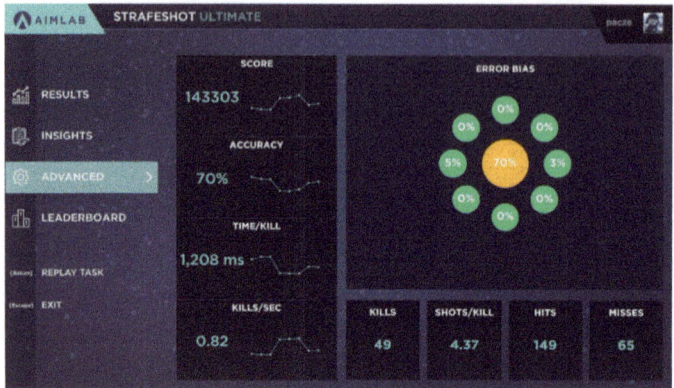

Fig. 13. Score and Error Bias by first test subject in StrafeShot

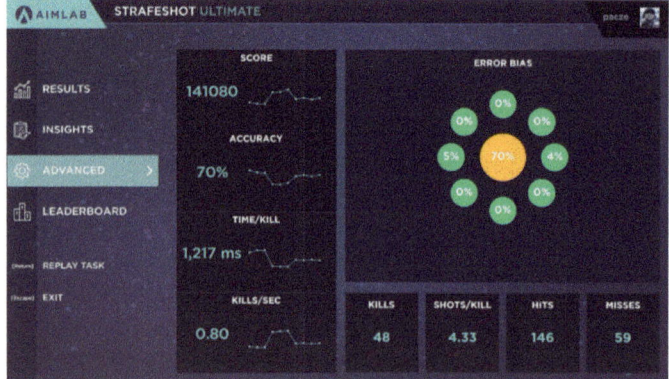

Fig. 14. Score and Error Bias by first test subject in StrafeShot

Each test subject wear Emotiv EPOC + NeuroHeadset device during training. Informations about brainwave activity was obtained from an application inte-grated with the Emotiv device called Emotiv Xavier TestBench, 3D Brain Visualizer and Brain Activity Map. After each exercise AimLab shows statistics and generates analyse. Having data from Emotiv EPOC + device and AimLab, it is possible to compile this data and conduct an analysis.

5 Summary

In conclusion, with all the information gathered from these exercises, we could come to some conclusions. Subjects having very different results from exercises shows how important is deep data from Emotiv EPOC + NeuroHeadset to show differences and similarities. Data was obtained with precision and payed attention to details. There was a further desire to play and punch higher scores among the subjects. Very similar to those that was mentioned in "Brain activity and desire for Internet video game play" article from Comprehensive Psychiatry. Both test subjects improved significantly when seeing their scores and analyses. Muscle memory adapts to the current equipment, settings. When the mouse and its DPI change, you have to adapt and learn how to move your hand again. Muscle memory is also the positioning of fingers on the keyboard, the ability to move the character, perform a jump which is as crucial as shooting, but, unlike shooting skills, there is no better way to position fingers on the keyboard. A player who learns certain habits stays with them, incorrectly remembered basics are extremely difficult to fix, this results in a reduction of quality and performance of the game. This is called bad habits. The higher the skill level, the harder it is to overcome the barrier of learned movements and overall skills. It turns out in the results. A player with higher skill level has very similar and close results and does not have much progress or better scores. The player who has less skills and does not have such knowledge of the program or generally games in this type assimilates and overachieves quickly. With each attempt, it definitely increases its results. One could say "one can't forget how to ride a bike" but one is going to lose aim skills if is not using it for some time, however it's simply to restore these skills.

References

1. Han, D.H., Bolo, N., Daniels, M.A., et al.: Brain activity and desire for internet video game play. Compr. Psychiatry **52**(1), 88–95 (2011)
2. Baglin, S.: Random numbers and gaming. Art and art History & Design Departments, ART 108: Introduction to games studies (2017)
3. Huang, J., Yan, E., Cheung, G.: Understanding gaming skill through practice and habit from gameplay behavior. Game XP: Action Games as Experimental Paradigms for Cognitive Science (2017)
4. Paszkiel, S., Sikora, M.: The use of brain-computer interface to control unmanned aerial vehicle. In: Szewczyk, R., Zielinski, C., Kaliczynska, M. (eds.) Automation 2019: Progress in Automation, Robotics and Measurement Techniques. Advances in Intelligent Systems and Computing, vol. 920, pp. 583–598 (2020). https://doi.org/10.1007/978-3-030-13273-6_54

5. Paszkiel, S.: Using the Raspberry PI2 module and the brain-computer technology for controlling a mobile vehicle. In: Szewczyk, R., Zielinski, C., Kaliczynska, M. (eds.) Automation 2019: Progress in Automation, Robotics and Measurement Techniques. Advances in Intelligent Systems and Computing, vol. 920, pp. 356–366 (2020). https://doi.org/10.1007/978-3-030-13273-6_34

6. Paszkiel, S., Hunek, W.P., Shylenko, A.: Project and simulation of a portable device for measuring bioelectrical signals from the brain for states consciousness verification with visualization on LEDs. In: Szewczyk, R., Zielinski, C., Kaliczynska, M. (eds.) Challenges in Automation, Robotics and Measurement Techniques. Advances in Intelligent Systems and Computing, vol. 440, pp. 25–35 (2016). https://doi.org/10.1007/978-3-319-29357-8_3

7. Paszkiel, S.: Augmented reality of technological environment in correlation with brain computer interfaces for control processes. In: Szewczyk, R., Zielinski, C., Kaliczynska, M. (eds.) Recent Advances in Automation, Robotics and Measuring Techniques. Advances in Intelligent Systems and Computing, vol. 267, pp. 197–203 (2014). https://doi.org/10.1007/978-3-319-05353-0_20

8. Paszkiel, S.: Using neural networks for classification of the changes in the EEG signal based on facial expressions. In: Analysis and Classification of EEG Signals for Brain-Computer Interfaces. Studies in Computational Intelligence, vol. 852, pp. 41–69 (2020). https://doi.org/10.1007/978-3-030-30581-9_7

9. Aim training: How to improve your FPS aim. https://www.pcgamer.com/aim-training/. Accessed 05 Jan 2021

10. Paszkiel, S.: Characteristics of question of blind source separation using Moore-Penrose pseudoinversion for reconstruction of EEG Signal. In: Szewczyk, R., Zielinski, C., Kaliczynska, M. (eds.) Automation 2017: Innovations in Automation, Robotics and Measurement Techniques. Advances in Intelligent Systems and Computing, vol. 550, pp. 393–400 (2017). https://doi.org/10.1007/978-3-319-54042-9_36

11. Paszkiel, S.: Data acquisition methods for human brain activity. In: Analysis and Classification of EEG Signals for Brain-Computer Interfaces. Studies in Computational Intelligence, vol. 852, pp. 3–9 (2020). https://doi.org/10.1007/978-3-030-30581-9_2

12. Paszkiel, S., Szpulak, P.: Methods of acquisition, archiving and biomedical data analysis of brain functioning. In: Hunek, W.P., Paszkiel, S. (eds.) Biomedical Engineering and Neuroscience. Advances in Intelligent Systems and Computing, vol. 720, pp. 158–171 (2018). https://doi.org/10.1007/978-3-319-75025-5_15

13. Paszkiel, S.: The use of facial expressions identified from the level of the EEG signal for controlling a mobile vehicle based on a state machine. In: Szewczyk, R., Zielinski, C., Kaliczynska, M. (eds.) Automation 2020: Towards Industry of the Future. Advances in Intelligent Systems and Computing, vol. 1140, pp. 227–238 (2020). https://doi.org/10.1007/978-3-030-40971-5_21

Privacy and Security in Brain-Computer Interfaces

Sebastian Słaby$^{(\boxtimes)}$ (iD)

Faculty of Electrical Engineering, Automatic Control and Informatics,
Opole University of Technology, Prószkowska 76, 45-758 Opole, Poland
`s.slaby@student.po.edu.pl`

Abstract. Brain-computer interfaces (BCIs) have significantly improved the quality of life of many people through, but not limited to, helping with various medical conditions. Given the recent development of BCIs to enable brain-to-internet and brain-to-brain communication a possibility for cybercriminals is generated, posing tremendous risk to the personal information and health of an individual. This paper presents characterizations of security attacks described in literature, along with their impact and countermeasures.

Keywords: BCI · Security · Privacy · Cybersecurity

1 Introduction

The purpose of brain-computer interfaces is to collect data related to a user's brain activities with the help of sensors and transfer the data to computers. This allows the skipping of peripheral nerves. The signals are captured by the BCI directly instead [1]. Over the years this technology has evolved to allow neural stimulation alongside neural recording with the use of electrical, optical, magnetic and ultrasonic methods and devices [2]. BCI communication can be bidirectional and unidirectional. The phases illustrated are not standardized in any way, so the most commonly mentioned in literature are presented. During phase 1 neurons interact with each other, the result of this is neural activity, which is based on previously agreed actions, like pressing a button or entirely spontaneous. In phase 2 this activity is gathered, transformed into digital data and analyzed by a BCI data processing system to determine what the user intended action is [3, 4].

The most widely used technology in BCI systems is EEG. Those devices work by measuring electrical potentials produced by the brain's neutral synaptic activities. EEG devices capture five waves from the human brain activity, Table 1 represents the waves, their frequency and related emotions or activities.

The security aspect of Brain-Computer technology has gained some traction in the recent years. A real, albeit small risk of BCI devices used against their users. Unauthorized access to BCI devices that are capable of impacting the user could potentially cause serious harm to the mentioned user [4]. This obviously depends on the type of the BCI system, which can belong to one of three groups: 1) invasive, 2) partial invasive, 3)

S. Paszkiel (Ed.): ICBCI 2021, AISC 1362, pp. 187–195, 2021.
https://doi.org/10.1007/978-3-030-72254-8_19

Table 1. Characteristics of waves captured by EEG devices.

Wave	Frequency [Hz]	Activity/Emotion
Gamma	31+	Excitement, arousal
Beta	12–30	Action, concentration
Alpha	8–12	Relaxation, disengagement
Theta	4–7	Inefficiency, daydreaming
Delta	0.5–4	User in hypnoidization

non-invasive. An invasive system requires physical implants of electrodes into the grey matter of the brain by surgery to function. A partially invasive system is applied inside of the skull but outside the grey matter. The last group: non-invasive systems, are applied outside the skull, just on the scalp [1].

Despite the recent interest in the security aspects of BCIs, it's security is still relatively underdeveloped and in its early stages. Because of the importance of private information in the brain of a person and the possible impact of BCIs on the individual the security of such devices is of the utmost importance. While gathering some data might prove more or less harmless, obtaining private information and gaining the ability to severely impact an individual's health is a serious issue. Already existing BCI systems would without a doubt strongly benefit from the implementation of robust security solutions and the focus of the device's manufacturers should shift their focus from device longevity and portability to its security [3]. The expansion of BCIs into fields like entertainment and video games introduces new risks to user privacy and data confidentiality [1, 5]. Research findings have already shown that it's possible to capture, read and manipulate information with basic hardware tools. On top of the already mentioned attacks, more sophisticated ones are possible with the use of eavesdropping, interception, denial of service or data modification. This opens the possibility for attackers to perform malicious actions on BCI systems without the users consent or knowledge [6].

2   Threat Models

In this section the various threat models will be listed and explored. The possible effects of various attacks will also be touched on in this section.

2.1   Man-in-the-Middle Attack

In this scenario the data, which is supposed to only travel between the BCI system to its paired receiver or another system over the network, gets intercepted by another device making it possible to modify the information within. While the attacker is able to read and influence the traffic between two parties, the parties themselves are unaware that such a thing is happening. This can occur even if the data is encrypted [7, 8]. The only prerequisite for the potential attacker is to have access to the same network in which the data is sent through or to be within the reach of the transmission of the BCI system, although the second method would work only when the system and the receiver aren't synchronized yet and the attacker can spoof their device to act as a receiver or has gained access to the receiver and is able to spoof the already synchronized one (Fig. 1).

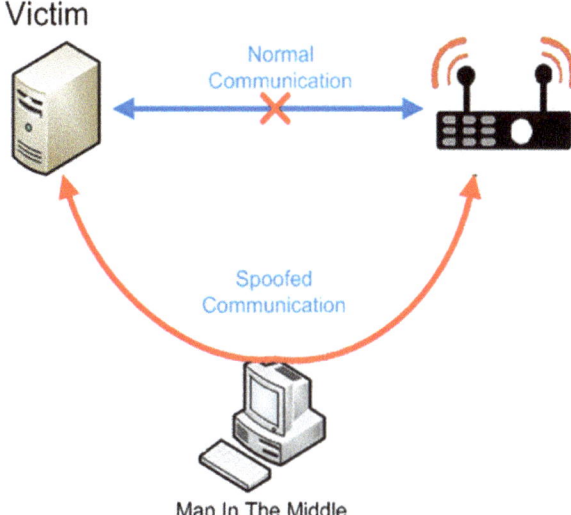

Fig. 1. Visualization of a man-in-the-middle attack.

The effects of such an attack vary substantially based on the type of BCI system in question. For neuro-medical application, which make it possible for patients with neurological disorders to interface with prosthetic limbs, the attacker would be able to control prosthetic limbs and invoke dangerous movements. The patients themselves might potentially take on the role of an attacker by trying to override limits imposed on them by security mechanisms or other systems in order to gain extra strength or eliminate the ability to feel pain [1].

2.2 Rootkit

A rootkit makes it possible for an attacker to gain access to an area of a system or software that is not supposed to be accessed at all by anyone, with the exception of its manufacturer. The term 'rootkit' is derived from the words 'root' and 'kit', where 'root' is the level of privileges possessed only by a user that is supposed to have full access to every part of a software or system, whereas 'kit' refers to the software component of the tool. Rootkits have proven themselves to be relatively hard to detect [11]. The effects of such an attack can range from harmless ones like adding functionality to a system by the user, through ones with medium impact like intercepting data up to potentially fatal ones where full control over a system makes it possible for the attacker to influence the behavior of the system and invoke dangerous movements or, in case of a more vital system, a complete shutdown. With root access there are basically no boundaries to what an attacker might be able to do from a privilege standpoint as he is in full control of the device.

2.3 Backdoor

This scenario is possible when the device has been tampered with by an attacker or the manufacturer themselves included such a backdoor. The aim of a backdoor is to bypass the regular authentication and authorization method already in place in order to gain access to a system or device.

2.4 Service Disruption/Physical Damage

There exist multiple reasons for a service disruption: hardware or software failure, power outage, connectivity loss or physical damage. Any of the above mentioned scenarios might happen by chance or by deliberate intent. In case of a service disruption a part of or all of the BCI system is rendered non-working. The effects of this outage will naturally depend on the type of BCI system in use and might be temporary or permanent, depending of the type of the outage and the severity of the damages. This is a case where the attack is possible on every component of the system, thus making it crucial to provide adequate security to all of its components as only one breach might be enough to shut down an entire system.

2.5 Eavesdropping

As user authentication and authorization have been used in conjunction with BCI systems and proven to be effective, a risk of using that mechanism to obtain private and sensitive data about a person manifests itself. Such data could potentially be intercepted with even less work than in a man-in-the-middle attack as there is no need to modify it. Encryption of the data might not be enough to prevent such actions. With such data an unauthenticated person may gain access to systems they shouldn't be able to get access to. Besides authentication and authorization the interception of such data may lead to the compromise of sensitive data not secured behind a computer system, like PINs or location of residence, especially if the attacker would be able to present images, videos or other stimuli to the victim which are designed to extract as much valuable information as possible [1].

3 Countermeasures

In this section possible countermeasures against the above described threats are listed and explored. As BCI systems are seeing wider adoption it is important for the creators and developers to take into account the importance of a secure framework and architectural design of their systems and software.

3.1 Certificate-Based Encryption

Digital certificates are a signature given by a trusted certificate authority that securely binds together several quantities. The most important parts of a certificate are the author-ity, user, public key and private key or secret. The computational overhead of encrypting and decrypting data using certificates or keys is negligible given modern processor architectures and speeds. The purpose of a certificate is to attest that a user is who he is claiming to be in terms of authentication and authorization. Encrypting data with the help of certificates makes it practically impossible to decrypt the data given certain security measures were taken beforehand. Basically only two attacks are taken into consideration when discussing attacks on encryption based on certificates: an uncertified client and the certifier itself. Any requests from an uncertified client should be instantly denied. The second is a bit trickier, if possible the devices should have corresponding keys baked into them out of the factory and the keys themselves should not be saved anywhere and not visible to any human. This is up to the manufacturer to decide, thus there might be no way to prove their claims. Another possible security breach is when a third person obtains the secret/private key to a pair of devices. This approach would work if two systems are linked together and not designed to work with other BCI systems, if they were to interface with multiple systems mutual certificate authentication could be used. This however requires another devices that would serve as an authentication server.

The keys should be stored in such a way that they are impossible to retrieve from a device even with root level access, preventing some effects of a possible rootkit or backdoor attack. This would limit the repairability of the devices in exchange for a more secure system. Another way of increasing security without limiting repairability is to implement a solution that makes it possible to revoke a certificate earlier than it's end of life date and create a new one. This would prove useful in a case when the private key is revealed. The certificate approach eliminates all of the network related attacks as the data is not readable by a third party. Given the potential risk of the manufacturer of the device having the private key and a possibility to include a backdoor in their product and the growth of BCI systems, customers should obtain the devices only from reputable and trustworthy vendors [9].

3.2 Timing

As evidenced in article [10] it is possible to detect tampering with the data based on the response time of the devices. This paper focuses on the importance of the distance between devices, a measurement which is also applicable to BCI systems. The basic premise of this countermeasure is that the second devices responds not only correctly, but within a given timeframe. If a message were to be intercepted and altered by a third party an increase in the response time would occur and trigger an appropriate reaction from the system. For BCI systems used for prosthetics, where the latency is low based on the physical proximity of the device of the BCI system. This approach has the potential to work for network based BCI systems, but in case of those such a solution would be harder to implement as it would have to take extra variables into account, like for example the load on the network in that point of time. Nonetheless such a method has already been proven to work and might prove invaluable in terms of security for the most vital BCI systems.

3.3 Access Control

Multi-factor authentication, which two-factor authentication is a part of, is a fundamental part of security in modern distributed systems. Most of, if not all, banks in the world have already adopted this solution and is actually mandated by the government in most countries. The basic premise of MFA is that a user is able to authenticate only after presenting multiple pieces of evidence to an authentication mechanism. The pieces include: knowledge (a password), possession (security token, like a USB stick) and inherence (biometrics). As in other modern computer systems it is very important to verify that a user is really who he claims to be and MFA makes it less likely for an attacker to gain access to a system in case they manage to obtain a user's login credentials (username and password). This not only applies to systems connected to a network but also ones that are only accessible physically, such devices should only be accessible behind appropriate security measures like biometric or identifier checks. The mentioned security measures reduce the risk of many potential attacks including but not limited to physical damage, a rootkit and backdoors. Access of an individual should be limited to only allow the operations that are required within the scope of the jobs of the individual in question, this method of privilege granting is names role-based access control, or RBAC. The advantages of this solution are that no user has complete ownership over everything and the permissions are divided between the users. This limits the work of a user to a single area, which makes it easier to educate and train that individual to perform his work properly and without errors.

3.4 Monitoring Solutions

It is possible to detect anomalies, attacks and avoid downtime altogether using various monitoring solutions. Such solutions provide data about the systems in use, access time and usage statistics. This can be linked with an alerting solution which will alert users to anomalies, violations and downtime in order for them to respond in a timely manner or trigger an automation that will try to resolve the issue by itself.

3.5 Personnel Training

Human error or social engineering are two possible methods of attack that can be prevented with the proper education and training. All the personnel with access to important systems and information should be properly trained and educated to detect and properly react to attempts at obtaining sensitive information through social engineering. Companies should also contact software and hardware vendors in order to provide their employees with adequate level of expertise in terms of operating their products. For any change done within a system or set of systems a procedure should be in place that requires the approval of multiple people, this will reduce the risk of human error.

3.6 High Availability

This one applies more to the computer and network systems used in conjunction with BCI. As outages are unavoidable in the long term it's important to mitigate their effects.

The premise of high availability is that the system keeps functioning despite a failure of one of its components, whether it's a network component or server failing. This induces the need for redundant infrastructure which is more costly and not always applicable to BCI, as it's simply not a viable solution to use multiple units of a single prosthetic at once. It's also not needed for each and every system in use, only for those where downtime is potentially very costly or dangerous like medical applications. Nonetheless redundancy in the components of BCI is recommended if possible.

3.7 General Computer, Network and Mobile Security

As computers are a part of BCI systems it is crucial to ensure they are properly secured. They should be managed by qualified personnel and have all the security measures like firewalls, security daemons in place and configured according to the use case and regulatory requirements. Many BCI connect to mobile devices thus their security is important as well. If used in a professional setting the mobile devices should be monitored and secured appropriately. For personal use where the data is most likely not as important as for medical appliances one should still pay attention to the security of their device, albeit in a less restrictive manner.

3.8 Testing

Companies that deliver BCI hardware and software should test their products rigorously to ensure its security and functionality. Both manual and automatic tests are advised. Once an issue is reported either by the customer or a tester the developers and engineers should work to address this issue. The reports of issues should not be publicly accessible until fixed as it makes it easier for potential attackers to exploit the devices or software. Ideally most of the issues should come out in development and in testing environments, but in reality this is not possible with complex architectures and software. Because of this a procedure that enables customers to report issues should exist.

3.9 Penetration Tests

This again is more relevant to networks and not strictly BCIs, although when a wireless connection is a method of communication for them it should be tested against possible exploits before release. In cases where BCIs connect over networks or the internet to another systems it's important to keep the network infrastructure safe from potential attacks. If possible the networks in use should not be accessible from the internet and if remote access is needed a VPN tunnel should be used.

3.10 Scalability and Load Balancing

In a system that is used by multiple users, like in multiplayer video games and systems that collect and process data from multiple users it is important to scale the performance of the infrastructure according to the load put on it by its userbase. Scalability can be automated and react dynamically to user load and previously recorded timeframes where the usage spiked. Load balancing is able to distribute the load to multiple servers. In case the servers hit a set threshold the number of instances rises to meet demand. In similar fashion the number of computational instances is reduced when demand is low, which often occurs at night when most users are asleep. This is true in case the users are in similar time zones, otherwise there should be running computational instances in multiple geographical regions to minimize latency. The concept of load balancing is illustrated in the figure below (Fig. 2).

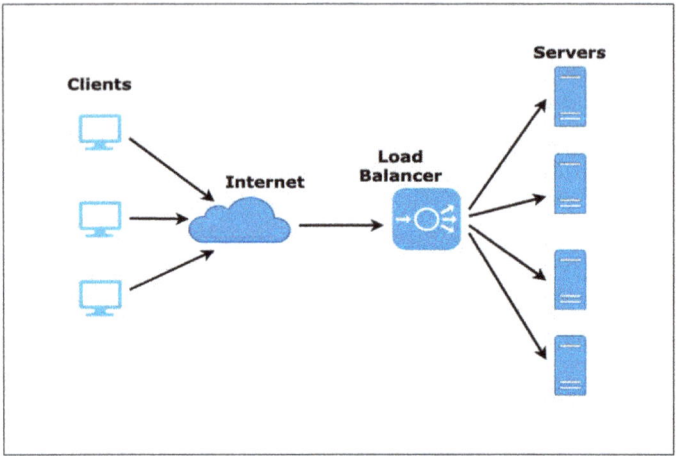

Fig. 2. Load balancer schematic.

4   Summary

This chapter performs a comprehensive analysis of potential threats and countermeasures to prevent them from succeeding. The importance of security of both BCI themselves and the surrounding infrastructure is growing constantly and the systems suffer from architectural design and implementation perspectives [15]. In this chapter various threats and attack models have been determined and described. This served as a starting point in determining the possible ways of attack and defining countermeasures that should prove effective against such threats. Every aspect of a BCI system, that is the BCI itself, computer networks, mobile devices and servers or computational nodes, has been taken into account when determining security measures to implement. Given the steady growth of BCIs more and more security concerns and challenges arise and should be the focus of developers and engineers. The countermeasures described in this paper should provide a foundation for further research and the ability to design and implement secure and robust BCI systems.

References

1. Li, O., Ding, D., Conti, M.: Brain-computer interface applications: security and privacy challenges. In: IEEE Conference on Communications and Network Security (CNS) (2015). https://doi.org/10.1109/cns.2015.7346884
2. Tyler, W.J., Sanguinetti, J.L., Fini, M., Hool, N.: Non-invasive neural stimulation. In: Micro-and Nanotechnology Sensors, Systems, and Applications IX (2017). https://doi.org/10.1117/12.2263175
3. Bernal, S., Celdrán, A.H., Pérez, G.M., Barros, M.T., et al.: Security in brain-computer interfaces: state-of-the-art, opportunities, and future challenges (2021). https://doi.org/10.1145/3427376
4. Rao, R.P.N.: Towards neural co-processors for the brain: combining decoding and encoding in brain-computer interfaces (2018). arXiv:1811.11876
5. Paszkiel, S., Szpulak, P.: Methods of acquisition, archiving and biomedical data analysis of brain functioning, biomedical engineering and neuroscience. In: Hunek, W.P., Paszkiel, S. (eds.) Book Series: Advances in Intelligent Systems and Computing, vol. 720, pp. 158–171 (2018). https://doi.org/10.1007/978-3-319-75025-5_15
6. Paszkiel, S., Dobrakowski, P., Lysiak, A.: the impact of different sounds on stress level in the context of EEG, cardiac measures and subjective stress level: a pilot study. Brain Sci. **10**(10), Article Number: 728 (2020). https://doi.org/10.3390/brainsci10100728
7. Paszkiel, S., Sikora, M.: The use of brain-computer interface to control unmanned aerial vehicle, automation 2019: progress in automation, robotics and measurement techniques. In: Szewczyk, R., Zielinski, C., Kaliczynska, M. (eds.) Book Series: Advances in Intelligent Systems and Computing, vol. 920, pp. 583–598 (2020). https://doi.org/10.1007/978-3-030-13273-6_54
8. Paszkiel, S.: Augmented reality of technological environment in correlation with brain computer interfaces for control processes, recent advances in automation, robotics and measuring techniques. Szewczyk, R., Zielinski, C., Kaliczynska, M. (eds.) Book Series: Advances in Intelligent Systems and Computing, vol. 267, pp. 197–203 (2014). https://doi.org/10.1007/978-3-319-05353-0_20
9. Paszkiel, S.: The use of facial expressions identified from the level of the eeg signal for controlling a mobile vehicle based on a state machine, automation 2020: towards industry of the future. In: Szewczyk, R., Zielinski, C., Kaliczynska, M. (eds.) Book Series: Advances in Intelligent Systems and Computing, vol. 1140, pp. 227–238 (2020). https://doi.org/10.1007/978-3-030-40971-5_21
10. Aziz, B.: Detecting Man-in-the-Middle Attacks by Precise Timin, DBLP (2009)
11. McAfee White Paper April 2006, Rootkits, Part 1 of 3: The Growing Threat (2006)
12. Paszkiel, S.: Characteristics of question of blind source separation using moore-penrose pseudoinversion for reconstruction of EEG signal, automation 2017: innovations in automation, robotics and measurement techniques. In: Szewczyk, R., Zielinski, C., Kaliczynska, M. (eds.) Book Series: Advances in Intelligent Systems and Computing, vol. 550, pp. 393–400 (2017). https://doi.org/10.1007/978-3-319-54042-9_36
13. https://www.researchgate.net/figure/Man-in-the-middle-attack-model_fig3_342618239. Accessed 05 Jan 2021
14. https://www.codeproject.com/Articles/326574/An-Introduction-to-Mutual-SSL-Authentication. Accessed 05 Jan 2021
15. Paszkiel, S.: Using BCI in IoT implementation, analysis and classification of EEG signals for brain-computer interfaces. In: Book Series: Studies in Computational Intelligence, vol. 852, pp. 101–110 (2020). https://doi.org/10.1007/978-3-030-30581-9_12
16. https://codeburst.io/load-balancers-an-analogy-cc64d9430db0. Accessed 05 Jan 2021
17. https://www.ekransystem.com/en/blog/rbac-vs-abac. Accessed 05 Jan 2021

Brain-Computer Interface Use to Control Military Weapons and Tools

Adrian Czech$^{(\boxtimes)}$

Faculty of Electrical Engineering, Automatic Control and Informatics,
Opole University of Technology, Prószkowska 76, 45-758 Opole, Poland

Abstract. The scope of this article will take into consideration a couple of factors which could provide military benefits established by BCI technology. BCI technology is still in development state. That is why this article will mostly provide predictions for near future. Assumptions made for future of this technology are strongly based on current development state of BCI in military field of its design. All military devices described in this article are currently at development state. Defense Advanced Research Projects Agency tries to apply BCI technology in military equipment. Their work is also focused on providing help to soldiers who come back from missions with combat injuries. Article will take in consideration safety factor during combat missions, which is essential on battlefield. Most of devices described in this article are being developed to increase this factor.

Keywords: BCI · Military · DARPA

1 Introduction

Human brain represents human's primary evolutionary advantage. It is still undiscovered area of humans body which can hide untapped potential for further evolution. Currently brain is used within our bodies to send electrical currents through the nervous system to perform certain kind of tasks. Research over BCI technology has shown that it could be automatized by linking human brain with a computer. This approach can free brain of its corporeal confine and provides an ability to control machines directly. Neurotechnological advances allow us already to perform some basic operations with only our brain. There is possibility, that human body will become a constraint that could be circumvented with appropriate neurotechnology [1]. BCI technology can help to extend use of brain outside of human body. It allows to control external device with bidirectional information flow (between the brain and a device) [2]. Potential impact of this technology is broad and far reaching, and policies on how to develop and manage such technology should be proactive, not reactive. Progress of BCI technology highlights the need to assess current and potential applications, and to ensure that the technology responds to actual needs in addition to the intentions of developers. As BCI transitions from basic research to more operational and commercial applications [3, 31], it will be important to devote early attention to the broader implications, to consider what policies and guidelines might maximize its benefits while mitigating potential downsides.

S. Paszkiel (Ed.): ICBCI 2021, AISC 1362, pp. 196–204, 2021.
https://doi.org/10.1007/978-3-030-72254-8_20

Developing such technologies as AI, data analytics, and robotics have captured headlines and fostered public discussion regarding potential benefits and risks. Limited comparable conversation has, as of yet, evolved for BCI. When compared to other prominent emerging technologies, BCI is relatively immature; few capabilities has been deployed commercially. However it has the potential to be no less influential.

There are many research laboratories working on BCI technology around the World. In places like these, BCI technology is constantly evolving to change the vision of future World. Further development of this technology will result in safe, practical and robust BCI capability. It will revolutionize health-care and communication. On the matter of BCI communication, researchers have already succeeded to facilitate brain-to-brain communication between two humans. This kind of technology will provide utility, relevance and tremendous strategic advantage on a future battlefield [4]. It is believed that BCI technology will dominate military systems in early 2030s [5]. Beside communication improvement, BCI technology can give advantage in common situational awareness, allow operators to control multiple technological platforms simultaneously. It would allow to remotely pilot air vehicles, sea vehicles or ground weapons system with an array of sensors and weapon systems that transmits and receives data input directly to and from the human operator's brain at an unprecedented rate, eliminating the delay of human sensory and muscle movement information processing. The operator's brain-computer interfacing could provide a multi-dimensional view of the battle space real-time with constant analysis and cognitive intuition. A fundamental difference between the brain and the computer is the brain uses billions of cells in parallel organization, whereas computer transistors are organized sequentially, and electronic computers operate at speed millions of times faster than brain computing. The brain has significant computational breadth, but limited depth. In contrast, computers have the depth to run an algorithm at high speed with significant data storage, but have limits on running multiple algorithms simultaneously6. Currently this mismatch limits BCI effectiveness, but the pace of computing and neuroscience research will solve this mismatch in the coming decades [7].

2 State of BCI Technology at Present

To perform properly, BCI has to translate the user's intentions accurately into actions. This requires proper correlation between brain and computer. User must modulate his brain signals to improve performance of the BCI, while the device must identify, interpret, and adapt to the neural signals that are most predictive of the desired output. This is achieved through feedback and fine-tuning mechanisms that are similar to those used when learning a new motor task [8]. Some BCI designs rely on a training phase in which the subject performs a designated task and a computational algorithm is employed to select the neuronal signals that best correlate with execution of that task. One of devices using this approach is Emotive EPOC+ NeuroHeadset, which is non-invasive type device to perform brain-computer connection (Fig. 1).

Fig. 1. Emotiv EPOC+ NeuroHeadset.

A code is generated for each command that can subsequently be used for control of an external device [9]. Alternatively, a real-time adaptive algorithm can be employed during the learning phase to concurrently select for the signals that are most predictive of the user's intentions by continuously refining them based on comparisons of past and intended trajectories. Recordings of larger populations of neurons, or neuronal ensembles, are generally the preferred source for extracting useful and relevant information to guide appropriate activity [9]. Although input form a single neuron can result in successful BCI control, averaging neuronal signals over many trial sessions is often necessary for predicting behaviour and therefore synthesizing the electrical firing of neuronal ensembles can remove the variability associated with using the input of a single neuron [9]. As the user learns to operate the BCI, neuronal plasticity leads to a tuning of ensemble signals such that activity in more discrete populations of neurons becomes the best determinant of action commands [10].

BCIs can be classified into those that use non-invasive, invasive, and partially invasive platforms. EEG, which obtains electrical signals form the scalp, has been the dominant method of recording used for non-invasive BCIs due to its relative safety and practical technical requirements. Invasive BCIs retrieve signals from single-neuron recordings via microelectrodes implanted in the cortical layers. Partially invasive BCIs use ECoG readings that come from sensors at the cortical surface placed either above or below the dura mater [9]. Platforms that belong to the latter 2 classes require neurosurgical implantation.

Using an EEG-based system, humans with motor debilities, including those that result form spinal cord injury or amyotrophic lateral sclerosis, have been able to control a computer cursor in 2 dimensions [11]. This technology has also been used by motor-intact individuals to command robots to manipulate objects, and has the potential to be applied in operating limb prosthetics [12]. The EEG devices, however, are fundamentally limited by their signal content, which does not convey information about components of movement such as position and velocity, and recordings are prone to interference from the electromyographic activity of cranial musculature [8].

Invasive BCIs, on the other hand, can acquire more informative signals that enable higher performance limits. For instance, human patients with locked-in syndrome are able to move cursors on a 2D keyboard to communicate using typed messages after undergoing implantation of electrodes that attract growth of myelinated nerve fibres [13]. With the aid of 96-microelectrode array that records signals from primary motor cortex, tetraplegic patients have been able to move a 2D cursor as well as to execute basic control over robotic devices, such as opening and closing a prosthetic hand, years after their initial spinal cord injury. Subsequent reports on tetraplegic patients who were enrolled in a pilot clinical trial of BCIs have demonstrated modest improvements in cursor control, thereby achieving greater functionality for practical tasks [14].

More recently, ECoG has proven to be useful tool in detecting input signals for BCIs. Unlike EEG, ECoG can detect high-frequency gamma wave activity that is the product of smaller cortical ensembles and correlates with discharge of action potentials from cortical neurons [8]. Because they are not embedded in brain parenchyma, ECoG electrodes inflict less damage to the cortex and also experience less signal deterioration than invasive electrodes. In patients with intractable epilepsy who required invasive monitoring. ECoG signals have been used for 2D movement control at a level of performance similar to that achieved with invasive BCIs. Although this approach has not been tested in patients with motor impairment, it can be applied more safely than invasive electrodes, and produces greater information content than EEG systems [8].

Although invasive and partially invasive BCIs hold great potential for functional recovery, the current risks and limitations associated with device implantation prevent BCIs from widespread use. One of the major shortcomings of the current technology is associated with the loss of signal reliability over time. In response to damage incurred by microelectrode penetration of the cortex, microglia and astrocytes begin a reactive process that unsheathes the prosthesis and disrupts its initial impedance properties. Neural and vascular damage at the site of insertion can also lead to development of infections [15]. As a result, the length of time over which an implantable electrode is able to produce signals is measured in months, and typically does not make it over 1 year without significantly losing quality. Moreover, motion between the implanted electrode and brain parenchyma, as is often caused by changes in brain volume or physical activity, can influence signal production [16].

3 Silent Communication

One way to gain advantage on a battlefield is to be able to send information as fast as possible. Even during reconnaissance, when quiet approach to the mission is crucial, giving word commands is at most cases impossible due to stealth nature of this type of mission. That's where silent communication may be a sufficient way to improve accomplish this type of mission with success. Researchers have already been working on this matter in the past. Some of them gave good overviews of classification algorithms. More exemplary article describing use of some current BCIs are the "Thought Translation Device" [18] and the "Berlin Brain Computer Interface" [19].

The aim of a BCI is to translate the thoughts or intentions of a subject into a control signal suitable for operating devices. When speaking about silent communication, "unspoken speech" refers to the process in which user imagines speaking a given word without actually producing any sound or performing any movement of the articulatory muscles at all. This research were provided mostly for people with disabilities, which caused inability to communicate, for example locked-in syndrome.

During the study, 16 channel EEG data were recorded using the International 10–20 System; results indicated that the motor cortex, Broca's and Wernicke's areas were the most relevant EEG recording regions for the task. The system was able to recognize unspoken speech from EEG signals at a promising recognition rate – giving word error rates on average 4 to 5 times higher than chance on vocabularies of up to ten words. In a follow-up study, Porbadnigk et al. (2009) [20] discovered that temporally correlated brain activities tend to superimpose the signal of interest, and that cross-session training (within subjects) yields recognition rates only at chance level. These analyses also suggested several improvements for suture investigations: using a vocabulary of words with semantic meaning to improve recognition results, increasing the number of repetitions of each word (20 were used in the study) and normalizing phrase length, in order to improve the model training, providing the subject with feedback on whether words were correctly recognized. Birbaumer (2000) [18] showed that subjects can be trained to modify their brain waves to enable words to be recognized more easily.

Another study proposed a control scheme for a silent speech BCI using neural activities associated with vowel speech imagery (DaSalla et al. (2009) [21]. They recorded EEG signals in three healthy subjects performing three tasks: unspoken speech of the English vowels /a/ and /u/; and a no-action state as a control. Subjects performed 50 trials for each task, with each trial containing two seconds of task-specific activity. To discriminate between tasks, the authors designed spatial trials from each of two tasks, the EEG time series data were decomposed into spatial patterns which were both common between, and optimally discriminative for, the two tasks. Applying these spatial filters to new EEG data produced new times series optimized for classification. Since the CSP method is limited to two-class discriminations, spatial filters for all pair-wise combinations of the three tasks were designed. Resultant spatial patterns showed mostly symmetrical activations centred at the motor cortex region, specifically the Cz and Fz positions in the International 10–20 system. After spatially filtering the EEG data, the authors trained a nonlinear support vector machine using the previously selected 20 trials per task, and classified the remaining 30 trials per task. This randomized training and testing procedure was repeated 20 times to achieve a 20-fold cross validation. Accuracies and standard deviations (in %) obtained for three subjects were 78 ± 5, 68 ± 7 and 68 ± 12. The study thus shows that motor cortex activations associated with imaginary vowel speech can be classified, with accuracies significantly above chance, using CSP and a nonlinear classifier. The authors envision the proposed system providing a natural and intuitive control method for EEG-based silent speech interfaces [22].

Researches described above give an amazing example of how significant BCI technology can be in near future. Silent speech can be a tremendous advantage on battlefield. Especially during reconnaissance when silent approach to mission is crucial to its success. Currently soldiers use hand gestures to communicate during stealth missions, which requires to looking at an ally and not on a field where soldiers are moving to. With silent speech this would not be necessary because commands could be given directly to brain where it would be interpreted. This approach appears to be faster, more sufficient and most of all more safer than current methods of communication. Americas research laboratory DARPA has already started working on this matter. They try to develop non-acoustic sensors for speech encoding in acoustically hostile environment, such as military vehicle where loud noise can be heard beside of commands given by soldiers. They also work on a shown above technology of "Silent Talk".

4 Further Military Enhancement with BCI Technology

DARPA did not stop on developing "Silent Talk" technology to improve soldiers safety and sufficiency during combat missions. They are also working on improving soldiers' perception. It's proposed under a name "Cognitive Technology Threat Warning System". Proposed by DARPA improvement of soldiers focuses on a visual neuroprosthetic to use on the battlefield to provide automated detection of environmental threats. This technology will be provided in form of binoculars enhanced with BCI technology. This devise will provide high resolution vision with quick threat detection. Such device can have detection range of up to 10 km against dismounts and vehicles and can expand soldiers' field of view to 120° [23].

Another concept presented by DARPA named "Telepresence" covers idea of remotely controlled vehicle by soldier using BCI technology. In this concept soldier is controlling vehicle placed on a dangerous area, being completely safe in a base or other concealed location [17]. This soldier still has the ability to sense and interact with hostile area without being at any danger. This concept gives opportunity to substitute human soldiers with BCI controlled machines in the future. In article written in 2020, authors managed to control humanoid robot P300 with BCI technology. To implement telepresence application they used two major techniques: Programming by demonstration in which the robot learns a task by observing someone performing it and BCI-based control in which the brain signal generated by the visual stimuli is converted to control signals by classifying the P300 signal generated.

During calibration session robot was learning the functions which were stored in the buffer. During real-time operation classified signals were sent to P300 which were interpreted as control commands saved in buffer. The system achieved a real-time accuracy of 78% on average [24]. Further development of telepresence can provide modernization of military on a bigger scale. Evolution of this concept can bring remotely controlled ships, submarines, drones, artillery. This would result in significant decrease of casualties between soldiers caused by combat during wars. Reconnaissance drones would give real-time information without or with minimal delay.

5 Function Restoration After Combat Injury – BCI Prosthetics

DARPA created BCI control of prosthetics which provides an innovation in prosthetics development. With this technology, prosthetic operates nearly the same as normal limb. This line of research and development involved real-time recording and decoding of motor cortical signals to provide research participants with tetraplegia the ability to control up to ten DOF (degree of freedom) with the prosthetic arm systems. To obtain brain signals responsible for control of the limbs, two intracortical microelectrode arrays were implanted in the participant's motor cortex (M1) and connected percutaneously through the use of two head-mounted pedestals. This allowed to record brain signals which were translated by complex algorithms into commands for the motors throughout the prosthetic arm system. Arrays were implanted 14 mm apart to additionally decode grasping behaviour [25]. Signal from both arrays allowed to record over 250 unique single units, which were then processed in real-time to ultimately send commands to move the prosthetic limb. The recorded signals passed through a Blackrock Microsystems NeuroPort data acquisition system, which converted neuronal firing rate (30-ms bins) into a functional mapping for prosthetic limb commands in endpoint velocity space. This enabled closed-loop control. This system allowed the participant to achieve control of the arm in 3 DOF (endpoint of the wrist) within two weeks of implantation, and 7 DOF within five weeks [26]. Different types of signals acquired from the brain (single neuron vs ECoG) required new population vector decoding methodologies and shared control architectures needed to be developed. This allowed users to initialize their control of the system and then learn to obtain control over a greater number of DOF. Applying such methodologies have further elucidated the neural mechanisms underlying human-tool interaction. All this effort allowed to understand better how the brain represents map control, environmental cues, object interaction and perception of neuroprosthetic control [26].

DARPA researchers have laid the foundation for adding the next generation of neuroprosthetic control by exploring the restoration of the sense of touch. To give the ability to identify objects, manipulate them and even grasp object without looking at them, researchers had to investigate how the non-human primate brain encodes sensory information provided via natural means and comparing that to the psychometric evaluation of the encoding of sensory information delivered through cortical stimulation [27]. Investigation explored simple percepts of touch as well as complex encoding of slip and texture by a series of experiments to demonstrate safety of chronically implanted stimulating electrode arrays in the somatosensory cortex of non-human primates. The implant consisted of two 100-channel electrode arrays connected via Cereport connectors to a CereStim R96 stimulator. During a six months stimulation, 300 Hz was delivered to three non-human primates across a range of charge amplitude, duty cycles and interval durations. The result revealed no deficits in fine motor control and demonstrated safety of the electrode-tissue interface [28]. Second study took place at the same time at the University of Chicago. This study was performed to characterize relationship between mechanical and electrical stimulation on tactile tasks [29]. During this study, two 96-electrode was implanted in the hand representation of Brodmann's area 1 and two 16-electrode Floating Microelectrode Arrays targeting the hand region in Brodmann's area 3b. This set was used to begin stimulation of via a CereStim stimulator. Then two types of detection tasks (electrical and mechanical) were compared. Study presented equivalent detection performance to mechanical stimulation of the native finger, with a psychometric curve function created during the stimulation test. Both of these studies were crucial in the support of FDA Investigational Device Exemption (IDE) approval for testing in human clinical populations. The main aspect of this studies and clinical use is to enable signals from sensors on prosthetic fingers to be translated into stimulation signals delivered directly to the sensory cortex, enabling patients to "feel" when they touch objects with their prosthetic arm [30].

6 Conclusion

BCI rapid evolution will have significant impact on military equipment. Development projects, which are already in progress in USA, can bring new weapons and equipment based on this technology even in the next decade. It will revolutionize the way of executing any military mission. Soldiers equipped with BCI technology will execute missions faster. Thanks to silent speech or Cognitive Technology Threat Warning System soldiers will be safer on hostile areas. BCI allows to restore near normal functionality after combat injuries where soldier lost one of his limbs. Prosthetics can be moved by this technology. DARPA went even further with research on this matter and created prosthetic which can restore even sense of touch for injured patient. All of this can be acquired with BCI technology.

References

1. Paszkiel, S.: Data acquisition methods for human brain activity, analysis and classification of EEG signals for brain-computer interfaces. In: Book Series: Studies in Computational Intelligence, vol. 852, pp. 3–9 (2020). https://doi.org/10.1007/978-3-030-30581-9_2
2. Paszkiel, S.: Using BCI in IoT implementation, analysis and classification of EEG signals for brain-computer interfaces. In: Book Series: Studies in Computational Intelligence, vol. 852, pp. 101–110 (2020). https://doi.org/10.1007/978-3-030-30581-9_12
3. Paszkiel S.: Using neural networks for classification of the changes in the EEG signal based on facial expressions, analysis and classification of eeg signals for brain-computer interfaces. In: Book Series: Studies in Computational Intelligence, vol. 852, pp. 41–69 (2020). https://doi.org/10.1007/978-3-030-30581-9_7
4. Major Cutter, P.A.: The shape of things to come: the military benefits of the brain-computer interface in 2040, accession number: AD1012768 (2015)
5. Shah, H.: Brain Computer Interface Technology: Interactive Applications You Probably Never Thought About, Futurista (2014)
6. Schalk, G.: Brain-computer symbiosis. J. Neural Eng. (2008)
7. Moore, B.E.: The brain computer interface future: time for a strategy, accession number: AD1018886 (2013)
8. Leuthardt, E.C., Schalk, G., Roland, J., Rouse, A., Moran, D.W.: Evolution of brain-computer interface: going beyond classic motor physiology (2009). https://doi.org/10.3171/2009.4.FOCUS0979
9. Nicolelis, M.A., Lebedev, M.A.: Principles of neural ensemble physiology underlying the operation of brain-machine interfaces. Nat. Rev. Neurosci. **10**, 530–540 (2009)
10. Carmena, J.M., Lebedev, M.A., Crist, R.E., O'Doherty, J.E., Santucci, D.M., Dimitrov, D.F., et al.: Learning to control a brain-machine interface for reaching and grasping by primates. PLoS Biol. **1**(2), E42 (2003)
11. Wolpaw, J.R., McFarland, D.J.: Control of a two-dimensional movement signal by a noninvasive brain-computer interface in humans. Proc. Natl. Acad. Sci. U.S.A. **101**, 17849–17854 (2004)
12. Bell, C.J., Shenoy, P., Chaldhorn, R., Rao, R.: Control of a humanoid robot by a noninvasive brain-computer interface in humans. J. Neural. Eng. **5**, 214–220 (2008)
13. Kennedy, P.R., Bakay, R.A.: Restoration of neural output from a paralyzed patient by a direct brain connection. NeuroReport **9**, 1707–1711 (1998)
14. Kim, S.P., Simeral, J.D., Hochberg, L.R., Donoghue, J.P., Black, M.J.: Neural control of computer cursor velocity by decoding motor cortical spiking activity in humans with tetraplegia. J. Neural Eng. **5**, 455–476 (2008)
15. Bjornsson, C.S., Oh, C.J., Al-Kofahi, Y.A., Lim, Y.J., Smith, K.L., Turner, J.N., et al.: Effects of insertion conditions on tissue strain and vascular damage during neuroprosthetic device insertion. J. Neural Eng. **3**, 196–207 (2006)

16. Santhanam, G., Linderman, M.D., Gilja. V., Afshar. A., Ryu. S.I., Meng, T.H., et al.: Hermes B: a continuous neural recording system for freely behaving primates. IEEE Trans. Biomed. Eng. **54**, 2037–2050 (2007)

17. Kotchetkov, I.S., Hwang, B.Y., Appelboom, G., Kellner, C., Connolly, E.: Brain-computer interfaces: military, neurosurgical, and ethical perspective (2010). https://doi.org/10.3171/2010.2.FOCUS1027

18. Birbaumer, N., et al.: he thought translation device (TTD) for completely paralyzed patients. IEEE Trans. Rehabil. Eng. **8**(2), 190–193 (2000). https://doi.org/10.1109/86.847812

19. Blankertz, B., Dornhege, G., Krauledat, M., Müler, K.R., Kunzmann, V., Losch, F., Curio, G.: The Berlin brain–computer interface: EEG-based communication without subject training (2006). https://doi.org/10.1109/TNSRE.2006.875557

20. Porbadnigk, A., Wester, M., Calliess, J., Schultz, T.: EEG-based speech recognition – impact of temporal effects, Biosignals 2009, Porto, Portugal (2009)

21. DaSalla, C.S., Kambara, H., Koike, Y., Sato, M.: Spatial filtering and single-trial classification of EEG during vowel speech imagery (2009). https://doi.org/10.1145/1592700.1592731

22. Denby, B., Schultz, T., Honda, K., Hueber, T., Gilbert, J.M., Brumberg, J.S.: Silent Speech Interfaces (2010). https://doi.org/10.1016/j.specom.2009.08.002

23. Williams, E.: JASON Defense Advisory Panel Reports: Human Performance, JSR-07-–625 (2008)

24. Chamola, V., Vineet, A., Nayyar, A., Hossain, E.: Brain-computer interface-based humanoid control: a review (2020). https://doi.org/10.3390/s20133620

25. Collinger, J.L., Kryger, M.A., Barbara, R., Betler, T., Bowsher, K., Brown, E.H.P., Clanton, S.T., Degenhart, A.D., Foldes, S.T.,Gaunt, R.A.,Gyulai, F.E., Harchick, E.A., Harrington, D.,Helder, J.B.,Hemmes, T., Johannes, M.S., Katyal, K.D., Ling, G.S.F.,Mcmorland, A.J.C., Palko, K., Para, M.P., Scheuermann, J., Schwartz, A.B., Skidmore, E.R., Solzbacher, F., Srikameswaran, A.V., Swanson, D.P., Swetz, S., Tyler-Kabara, E.C.,Velliste, M., Wang, W., Weber, D.J., Wodlinger, B., Boninger, M.L.: Collaborative approach in the development of high-performance brain-computer interface for a neuroprosthetic arm: translation from animal models to human control (2013). https://doi.org/10.1111/cts.12086

26. Collinger, J.L., Wodlinger, B., Downey, J.E., Wang, W., Tyler-Kabara, E.C., Weber, D.J.: High performance neuroprosthetic control by an individual with tetraplegia (2013). https://doi.org/10.1016/S0140-6736(12)61816-9

27. Zaaimi, B., Ruiz-Torres, R., Solla, S.A., Miller, L.E.: Multi-electrode stimulation in somatosensory cortex increases probability of detection (2013). https://doi.org/10.1088/1741-2560/10/5/056013

28. Chen, K.H., Dammann, J.F., Boback, J.L., Otto, K.J., Gaunt, R.A., et al.: The effect of chronic intracortical microstimulation on the electrode-tissue interface (2014). https://doi.org/10.1088/1741-2560/11/2/026004

29. Berg, J.A., Dammann, J.F., Tenore, F.V., Tabot, G.A., Boback, J.L., Manfredi, L.R.: Behavioral demonstration of a somatosensory neuroprosthesis (2013). https://doi.org/10.1109/TNSRE.2013.2244616

30. Miranda, R.A., Casebeer, W.D., Hein, A.M., Judy, J.W., Krotkov, E.P., Laabs, T.L., Manzo, J.E., Pankratz, K.G., Pratt, G.A., Sanchez, J.C., Weber, D.J., Wheeler, T.L., Ling, G.S.F.: DARPA-funded efforts in the development of novel brain-computer interface technologies (2014). https://doi.org/10.1016/j.neumeth.2014.07.019

31. Paszkiel, S., Dobrakowski, P., Lysiak, A.: The impact of different sounds on stress level in the context of EEG, cardiac measures and subjective stress level: a pilot study. Brain Sci. **10**(10), Article Number: 728 (2020). https://doi.org/10.3390/brainsci10100728,

Role of Brain-Computer Technology in Synthetic Telepathy

Krzysztof Hanczak[✉] [iD]

Faculty of Electrical Engineering, Automatic Control and Informatics,
Opole University of Technology, Prószkowska 76, 45-758 Opole, Poland

Abstract. The objective of this work is to explore the potential use of electroencephalography (EEG) as a means for silent communication by way of decoding imagined speech from measured electrical brain waves. Communication using BCI can be used for medical and non-medical application. Worldwide a large number of people suffer from disabilities which impair normal communication. Communication BCI's are excellent tool which helps the affected patients communicate with others. BCI technology is used can be used in many different areas of life, beginning on most important field which is help people, through improving or making life easier and ending with entertainment. Also BCI technology can be used in military in different aspects, it may be used to monitor a soldier's cognitive workload, control a drone swarm, or link with a prosthetic, among other examples.

Keywords: BCI · Synthetic telepathy · EEG

1 Introduction

Humans use computers through interfaces such as keyboard, mouse, touch screen, digital camera or a data glove (Fig. 1). These interfaces have one thing in common they need physical movement of the user [4].

Fig. 1. Conventional human computer interface.

Using Brain-Computer technology humans use only brain. The human brain is a complex organ. The brain is divided into different areas with each area performing a specialized function. The different lobes and structures of the brain are frontal lobe, parietal lobe, occipital lobe, temporal lobe, cerebellum limbic system and brain stem.

S. Paszkiel (Ed.): ICBCI 2021, AISC 1362, pp. 205–211, 2021.
https://doi.org/10.1007/978-3-030-72254-8_21

The brain is made up of cells called neurons. These work by sending electrical signals among themselves which is responsible for generation of brain waves. Brain waves or brain rhythms are divided into bandwidths to describe their functions. The different brain rhythms are as follows: 1. Delta wave (0.5–3 Hz) – Delta brainwaves have the least frequency but highest amplitude. They are generated in dreamless sleep and deepest meditation. Delta waves occur when external awarness is suspended. Regeneration and healing takes place in this state and this is why it is often said that deep restorative sleep is so essential to the healing process; 2. Theta wave (3–8 Hz) – Theta brainwaves also occur in the sleep but are also dominant in the deep meditation. It acts as gateway to learning and memory. It is that state which is normally experienced fleetingly as one wake up or drifts off to sleep. Theta is a state of intuition, vivid imagery and information beyond one's normal conscious awarness; 3. Alpha wave (8–13 Hz) – Alpha waves occur durning quietly flowing thoughts, but not quite meditation. Alpha is the brain's resting state. Alpha waves aid overall, calmness, alertness, mental coordination, mind/body integration and learning; 4. Beta wave (13–30 Hz) – Beta brainwaves are present in one's normal waking state of consciousness when attention is focussed towards cognitive tasks and the outside world. Beta is a 'fast' activity, present when a person is attentive, alert, engaged in problem solving, judgment, decision making or engaged in focused mental activity. Beta brainwaves can further divided into three bands namely Low Beta (Beta1, 12–15 Hz), Beta (Beta2, 15–22 Hz), High Beta (Beta3, 22–30 Hz) 5. Gamma wave (30–80 Hz) – Gamma brainwaves are the highest frequency brain waves and associated with simultaneous processing of information from different brain areas. It passes information rapidly. Earlier gamma waves were more often dismissed as 'spare brain noise' untill reaserchers discovered it was highly active when it states of universal love, altruism and the 'higher virtues'. Gamma rhythms modulate perception and consciousness and are found to disappear under anaesthesia. Gamma is observed to be above the neuronal firing frequency, so how it is generated remains a mystery and continues to arouse a great interest in the minds of scientists. A brain-computer technology (BCI) is a device in which person uses his brain to control the machine to be used. BCI is a technology that provides a direct communication between brain and computer for conveying messages to the external world from one's thoughts, without using any of the appendages. Each time we do a task or think about performing one, our brain generates distinct signals. A BCI picks signals from the brain of an user in the form of Electroencephalography (EEG). EEG is a technique to measure the electronic fields produced by brain activity. EEG can be recorded in different forms depending on its use. Scalp EEG recording is the most common and preffered way to acquire EEG data because it involves placing electrodes on scalp and hence it is non-invasieve and easy to acquire [1]. From EEG measurements, it may be possible to extract information and decipher this internal thought, thus resulting in some form of synthetic telepathy. Classifying of EEG signals can be provided by experiments where subject performed by identify imagined speech syllables or covert actions such as imagined movement, mental tasks, imagined sylabic and vowel speech.

Several BCI techniques evolved over the past decade restoring communication to persons with severe paralysis. These assistive devices range from simple binary (yes/no) communication device, the speller device, a virtual keyboard to imagined speech com-munication to name a few [4]. Birbaumer et al. [5] and Perelmouter et al. [6] have developed a speller device for a "locked-in" person to compose letters. Binary tree structure decision of the BCI is used, dividing the alphabet into succesive halves until the desired letter is selected. A similar kind of speller is portrayed by Wolpaw et al. [7] where the alphabets iteratively divide into fourths instead of halves. Donchain et al. [8] has developed a method based on the P300 component of event-related potentials. The rows and columns of a two dimensional alphabet grid are illuminated in sequence, and allow the user to select the desired letter. A 2-D cursor navigation to select letters from a WiViK virtual keyboard for "locked -in" subjects suggested by Kennedy et al. [9]. BCI is used in medical and non-medical field. The neuroimaging techniques used in BCI can be broadly classified into invasive and non-invasive methods. Invasive BCI's involve implanting electrodes inside the brain and the non-invasive ones include haptic controllers and EEG scanners.

Hemispheric assymetries can be detected in the EEG data during imagined left or right hand movements [10]. Gevins et al. [11] have also investigated such hemispheric asymmetry in EEG data, but for cognitive (non-motor) verbal, logical and spatial tasks using pattern recognition techniques. The authors concluded from their data analysis that there seems to be no evidence of lateralzation during cognitive tasks, as large, bilateral areas of the cerebral cortex appear to be involved in the performance of higher cortical functions. There has been little research to-date, however on classifying EEG signals associated with imagined speech. DaSalla et al. [12] performed experiments where EEG data was recorded while subjects imagined vocalizing and mouthing the vowels /a/ and /u/. These vowels were chosen specifically for their differing mouth formations. Features were extracted using the Common Spatial Patterns method. The resulting spatial patterns from comparing /a/ and /u/ showed enhanced activity located symmetrically over the motor cortex region for both vowels, which was actually to be expected since speech musculature is innervated from both the left and right motor cortices. This suggests that the subsequent classification of these EEG signals may have picked up mostly on the imagined speech muscle movements rather than the imagined speech itself. D'Zmura et al. [2] have also performed speech imagery experiments for EEG signal classification, but with subjects imagining syllabic speech without the associated muscle movements. Either one of two syllables, /ba/ or /ku/, was covertly spoken at a specific rhythm, provided by audio cues. D'Zmura et al. classified this EEG signals by computing the Hilbert envelope of each electrode waveform and averaging signal envelopes across each electrode separately to form templates for each class. Matched filters were then employed to classify the imagined syllables. These works present different approaches to classifying different types of mental activites. EEG data captured during motor imagery exhibit hemispheric specialization, and classification methods are typically built with this in mind. However with covertly performed cognitive tasks, it is not clear if such hemispheric asymmetry is present in EEG data, especially with imagined speech. Thus there is no clear answer on how to approach EEG signal classification for general cognitive tasks [13].

2 Previous Research

Experiment conducted by University in California [2] in which volunteer subjects were fitted with EEG equipment and instructed to simply imagine speaking two different syllables, /ba/ and /ku/. These syllables are imagined once every 1.5 s over 6 s period, resulting in 3 imagined syllables per trial. In each trial subject is cued by spoken syllable (/ba/ or /ku/), followed by series of clicks at desired rhythm for imagined speech. Approximately 1.5 s after last click, subject is supposed to begin to imagine speaking the spoken syllable at the given rhythm.

As described in [2], EEG data were recorded using 128 Channel Sensor Net by Electrical Geodesics and sampled at 1024 Hz. A single experimental session was comprised of 20 trials per syllable and data were recorded over 6 separate session for a total of 120 trials per syllable per subject. During recording subjects were seated in a dimly lit room and instructed to keep their eyes open and to fixate on certain point while avoiding any eye blinks and muscle movement. Classification of these EEG signals representing imagined speech is significantly challenging due to number of factors including inherently low signal-to-noise ratio (SNR) and presence of artifacts such as signal spikes caused by eye blinks and electromyographic activity, which can dominate and obscure actual cortical signal. Additionally in some cases these artifacts can be fairly predictive. This results in deceptively high recognition rates since a classifier will succeed by identifying these artifacts as opposed to portions of signal that reflect true imagined speech. EEG signal classification is further complicated by inter-subject variation as the anatomical details of people's heads will differ. As such classification experiments are typically performed for each subject separately to minimize error caused by forcing classification method to generalize over inherently different sets of data. There is also intra-subject variation since from session to session electrodes are not placed in exact same locations. This makes it difficult for classification method to generalize using individual subject data as well. However significant complication, particularly for imagined speech EEG data, is the potential for trials to be devoid of any useful information. If and how imagined speech and its characteristics manifest in EEG data and what kind of impact each subject's particular variation of covert speech may have on these signals are still open research questions [3].

3   Data Pre-processing

Common model for EEG signals is to consider them as a linear combination of a number of current sources and noise. The premise behind this assumption is that at the frequencies of interest (<100 Hz), tissue is primarily a resistive medium and therefore the contribution of single current source to the surface potentials is linear [14]. Parra et al. [14] maintain that the linear model shown in (1) is an accurate representation of the electro-physics of EEG.

$$x(t) = As(t) + n(t) \tag{1}$$

where $x(t) = [x_1(t)x_2(t) \ldots x_N(t)]^T$ is a vector of observed noisy sensor signals from N sensors, A is the forward model relating the source activity to the sensor activity, $s(t) = [s_1(t)s_2(t) \ldots s_M(t)]^T$ is a vector of M unknown sources with $M \leq N$ and n(t) represents background activity that would be considered noise.

Finding the forward model A to recover s(t) in (1) is a difficult inverse problem, as there are many possible configurations of current sources inside the brain that may "explain" the electric potentials seen on the scalp. However one may attempt to recover some form of brain "source" by estimating the mixture matrix A using Independent Component Analysis (ICA) to find sources that are statistically independent. Makeig et al. [15] suggest that these sources or ICA components, might represent activity in a spatially broadly distributed set of cortical domains, if their activities were somehow synchronized. It is assumed that different biological processes operate independently, so therefore ICA could be used to separate components that are generated from different biological sources. With this, one may attempt to remove artifacts from EEG data by identifying and discarding sources that appear to have been generated by extraneous movement related activity.

Electromyographic (EMG) artifacts are first considered for removal using the same preprocessing steps performed by D'Zmura et al. in [2]. EEG signals from 18 of the 128 electrodes that are closest to the neck, eyes and temple are discarded since they are most prone to EMG artifacts. Also certain number of trials (depending on subjects) that contain large number of electrodes exceeding the thresholds of $\pm 30\ \mu V$ are discarded since they are most likely heavily contaminated by these artifacts. Since EMG artifacts are typically present in frequencies greater than 25 Hz, remaining EEG signals are filtered to frequency range 4–25 Hz which additionally removes 60 Hz line noise from these signals. Data is then detrended to remove baseline drift and downsampled to a more manageable sampling rate of 256 Hz.

After initial preprocessing steps, ICA is performed to remove additional artifact components. Independent components are found using following "demixing" model:

$$y(t) = Wx(t) \tag{2}$$

where: $y(t) = [y_1 y_2 \ldots y_M(t)]^T$ is a vector of M idependent components, $x(t) = [x_1 x_2 \ldots x_N(t)]^T$ is a vector of observed noisy sensor signals from N sensors and W is the separation matrix [16]. For source separation, Equivalent Robust ICA (ERICA) algorithm is employed due to its roubstness to addictive noise. In ERICA [17], W is iteratively calculated until convergence using following update equation:

$$W_{l+1} = (I + \mu(C_{y,y}^{1,3} S_y^3 - I))^{-1} W_t \tag{3}$$

where: W_t is the estimate of separation matrix at step t, μ is the learrning-rate parameter, $C_{y,y}^{1,3} = M_{y,y}^{1,3} - 3M_{y,y}^{0,2} M_{y,y}^{1,1}$ is a 4th order cumulant matrix with $M_{y,y}^{k,1} = E[y^k(y^l)^T]$ and S_y^3 is diagonal matrix of cumulant signs (i.e. $[S_y^3]_{ii} = sign([C_{y,y}^{1,3}]_{ii})$. The mixture matrix A is subseqently calculated as Moore-Penrose pseudo-inverse [18] of the separation matrix W. Typically artifact components are manually identified by visual inspection. However this approach may be impractical for datasets. Vorobyov and Cichocki [16] developed method to automate artifact removal by using Hurst exponent to identify components that contain artifacts. Hurst exponent, which measures predictability of a time series, is a metric that ranges from 0 to 1, where value of 0 indicates signal in unpredictable and value with 1 indicates signal is completely predictable. Hurst exponent of sources containing heartbeat and eye blink artifacts appear to fall within range 0.58–0.69. Therefore by discarding independent components that have Hurst exponent valeus within range, one may remove effects of these artifacts on EEG data. Vorobyov and Cichocki also found that the Hurst exponent can be used to detect "useful" EEG components. They note that many researchers have found that for numerous natural, economic and biological phenomena, Hurst exponent typically falls within range of 0.70–0.76, which authors also deem to be an appriopriate range for characterizing "interesting" EEG sources. So components that have Hurst exponent values outside range 0.70–0.76 are discarded as well. Remaining source components are mixed back to reconstruct sensor data using mixture matrix A. If trial does not contain any independent components, trial is discarded. Following data sensor data reconstruction, EEG signals re filtered with a subspace-based Wiener filter (described in [9]) for further noise reduction. Due to typically sparse structure of EEG signals, low-rank filter can be built in place of full-rank filter. If filter were designed using whole signal, filter would likely describe both signal and noise structure. By applying Principal Component Analysis (PCA), signal and noise subspace can be partially separated. For each EEG signal, only first principal component of subset of signal was used to construct its Wiener filter [3].

4 Conclusion

BCI technology can be used in almost any area [19–21]. BCI can be applied in medicine, military, gaming and many other aspects of human life. Primary goal of BCI is to restore communication in severely paralyzed population, the speech communication has expanded its application in silent speech communication, synthetic telepathy and cognitive biometrics. Method of BCI used are invasive ECOG and non-invasive EEG, MEG, MRI, NIRS. EEG is a non-invasive, non-injurious method for probing cortical activity which has a high temporal resolution, moderate spatial resolution, relatively low cost and which is increasingly portable. Classifying EEG data from imagined speech is a difficult task, as it is currently unknown if and how imagined speech manifests in EEG data. The data is pre-processed through several stages in an attempt to extract any information that may be present in the signals since they are typically buried in noise and dominated by the artifacts. Even after pre-processing, there seem to be trials that do not contain discriminative information. However, discarding trials based on measurements of signal persistence appears to be able to filter out trials that may not effectively contribute information for imagined syllable discrimination. Final classification results demonstrate that there is indeed some discriminative information present in imagined speech EEG signals, and proposed method is able to classify imagined syllables for several subjects [3].

References

1. Iqbal, S., Muhammed Shanir, P.P., Khan Uzzaman, Y., Farooq, O.: EEG analysis of imagined speech (2016). https://doi.org/10.4018/978-1-5225-9273-0.ch033
2. D'Zmura, M., Deng, S., Lappas, T., Thorpe, S., Srinivasan, R.: Toward EEG sensing of imagined speech (2009). https://doi.org/10.1007/978-3-642-02574-7_5

3. Brigham, K., Vijaya Kumar, B.V.K.: Imagined speech classification with EEG signals for silent communication: a preliminary investigation into synthetic telepathy (2010). https://doi.org/10.1109/ICBBE.2010.5515807
4. Mohanchandra, K., Saha, S., Lingaraju, G.M.: EEG based brain computer interface for speech communication: principles and applications (2015). https://doi.org/10.1007/978-3-319-10978-7_10
5. Birbaumer, N., Hinterberger, T., Kubler, A., Neumann, N.: The thought-translation device (TTD): neurobehavioral mechanisms and clinical outcome (2003). https://doi.org/10.1109/TNSRE.2003.814439
6. Perelmouter, J., Birbaumer, N.: A binary spelling interface with random errors (2000). https://doi.org/10.1109/86.847824
7. Wolpaw, J.R., McFarland, D.J., Vaughan, T.M.: Brain-computer interface research at the Wadsworth Center (2000). https://doi.org/10.1109/86.847823
8. Donchin, E., Spencer, K.M., Wijesinghe, R.: The mental prosthesis: assessing the speed of a P300-based brain-computer interface (2000). https://doi.org/10.1109/86.847808
9. Kennedy, P.R., Bakay, R.A., Moore, M.M., Adams, K., Goldwaithe, J.: Direct control of a computer from the human central nervous system (2000). https://doi.org/10.1109/86.847815
10. Ramoser, H., Muller-Gerking, J., Pfutscheller, G.: Optimal spatial filtering of single trial EEG during imagined hand movement (2000). https://doi.org/10.1109/86.895946
11. Gevins, A.S., Zeitlin, G.M., Doyle, J.C., Yingling, C.D., Shaffer, R.E., Callaway, E., Yeager, C.L.: Electroencephalogram correlates of higher cortical functions (1979). https://doi.org/10.1126/science.760212
12. DaSalla, C.S., Kambara, H., Sato, M., Koike, Y.: Single-trial classification of vowel speech imagery using common spatial patterns (2009). https://doi.org/10.1016/j.neunet.2009.05.008
13. Paszkiel, S., Dobrakowski, P., Lysiak, A.: The impact of different sounds on stress level in the context of EEG, cardiac measures and subjective stress level: a pilot study. Brain Sci. **10**(10), 728 (2020). https://doi.org/10.3390/brainsci10100728
14. Makeig, S., Debener, S., Onton, J., Ddorme, A.: Mining event-related brain dynamics (2004). https://doi.org/10.1016/j.tics.2004.03.008
15. Vorobyov, S., Cichocki, A.: Blind noise reduction for multisensory signals using ICA and subspace filtering with application to EEG analysis (2002). https://doi.org/10.1007/s00422-001-0298-6
16. Cruces, S., Castedo, L., Cichocki, A.: Novel blind source separation algorithms using cumulants (2000). https://doi.org/10.1109/ICASSP.2000.861206
17. Golub, G.H., Van Loan, C.F.: Matrix Computations, 3rd edn. John Hopkins University Press, Baltimore (1996)
18. Kay, S.M.: Modern Spectral Estimation: Theory and Application, pp. 228–230. Prentice Hall, Englewood Cliffs (1988)
19. Paszkiel, S.: Augmented reality of technological environment in correlation with brain computer interfaces for control processes. In: Szewczyk, R., Zielinski, C., Kaliczynska, M. (eds.) Recent Advances in Automation, Robotics and Measuring Techniques. Advances in Intelligent Systems and Computing, vol. 267, 197–203 (2014). https://doi.org/10.1007/978-3-319-05353-0_20
20. Paszkiel, S.: Using BCI in IoT implementation, analysis and classification of EEG signals for brain-computer interfaces. In: Studies in Computational Intelligence, vol. 852, pp. 101–110 (2020). https://doi.org/10.1007/978-3-030-30581-9_12
21. Paszkiel, S.: The use of facial expressions identified from the level of the EEG signal for controlling a mobile vehicle based on a state machine. In: Szewczyk, R., Zielinski, C., Kaliczynska, M. (eds.) Automation 2020: Towards Industry of the Future. Advances in Intelligent Systems and Computing, vol. 1140, pp. 227–238 (2020). https://doi.org/10.1007/978-3-030-40971-5_21

Brain-Computer Interface in Intelligent Environment

Paweł Piróg[✉]

Faculty of Electrical Engineering, Automatic Control and Informatics,
Opole University of Technology, Prószkowska 76, 45-758 Opole, Poland
p.pirog@student.po.edu.pl

Abstract. Brain-computer interface (BCI) is a new form of communication between human and computer without using any physical means. Applicating BCI system into the intelligent environment can become very appropriable especially when we are stepping into times where efficiency and comfort is becoming top priority. This research paper focuses on application of different brain-computer mechanisms in smart homes and environments.

Keywords: BCI · Human and computer · Intelligent environment

1 Introduction

Brain computer interfaces use different signals from the user in order to control external devices. This can happen basing on muscle activity, eye movement, respiration or heart rate. The improvement of technology also allows researchers to use electrical brain activity as potential signal.

BCI is an approach that is not dependent on the brain normal channels or nerves and muscles. That is why it's most important research is to provide people with confined motor skills due to damaged central nervous system or impaired motor system but whose brain is intact. With increasing aged population, systems that will provide users with environment that is going to be more intelligent, easier too use and more secure is now even more needed than before for the elderly.

Thanks to the technological advancements BCI approach is possible without any invasive methods. Noninvasive methods are safe and easy to maintain if treated properly. There are several noninvasive approaches like electroencephalography (EEG), magnetoencephalography (MEG) and functional magnetic resonance imaging (fMRI) of which most effective one is EEG because of its probability. Several studies attempted developing BCI system for improvement of quality of life for people with different disabilities such as ASL, paralysis, brain stroke etc. [1–4].

Through the years various EGG-based BCI systems have been introduced. Those include slow cortical potential (SCP) [5], motor imagery (MI) [6], visual evoked potential (SSVEP) [8], the P300 potential [7] and hybrid BCI [9].

With the advancement of the technology ideas of smart environments is becoming reality. Research teams across the globe have achieved success in implementing the

S. Paszkiel (Ed.): ICBCI 2021, AISC 1362, pp. 212–218, 2021.
https://doi.org/10.1007/978-3-030-72254-8_22

BCIs into their smart environments. Although most of these research groups focused on using BCI in virtual environments. There are few studies that that implemented this technology into real environment. For example, Carabalona et al. used P300 potential while constructing a smart home system [10], same approach was done by Kosmya et al. With 77% task accuracy [11].

This presents first obstacle of implementing BCI technology into the intelligent environments and that is the compatibility and cost efficiency. Because of that most users cant benefit from the bulky and expensive EGG equipment. Another thing to mention is that it was only used to recognize the cognitive state. Due to this fact only a few possible applications were used, for example controlling light by reading alpha and theta rhythms.

Another obstacle of implementing the BCI into an intelligent environment is that the certain percentage of people is currently unable to operate specific BCI systems due to various reasons. With almost every BCI system around 20%–25% of users is unable to control one in satisfactory way. That's why the hybrid BCI approach might be appropriate to overcome this problem by using the output of somatosensory rhythm BCI along with the P300 or steady state visually evoked potential (SSVEP) based BCIs that will give subjects ability to choose different approach for optimal control [12].

Last obstacle in creating BCI based smart environment is a way of controlling different devices and powering them without losing signal while having reasonable cost efficiency. This is how the idea of supplying the devices by the Power over Ethernet was created. Power over Ethernet (POE) is a technology based on Cat.5 infrastructure, which allows for simultaneous transmission of data and current. This technology cuts costs of setting up whole independent power infrastructure. Thanks to this approach the maintenance of the network is only limited to network cable.

Basing on previous studies it is known that we can achieve lightning control using PoE system to control LED [13]. PoE technology can also be used in several other applications such as VoIP, home automation, access points and many others.

Goal of this research chapter is to research ways of implementing BCI technology into a intelligent environment. This paper is going to be mostly focused on creating a smart home system for paralyzed and elderly people.

2 Approaches

2.1 Hybrid Approach

The BCI system that can be used in creating the smart environment is combination of P300 and SSVEP BCIs which were implemented and proven working with the Emotiv EPOC NeuroHeadset [14, 15]. This headset is a good device as a base of research because it is a very compact and cost-efficient.

2.2 P300 System

P300 system is used when selection of many action is needed. This is possible thanks to a rectangular matrix layout of characters or icons displayed on a screen. Characters on the matrix are flashing in random order. When the figures are flashing user must concentrate on a character he want's to select. When the character starts flashing and the user is focused on it EEG amplitude is going to be reached after 300 ms after a flash. Signal response of P300 is more conspicuous for a single characters speller than for the whole row/column speller that is why it is easier to detect.

When training the BCI system EGG data will be acquired from user when he will focus on specific characters for previously preset amount of times the character will flash. As the user will progress through all the characters the evaluated EGG data is send to an linear discriminant analyser (LDA) for separation between target characters and non target characters. This will return the user weight vector for a real-time evaluation (All information about P300 speller setup can be found in [7]).

When it comes to the difference between using characters and icons the result is very similar. This subject was approached by Bayliss [17] and Holzner [18] who have already proven that usage of icons is as effective. It was tested on 12 different subjects of which each had select 7 icons of each control mask that was tied to position in smart home for training. Each icon was tested 15 times. During these test the flash time (when icon is highlighted) and dark time (time between highlights) is also an important parameter. Flashes must be calibrated in a way that the highlight time won't be too long so that the different responses are not overlapping but long enough for subject to recognise the highlight. The system was tested on different amount of flashes for each icon.

2.3 SSVEP System

SSVEP BCI system is using the flickering lights or symbols on a screen to stimulate the user with frequency between 5 and 25 Hz [19]. This system can be responsible for on and off switches of devices. While the light is flickering at a for example 14Hzwhile user is looking at it the EEG signal is created near occipital area and can be made visible on a power spectrum. The signal power is falling down as the frequency increases. For a fixed frequency threshold criterion is used for determination if user looks at the lights/symbols, if not the LDA is trained to find the optimal threshold. With increasing amount of light sources the multi-dimensional control can be reached.

There is small amount of fully integrated non-virtual smart homes but thankfully they exist [21]. In Helsinki was conducted experiment that was part of KogniHome Project. Goal of that project was to create a home that will support its residents in daily life. This process took 3 years and as part of this project multiple demonstrative uses of BCI were created and connected using Robotic Service Bus.

The apartment was equipped with a smart door, networked wardrobe and smart mirror at the entrance. Near the smart door delivery flap a mobile robot was stationed. Kitchen featured a lifted worktop, cooking assistant [22] and medicine dispenser. In living room a dialogue assistant was installed. Whole floor was integrated with guide lightning in the floor [23]. For a visual stimulation 3 devices were installed into a apartment. At the entrance door a LED stripes were installed to utilize 9 Hz flicker. These lights were used to turn on and off the lights in the apartment. Another device was a smart TV in a living room. It was used do display eight flickering boxes. On a TV at a 60 Hz refresh rate 8 different stimulation boxes were displayed with lights flickering at frequencies: 11.25 Hz, 11.75 Hz, 12.25 Hz, 12.75 Hz, 13.25 Hz, 13.75 Hz, 14.25 Hz and 14.75 Hz.

Last device was installed in the kitchen. It was a custom made monitor that displayed six fields contained separate LED lights. Thanks to this approach the monitor function as BCI interface and could be used for normal operations with overlaid flickering. Because the LED lights we separate from the monitor no approximation method was needed for setting up the frequencies. Frequencies of the fields were 11.5 Hz, 12.0 Hz, 12.5 Hz, 13.0 Hz, 13.5 Hz and 14.0 Hz.

All commands on the devices were streamed using Lab Streaming Layer [24] which is system for collection of measurement time series in research experiments. This program can handle networking and time-synchronisation at near real-time according to creator. Recorded action were mapped and sent through a bridge and then processed to the smart home actuators using RSB. For creation of the BCI in smart environment it is very important to transfer of data at the fastest rate because if we implement some kind of emergency command in our system we are going to need fastest possible reaction time.

2.4   Equipment Setup

For this research it is assumed that person using the BCI will in some degree movement impaired and has to move in a wheelchair. Wheelchair will be equipped with a portable power station capable to supply the BCI with enough power for a longer periods of time. Each room in the smart environment should be equipped with identical monitors that are connected into a network responsible for controlling the environment.

According to different studies a wireless transmission of EGG data can be implemented. Because Emotiv Headset is created to be a wireless device it is only needed to implement good signal amplifier.

2.5   Smart Home Preparation

For controlling the environment the virtual base structure should be developed. For example structure will consists of 3 rooms each equipped with PoE connected LED lights and different appliances like TV, automatic blinds, lamps, telephone, thermostat, door lock. Devices are will have different programmed functions and depending on the device P300 system will be implemented or not. If the P300 system is not implemented it means that the device can be only switched on and off using SSVEP.

All of the appliances are presented on the monitors. If user will select the appliance with SSVEP system implemented then he will be able to choose between its functions using P300 system (Table 1). The limitation of the amount of devices is dependent on the screens refresh rate and their size. Screens with higher refresh rate can be used to display more devices or have better accuracy. Although it might look like that higher refresh rate will give higher accuracy for the same frequencies used the difference it is not that significant to invest into higher frequency screens [16].

Table 1. Devices and their functions

Device	P300 system	Functions	SSVEP frequency
LED lights	No	On/Off	4
Automatic blinds	Yes	Move Blinds Up/Down, Lock/Unlock, Set transparency	8
Lamps	Yes	Select lamp to be turned on/off	12
Telephone	Yes	Call number, Call contact, Call emergency	16
Thermostat	Yes	Set temperature, Automatic On/Off, Turn On/Off	20
TV	Yes	Select channel, Volume Up/Down, Next/Previous channel, Turn On/Off	24
Door lock	No	On/Off	28

2.6 Neural Impulse Actuator (NIA)

NIA is a technology that was introduced by OCZ Technology Group Inc. This controller is a headband that is equipped with electrodes which capture the electrical potentials. Those potentials are: electromyogram, electroencephalogram and electrooculogram. NIA is mentioned because this is also a very compact device that is also easy to set up because it registered as USB device that can act as a mouse, keyboard or any other input device. After calibration NIA can be used as a joystick or for switching events triggered by electric potentials or muscle movements. It is possible for NIA to track: eye movement, heavy muscle movement of the forehead, movement of the jaw, heavy thinking and relaxing or closing eyes. For the most accurate results it is best to use only eye movement and forehead muscle movement because those type of signals are easy to use in the interface. One of the problems of the NIA is that is capable of performing signals at frequency 10 per minute which is very low.

This device was used by Rusher [20] while trying to implement the BCI into their Smart Lab. The functionalities that the interface should be able to do were: possibility to add new devices without fixing the environment, operation on the interface while doing other tasks. First problem that occurred in implementation of the BCI into their network was that you can't use the BCI in a high traffic environment because there is too many triggers. Another problem was that creation of compact graphical interface that will be able to contain every possible action of any added device is also very complex. After implementing the NIA the conclusion was that using the BCI is not good when trying to do other actions simultaneously. This happens because while the device is on, other actions are recorded at the same time. Rusher team concluded that using the BCIs in not fixed environment is currently really tough although not impossible. Even adding simple devices made a huge impact on the menus of their BCI. Rusher attempt of implementation of the interface is a good example of possible future of this kind of devices and how they should be created.

3   Conclusion

When creating an intelligent environment it is obligatory for creators to decide of what kind of approach will be appropriate. For small or less complex environments where the functionalities of devices are limited to the on and off switch SSVEP solution should be implemented. That method is really good because it operates on simple functions that will be easier to implement. For bigger infrastructures hybrid of P300 and SSVEP should be applied. The P300 will also work better when more items are on the control mask because the P300 response becomes more pronounced. This method might lower the transfer rate but enables control on almost any device. This will allow for the interface to have layered menus and that not all appliances will be shown on one specific panel. Creation of a VR system that will serve as a base before creation of an actual intelligent environment is very good step. Such system will serve in the future because it is much easier to reconfigure the virtual system than to make changes in an actual home. One of the variables that a creator of this kind of intelligent environment must account for is a type of device that is going to be used to get an EGG data. Use less advanced devices with limited set of function will create many problems. For example NIA is really compact device, unfortunately because it has a very limited amount of sensors it is unable to make fast and accurate readings. Very advanced intelligent environments controlled by BCI are unfortunately limited to the current technology although less advanced prototypes can already be used by people with disabilities to improve their quality of life.

References

1. Birbauer, N., Murguialday, A.R., Cohen, L.: Brain-computer-interface (BCI) in paralysis. Curr. Opin. Neurol. (2008).https://doi.org/10.1097/WCO.0b013e328315ee2d
2. Kübler, A., Kotchoubey, B., Kaiser, J., Wolpaw, J.R., Birbaumer, N.: Brain-computer communication: unlocking the locked in. Psychol. Bull. (2001). https://doi.org/10.1037/0033-2909.127.3.358
3. Riccio, A., Simione, L., Schettini, F., Pizzimenti, A., Inghilleri, M., Belardinelli, M.O., et al.: Attention and P300-based BCI performance in people with amyotrophic lateral sclerosis. Front. Hum. Neurosci. (2013).https://doi.org/10.3389/fnhum.2013.00732
4. Mao, X., Li, M., Li, W., Niu, L., Xian, B., Zeng, M., et al.: Progress in EEG-based brain robot interaction systems. Comput. Intell. Neurosci. **2017**, 1–25 (2017). https://doi.org/10.1155/2017/1742862
5. Birbaumer, N., Ghanayim, N., Hinterberger, T., Iversen, I., Kotchoubey, B., Kubler, A., Perelmouter, J., Taub, E., Flor, H.: A spelling device for the paralysed. Nature **398**, 297–298 (1999). https://doi.org/10.1038/18581
6. Sellers, E.W., Krusienski, D.J., McFarland, D.J., Vaughan, T.M., Wolpaw, J.R.: A P300 event-related potential brain-computer interface (BCI): the effects of matrix size and inter stimulus interval on performance (2006). https://doi.org/10.1016/j.biopsycho.2006.04.007

7. Guger, C., Daban, S., Sellers, E., Holzner, C., Krausz, G., Carabalona, R., Gramatica, F., Edlinger, G.: How many people are able to control a P300-based brain-computer interface (BCI)? (2009). https://doi.org/10.1016/j.neulet.2009.06.045

8. Allison, B., Luth, T., Valbuena, D., Teymourian, A., Volosyak, I., Graser, A.: BCI demographics: how many (and what kinds of) people can use an SSVEP BCI? IEEE Trans. Neural Syst. Rehabil. Eng. (2009). https://doi.org/10.1109/TNSRE.2009.2039495

9. Friman, O., Volosyak, I., Graser, A.: Multiple channel detection of steady-state visual evoked potentials for brain-computer interfaces (2006). https://doi.org/10.1109/TBME.2006.889160

10. Carabalona, R., Grossi, F., Tessadri, A., Castiglioni, P., Caracciolo, A., de Munari, I.: Light on! real world evaluation of a P300-based brain-computer interface (BCI) for environment control in a smart home (2012). https://doi.org/10.1080/00140139.2012.661083

11. Nataliya, K., Franck, T.B., Nicolas, B., Bertrand, R.: Feasibility of BCI control in a realistic smart home environment. Front. Hum. Neurosci. (2016). https://doi.org/10.3389/fnhum.2016.00416

12. Pfurtscheller, G., Allison, B.Z., Brunner, C., Bauernfeind, G., Solis-Escalante, T., Scherer, R., Zander, T.O., Mueller-Putz, G., Neuper, C., Birbaumer, N.: The hybrid BCI. Front. Neurosci. (2010). https://doi.org/10.3389/fnpro.2010.00003

13. Gao, Q., Li, X., Sui, Y., Gu, H., Yu, Z., Zhang, Q., et al.: Design and implementation of LED intelligent lighting system based on the technology of PoE. In: CCDC: The 27th Chinese Control and Decision Conference (2015). https://doi.org/10.1109/CCDC.2015.7162366

14. Liu, Y., Jiang, X., Cao, T., Wan, F., Mak, P.U., Mak, P.I., Vai, M.I.: Department of Electrical and Computer Engineering Faculty of Science and Technology University of Macau Implementation of SSVEP Based BCI with Emotiv EPOC (2012). https://doi.org/10.1109/VECIMS.2012.6273184

15. Paszkiel, S., Dobrakowski, P., Lysiak, A.: The impact of different sounds on stress level in the context of EEG, cardiac measures and subjective stress level: a pilot study. Brain Sci. **10**(10), 728 (2020). https://doi.org/10.3390/brainsci10100728

16. Nakanishi, M., Wang, Y., Wang, Y.T., Mitsukura, Y., Jung, T.P.: An approximation approach for rendering visual flickers in SSVEP-based BCI using monitor refresh rate (2013). https://doi.org/10.1109/EMBC.2013.6609966

17. Bayliss, J.: Use of the evoked potential P3 component for control in a virtual apartment (2003). https://doi.org/10.1109/TNSRE.2003.814438

18. Holzner, C., Guger, C., Edlinger, G., Groenegress, C., Slater, M.: Virtual smart home controlled by thoughts (2009). https://doi.org/10.1109/WETICE.2009.41

19. Paszkiel, S.: Using BCI in IoT implementation. In: Analysis and Classification of EEG signals for brain-computer interfaces. Studies in Computational Intelligence, vol. 852, pp. 101–110 (2020). https://doi.org/10.1007/978-3-030-30581-9_12

20. Ruscher, G., Krüger, F., Bader, S., Kirste, T.: Controlling smart environments using a brain computer interface (2011)

21. Adams, M., Benda, M., Saboor, A., Krause, A.F., Rezeika, A., Gembler, F., Stawicki, P., Hesse, M., Essig, K., Ben-Salem, S., Islam, Z., Vogelsang, A., Jungeblut, T., Rückert, U., Volosyak, I.: Towards an SSVEP-BCI controlled smart home (2019). https://doi.org/10.1109/SMC.2019.8914668

22. Neumann, A., Elbrechter, C., Pfeiffer-Leßmann, N., Koiva, R., Carlmeyer, B., Ruther, S., Schade, M., Uckermann, A., Wachsmuth, S., Ritter, H.J.: KogniChef: a cognitive cooking assistant (2017). https://doi.org/10.1007/s13218-017-0488-6

23. Esselmann, C., Gabel, K., Schwenzfeier-Hellkamp, E.: Realization of a test setup for a smart light guiding system with assistance functions for elderly people (2017)

24. Kothe, C.: Labstreaminglayer. https://github.com/sccn/labstreaminglayer

Research on the Influence of Noise on Concentration Using Brain-Computer Interface Technology

Adam Lukosik$^{(\boxtimes)}$ ⬤

Faculty of Electrical Engineering, Automatic Control and Informatics,
Opole University of Technology, Prószkowska 76, 45-758 Opole, Poland
adam.lukosik@student.po.edu.pl

Abstract. Noise has become inexorable stress due to the increase in urbanization, automobile usage, Noise based occupation, and lifestyle modifications such as recreation in day to day life of an individual. The noise has an influence on our perception, such disturbances can affect abilites like reading, calculating, learning. The scope of this study will cover noise impact issues on concentration and general performance in daily activities. This chapter proposes concepts of using Brain-Computer Interfaces (BCI) technology to study the impact of noise on concentrations.

Keywords: Noise and health · Brain-Computer Interfaces · Performance · Short-term memory task

1 Introduction

Noise not only affects the auditory system but can also debilitate other non-auditory systems as evidenced in animal and human models. Noise has negative effect on mental health [1]. Also the human performance is affected by several factors including occupational noise. This influence can even occur at low sound pressure levels, those far away from being dangerous. Listening is a part of learning, so the acoustic environment should not be distracting because this would cause lower performance at doing basic tasks.

The World Health Organisation provided values that should not be exceeded in classrooms, the one that applies to noise is 35 dB. Unfortunately those guidelines are often neglected when it comes to teaching in schools. Many classes are held in noisy places and this can be one of the problems when it comes to effective learning. Another issue when it comes to noise is the reverbation, together with the transmitted content can cause th voice to be hardly recognisable. This factor also provides students with not enough focus on subject.

2 Description of Measurement Device

On the purpose of this concept study the Emotiv EPOC + NeuroHeadset will be used. It is a commercially available device that allows people to control a computer using their facial

S. Paszkiel (Ed.): ICBCI 2021, AISC 1362, pp. 219–225, 2021.
https://doi.org/10.1007/978-3-030-72254-8_23

expressions or their thoughts. The Emotiv EPOC + NeuroHeadset is a high resolution, neuro-signal acquisition and processing wireless Emotiv EPOC + NeuroHeadset used to read brain waves. It uses a set of sensors to tune into electric signals produced by the brain to detect a persons thoughts, feelings and expressions and connects wirelessly to a computer. Set used for the purpose of this study contains: Emotiv EPOC + NeuroHeadset Assembly with battery already in, USB transceiver dongle, Hydration Sensor Pack, Saline solution, Battery charger, installation disk.

More detailed information about performance and features of the device: 14 channel EEG - helps to get more information from brain; saline based electrodes – wet sensors, no need to use gels; Bluetooth interface – Wireless connection to PC and mobile devices; battery life up to 12 h, 9 axis motion sensors – help with detecting head movements; professional grade signals – very high quality of recorded data; mobile Recording – transmission of wireless data at 128 or even 256 Hz, this means work can be done on phone with very good quality of recorded and transmited data; easy configuration – configuration takes up to 5 min and there is no need to use any kind of gels, saline is cleaner in use and doesn't make mess like gels.

Emotiv Control Panel comes with many useful functionalities that help with configuration, understanding the device and tools to manage and display gathered data. The Cognitiv Suite panel contains cube to display representation of the Cognitiv detection output. This cube is also used to assist the user in visualizing the intended action during the training process. The right section displays current state of detection, there is few actions to be done in order to complete training. The neutral action is focused on staying at passive mental state [1] (Fig. 1).

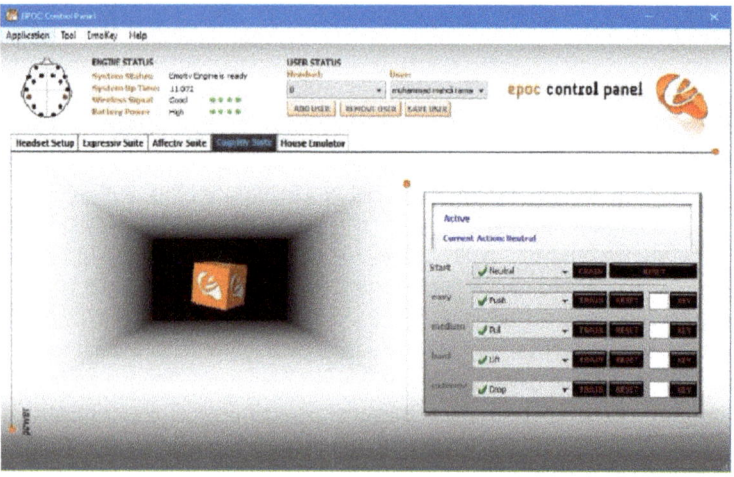

Fig. 1. Interface of The Cognitiv Suite panel.

The Expresive Suite Tab can distinguish facial expressions. The displayed avatar can help with adjusting and establishing mimic movements. Included gestures are horizontal eye movements to the left and right, normal eye blinks, left and right sided winks, clenching the teeth and smiling. Sliders on the right side allow us to adjust sensitivity of expressions so they are equivalent to ours [2] (Fig. 2).

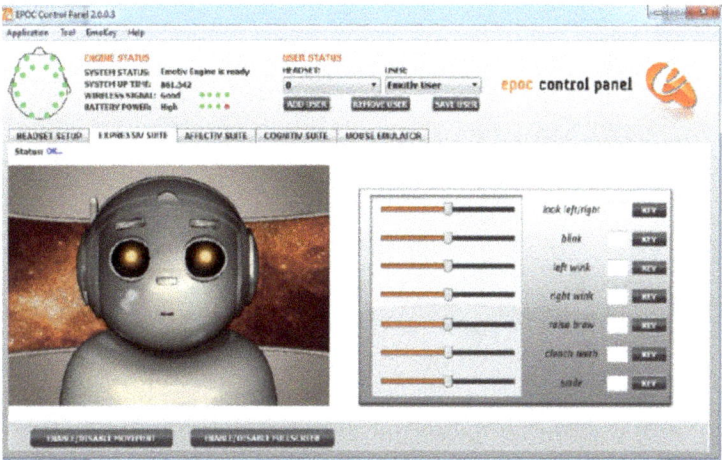

Fig. 2. Interface of the expressive suite tab.

EEG is a recording of the electrical activity of the brain from the scalp. EEG activity is quite small, measured in microvolts (μV). The recorded waveforms reflect the cortical electrical activity. 3D Brain Visualiser module allows us to see the activity ov given brain frequencies. Every frequency threshold can be adjusted separately using the slider on the left side. The 10–20 System is an international method of placing the electrodes on the head [3, 4]. Brain Activity Map module displays a real time map of brain activity. There are 4 wavebands we can monitor at a time: Delta (0.5–4 Hz) – indicating deep sleep, restfulness, and conversely excitement or agitation when delta waves are suppresse; Theta (4–8 Hz) – indicating deep meditative states, daydreaming and automatic tasks; Alpha (8–15 Hz) – indicating relaxed alertness, restful and meditative states; Beta (15–30 Hz) – indicating wakefulness, alertness, mental engagement and conscious processing of information [5–7]. The gain knob in 3D Brain Visualiser allows us to adjust the visual data output on relative strengths between different brain regions. The buffer knob can be modified to show instant response data or the average data based on longer periods of time. Once the settings are adjusted to our preferences, the colors will indicate brain activity strength.The more warm colors the higher brain activity occurs.

3   Measurement Methodology

First of all after selecting test subject the callibration of the device should be made. For this purpose USB Transceiver should be plugged in to computer, it will automatically install needed drivers. The next step is to place Emotiv EPOC + NeuroHeadset on subject's head keeping in mind that correct placement of sensors is crucial in the process. Now there should be ability to check the signal quality on the EmoEngine Status Panel.

It allows us to check all information about the engine and user status. Various Dots on the head graphic indicate the strength of the signal of specific electrodes: Black: No signal; Red: Very poor signal (Not Acceptable). Orange: Poor signal; Yellow: Fair signal; Green: Ideal signal [4] (Fig. 3).

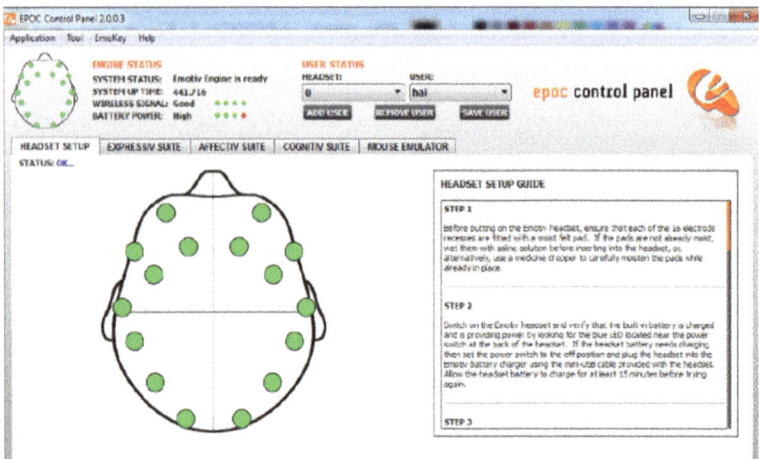

Fig. 3. Example of good connection of electrodes.

We should always aim at having ideal signal where all indicators are green. Situation where some of them are black wont give a reliable result of measurements and will not allow to make certain types of them. Things that may cause the measurements not reliable or impossible are as follows:not enough saline on the pads, weak battery, bad wireless connection, pressence of other power devicesthat may interfere with signal, not good enough adjustment of sensors.

It is advised to double check factors that could cause complications, measurements made without certainty of being reliable should not be taken into account, that could make all this experiment not valuable.

After all adjustements done, subject will participate in a simple test consisting of few mathematic equations to solve. The main goal of this part is to collect data of brain activity in different frequencies.

4   Implementation Concept

All attempts consists of one common part which is solving 10 equations with rising difficulty, different for each attempt and second part consisting in adding noise. The first attempt is to be done in a total silence, this will give us the results of normal brain activity for this task when rested. The second attempt causes the subject to wear headphones with classical music playing. Third is going to be metal music and fourth rock music. Fifth attempt is quite different, the noise is less pleasant and more distractive, loud sound of TV news. The last attempt is going to be most challenging, the subject will be the addressee of the spoken messages, consisting of mathematical equations, random questions, also there will be present unexpected loud sound of clapping. For all attempts the time of each will be measured (Fig. 4).

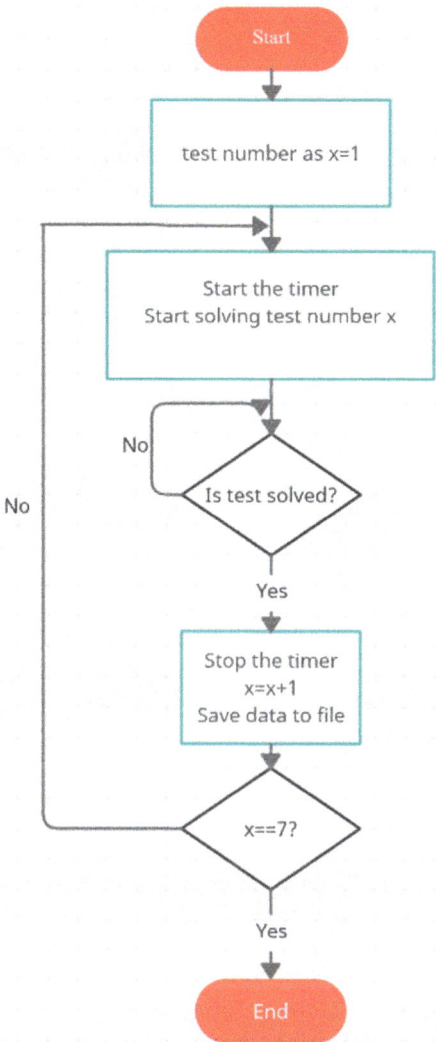

Fig. 4. Flowchart explaining data collection process.

For the first attempt, subject will sit down at the desk in empty classroom, and once the Emotiv EPOC + NeuroHeadset is placed and turned on, he will start solving 10 simple mathematical equations, at the sme time the timer will start. Once the first attempt is completed, the data from Emotiv EPOC + NeuroHeadset will be saved along with the time value. Second attempt will differ from previous one by adding headphones with classical music playing. Third attempt will change the music to metal. Fourth attempt will change themusic to rock. Fifth attempt will change the type of noise to TV news. Sixth attempt will change the type of noise to personal messages and unexpcted loud sounds.

After all data is collected, it will be inserted do database in MS Excel, bar charts will show the differences between brain activity in each attempt. Also there should be visible differencies in time, and with rising amount of time value some brain waves will be enhanced. Those data can help with estimating what kind of noise have positive or negative influence on concentration on tasks.

It is confirmed that music can have positive influence on concentration during simple tasks but it is highly dependent on someone's preferences. From observations, soft music in background may have positive influence on brain performance, for more complex tasks people prefer calm and silent environment. There is similiar experiment made by I. Voznenko Alexander A. Dyumin Evgeniya, V. Aksenova Alexander, A. Gridnev, Vladislav A. Delov with conclusion: "The impact of music listening on control performance through BCI was investigated in this work. The main finding is that different music styles can influence differently based on operator's profile" [8]. On the other hand there are activities that require from us being focused and there are intentionally placed sounds that will help achieving better results. One of such activities can be playing video games, it's not only about the visual output but sounds play an important role in terms of performance. MMORPG games can be cited as a good example, there are mechanics like killing bosses with class specified abilities, they are correlated with others and there is a percentage chance that one ability will be more appriopriate to use at given time. Such an occasion is spiced with a visual and audio effect, most of the times user cannot predict when will it happen so he relies on those sounds and pop ups. As a test while playing such a game we managed to disable sound, this caused my performance to drop drastically, we were able to see those visual effects but hearing no sound was distractive and caused slower reaction time.

5 Summary

Sound plays a very important role in the life. It cannot be clearly stated whether it helps or disturbs. It is highly dependent on the person experiencing it. For some people there are sounds that may help with concentration or be just neutral. On the other hand most sounds that people experience are rather distractive, annoying and not helpful. While some of them can be ignored, others are too overwhelming just to ignore them. People tend to focus on trying not to focus on them and this may be the main reason why those sounds cause such distraction and loss in performance. Brain responds to external stimuli even while we are not aware of it. In this case it's a neutral position where the performance is not affected at all, but once the focus is split among all those factors the sounds start to be distracting. The time or possibility of this is dependent on person's mood, age, sex and many other agents. Other similiar experiment is about ability to understand spoken language without occupational noise, with reverb and with background noise like renovation on the street. Maria Klatte, Thomas Lachmann, Markus Meis made similiar tests in their work and it showed that background sound has huge influence especially on children in schools: "Clearly, children who, due to poor interior acoustics, often lose the content of the teachers' instructions, are at risk of poor academic achievement". This is obviously negative aspect of background noise [9].

References

1. Paszkiel, S., Dobrakowski, P., Lysiak, A.: The impact of different sounds on stress level in the context of EEG, cardiac measures and subjective stress level: a pilot study. Brain Sci. **10**(10), 728 (2020). https://doi.org/10.3390/brainsci10100728
2. Paszkiel S.: The use of facial expressions identified from the level of the EEG signal for controlling a mobile vehicle based on a state machine. In: Szewczyk, R., Zielinski, C., Kaliczynska, M.: Automation 2020: Towards Industry of the Future. Advances in Intelligent Systems and Computing, vol. 1140, pp. 227–238 (2020). https://doi.org/10.1007/978-3-030-40971-5_21
3. Paszkiel S., Hunek WP., Shylenko A.: Project and simulation of a portable device for measuring bioelectrical signals from the brain for states consciousness verification with visualization on LEDs. In: Szewczyk, R., Zielinski, C., Kaliczynska, M.: Challenges in Automation, Robotics and Measurement Techniques. Advances in Intelligent Systems and Computing, vol. 440, pp. 25–35 (2016). https://doi.org/10.1007/978-3-319-29357-8_3
4. Emotiv EPOC + Neuroheadset User Manual. https://fccid.io/2ADIH-EPOC02/User-Manual/Manual-2596562
5. Paszkiel, S.: Using BCI in IoT implementation. In: Analysis and Classification of EEG Signals for Brain-Computer Interfaces. Studies in Computational Intelligence, vol. 852, pp. 101–110 (2020).https://doi.org/10.1007/978-3-030-30581-9_12
6. Paszkiel, S., Sikora, M.: The use of brain-computer interface to control unmanned aerial vehicle. In: Szewczyk, R., Zielinski, C., Kaliczynska, M. (eds.) Automation 2019: Progress in Automation, Robotics and Measurement Techniques. Advances in Intelligent Systems and Computing, vol. 920, pp. 583–598 (2020). https://doi.org/10.1007/978-3-030-13273-6_54
7. Paszkiel, S.: Data acquisition methods for human brain activity. In: Analysis and Classification of EEG Signals for Brain-Computer Interfaces. Studies in Computational Intelligence, vol. 852, pp. 3–9 (2020). https://doi.org/10.1007/978-3-030-30581-9_2
8. Voznenko, T.I., Dyumin, A.A., Aksenova, E.V., et al.: The experimental study of 'unwanted music' noise pollution influence on command recognition by brain-computer interface. National Research Nuclear University MEPhI (Moscow Engineering Physics Institute) Moscow, Russia. Moscow Institute of Physics and Technology Dolgoprudny, Russia. Mental Health Research Center Moscow, Russia (2018)
9. Klatte, M., Lachmann, T., Meis, M.: Effects of noise and reverberation on speech perception and listening comprehension of children and adults in a classroom-like setting. Noise Health **2010**(12), 270–282 (2010)

Brain-Computer Interface: A Possible Help for People with Locked-In Syndrome

Anna Nolte$^{(\boxtimes)}$ (iD)

Faculty of Electrical Engineering, Automatic Control and Informatics,
Opole University of Technology, ul. Prószkowska 76, 45-316 Opole, Poland
anna.nolte@student.po.edu.pl

Abstract. Brain-computer interface (BCI) is a communication system which enables the user to interact with a computer using brain signals recorded with an electroencephalograph (EEG). This feature is vital for persons with such severe form of paralysis as is displayed in locked-in syndrome, where patients experience total paralysis but they retain their consciousness. With the rise of affordable commercial EEG headsets there is a possibility of improving their quality of life (QOL).

Keywords: Brain-computer interface (BCI) · Locked-in syndrome (LIS) · Electroencephalography (EEG) · P300 · SSVEP

1 Introduction

Locked-in syndrome (LIS) is a rare condition in which a person cannot move their own body and they don't have the ability to communicate with other people [1, 2]. They are conscious, but their body is completely paralyzed. Amyotrophic lateral sclerosis (ALS) and other progressive diseases, brainstem stroke, or brain injury caused by an accident could all cause the syndrome. Individuals who lost their muscle control also lose the ability to communicate with the world around them which makes it hard to give them a proper diagnosis as they cannot show that they comprehend what other people are saying and what goes on around them. This results in significant rates of misdiagnosis, patients have been classified as either being in a state of a coma or a semi-coma [3]. Complete inability to move is a significant issue, but the inability to communicate poses an even bigger problem [4]. Nonverbal communication is usually achieved through the utilization of the only remaining movement, which is in the patient's eyes. There are three categories of LIS: Classic, where patients are quadriplegic and anarthric with preserved consciousness and upper eyelid and vertical eye movements; Incomplete, where voluntary movements are limited, not including eye movements, most frequently of fingers, toes and head; Total, where patients exhibit complete immobility including eye movements but they are still conscious [5].

The syndrome occurs equally in both males and females and the age range of patients is vast, the mean age at onset is 52 years. There is no cure for it, a standard treatment plan also doesn't exist [6]. A study report found that the life expectancy of patients reached

S. Paszkiel (Ed.): ICBCI 2021, AISC 1362, pp. 226–232, 2021.
https://doi.org/10.1007/978-3-030-72254-8_24

80% over 10 years [7]. Which illustrates the need for proper communication ways for people affected by LIS. They require tools that will enable them to connect to those around them, family and caregivers, which is crucial for survival and improving their quality of life (QOL). Using only their eye movements to indicate either yes or no is not sufficient in the long-term, as it can be tiring, frustrating and inefficient when more information is needed.

Brain-computer interface (BCI) is a device that enables direct communication between the brain and a computer, utilizing brainwaves, which is essential for people with limited motor skills. It can be either invasive or non-invasive.

The non-invasive approaches are more accessible and safer as they don't require surgery that a person who already may have gone through trauma which resulted in LIS may not be interested in. Obtaining informed consent about participating in BCI research trials from LIS patients is crucial, but challenging. For participating in any research, informed consent is a foundational principle which stems from the need to respect patients. The inability to give proper consent or decline participation when the person has full consciousness and is capable of forming their stance on the issue undermines the whole principle, since they simply cannot express their desires. Therefore, research into communication with LIS patients is needed, because in turn they can use BCI to give consent to other research [8].

Electroencephalography (EEG) is a recording of brain activity, an EEG signal is an electrical signal produced when brain cells send messages to each other, it consists of many waves with different characteristics. The differences in frequency ranges makes it possible to distinguish six brain rhythms in EEG, they are: delta (1–4 Hz), theta (4–7 Hz), alpha (8–12 Hz), mu (8–13 Hz), beta (12–30 Hz), and gamma (25–100 Hz) [9]. When an individual performs an activity their brain generates different kinds of EEG bands. Different degrees of consciousness are displayed when different EEG bands dominate, thus making it a suitable method for helping patients with LIS to improve their QOL.

Besides EEG there is magnetoencephalography (MEG) and functional magnetic resonance imagining (fMRI). EEG has been largely used in clinical and research applications due to the relatively low cost and portability of the required equipment. Over the years many BCI systems that base upon the EEG were developed, including motor imagery (MI) [10], slow conical potential (SCP) [11], steady-state visual evoked potential (SSVEP) [12], P300 [13] and hybrid SSVEP/P300 BCI [9].

Previously, BCI was less accessible because of the high cost of medical devices. In recent years, commercial devices such as the Emotiv EPOC+ NeuroHeadset (Fig. 1) became available to the public [14].

Researchers across the globe have achieved success in terms of BCI communication for LIS individuals. Some trials have used healthy individuals [15, 16], presumably since LIS is not so prevalent to have multiple patients in the vicinity of where each study takes place. They are still useful for LIS research, since the subjects are asked to remain still and only their brainwaves are recorded and evaluated. Trials involving only healthy individuals enable the researchers to evaluate the functionality of the proposed idea without mentally exhausting the disabled patients.

Fig. 1. Emotiv EPOC+ NeuroHeadset device.

In terms of research involving disabled subjects, a study using laboratory-grade EEG device in combination with a P300 speller [17] was done on a cohort of 20 severely disabled ALS patients. They reported high patient satisfaction and good recognition rates, having a mean satisfaction score of 8.7/10 for three metrics of usefulness, ease of use and comfort.

There are trails done using a mix of disabled and healthy individuals, a study [18] used two groups. The first consisted of 12 healthy subjects and the second group contained 12 healthy subject and 6 LIS patients. The study was a success for healthy individuals, but disabled patients had a low success rate, which can be explained by some of them stopping the test due to fatigue or persistent nystagmus.

2 Approaches

2.1 Steady-State Visually Evoked Potential (SSVEP) and P300 Hybrid System

Information transfer rate (ITR) and required training time are two vital features of each BCI system. SSVEP and P300 both have relatively low training time combined with high ITR which makes them good candidates for a hybrid system.

BCI can be used in combination with SSVEP and P300 to give LIS individuals a way to communicate with others around them. There are two ways to combine BCI systems. The first way is called a sequentially hybrid BCI, where each system has a separate input signal. The second way is a simultaneous hybrid BCI where both systems are processed in parallel [19]. The main purpose of the first system in sequential hybrid BCI is to act as a switch. To perform that task SSVEP is one of the most appropriate options, since it has high classification accuracy and high information transfer rate does not need training [9].

SSVEP-P300 hybrid asynchronous BCI system can be seen in [20], where this combination worked well as the stimuli for evoking both patterns that can be simultaneously shown on one screen. The P300 in this trial is a speller matrix based on the original P300 row/column paradigm and one frequency is allocated for the SSVEP paradigm. The background colour was flashing with a frequency slightly below 18 Hz. The change of the background colour enables SSVEP detection. A band pass filter separated P300 and SSVEP signals during the classification. The SSVEP was used for control state (CS) detection. The SSVEP was detected when the user was looking at the screen and there was an assumption that a command was intended to be send. The selection of the P300 target and the CS were both detected at the same time by the system [20].

2.2 Steady-State Visually Evoked Potential System

Steady-state visually evoked potential (SSVEP) BCI system uses flickering lights with a frequency above 4 Hz to stimulate the user, it arises mainly in the visual cortex during the test. The strongest response is obtained for stimulation frequencies in the range 5–20 Hz [21].

Higher information transfer rate (ITR) and higher accuracy are both crucial advantages of a SSVEP-based BCI system. Another positive is that it requires short or no training time and fewer EEG channels, as compared to other BCI approaches [9].

In one study [18] that used SSVEP-BCI in ALS patients with LIS only one out of four was able to communicate online. Such a low success rate was attributed to symptoms such as fatigue and persistent nystagmus. The system relies on the participant's ability to use their sight which is a significant obstacle for individuals with a loss of eye control and other clinical issues. This study demonstrated that online communication is possible using a covert SSVEP paradigm, which is fully independent of any neuromuscular activity. 24 participants were split into two groups, group A had 12 healthy participants and group B had 12 healthy participants and 6 LIS patients. In group A, a panel was placed at the 30cm distance from the person's face. It had a white fixation cross in the center and it was a 7×7 cm^2 'interlaced square' that had red and yellow 1×1 cm^2 (LED)-squares which emitted light. The yellow squares were set to flicker at 10 Hz and the red ones at 14 Hz during the experiments, the duty cycle equaled 0.5. This system has shown to be compact, convenient and simple to use. Each person went through 6 runs, the duration of every one of them was about 5 min. Every run had ten 7 s trials, in between which there was a space of 23 s, with 7 s dedicated for resting and 16 s for listening to instructions. The pattern steadily flashed during each run and the choices were presented randomly, although both were shown the same number of times. Each person was told to zero in onto the white cross in the center and to concentrate on one of the flashing colours [18]. In group B, an overt session was performed before the training session by 3 out of 12 healthy subjects and 4 out of 6 LIS individuals. The participants had about 30 cm of space between their faces and a block pattern that contained a white fixation cross in the center and two 2×2 cm^2 blocks made of 1x1 cm2 LED squares on either side of the cross, the blocks had 12 cm of space in between them. The one yellow square flashed at 10 Hz and one red square flashed at 14 Hz, the duty cycle equaled 0.5. Afterwards, the same training session was performed as the one done with group A. In order to train the classifier a short break was made and it was followed by an online communication session in which the individual was asked 33 closed questions (yes/no), an example of an asked question is "are you 25 years old?". The answers to these questions had to be known before the test and only questions with clear answers were asked. Throughout the 7 s window of time, each subject answered by focusing on one of the squares, the yellow was picked to answer 'yes' and red to answer 'no'. During the question/response time, the simulation panel was switched on in order to prevent the subjects from becoming fatigued [18].

2.3 P300 System

Due to there being no requirement for training or imagination of movement, P300-based BCI systems are a prime choice for LIS patients.

The P300 system allows the user to select multiple actions, one after the other. The most common version is an alphabet matrix where the characters flash in a random order. Patients select a letter from a matrix in which each letter is randomly illuminated. The user concentrates on a letter and when it flashes, a P300 ERP is elicited which triggers the selection of the letter. The biggest component of an ERP is called P3, or P300 [22]. A study [23] was done to evaluate the capability of the Emotiv EPOC+ NeuroHeadset to record the P300 wave. The Emotiv EPOC+ has 14 electrodes: AF3, AF4, F3, F4, FC5, FC6, F7, F8, T7, T8, P7, P8, O1, O2 plus two standard reference electrodes, CMS and DRL, arranged according to the 10–20 international electrode system (Fig. 2). The EEG signals were recorded from all 14 electrodes with sampling rate at 128 Hz and they were sent wirelessly to the OpenViBE software, which was used to run the P300 speller. Five healthy subjects participated in the study, they were shown a 6x6 or 3x3 matrix during the experiment [23].

A P300-BCI speller gives people with severe paralysis, like LIS patients, a communication way so that they can express their wishes and converse with others. This trial has demonstrated the capability of the Emotiv EPOC+ Neuroheadset to detect the P300 signals from two channels, O1 and O2. Moreover, the study investigated the usage of different factors (matrix size, flash duration, colours) on the P300 amplitude and accuracy. The results illustrate that when the matrix size or flash duration increases, the P300 amplitude also increases. Although, the P300 performance decreases with a larger inter-stimulus interval. Since the right hemisphere is responsible for colour processing and the O2 sensor is located on it, the effect of using the coloured matrix could be seen particularly clearly on the O2 channel [23, 24].

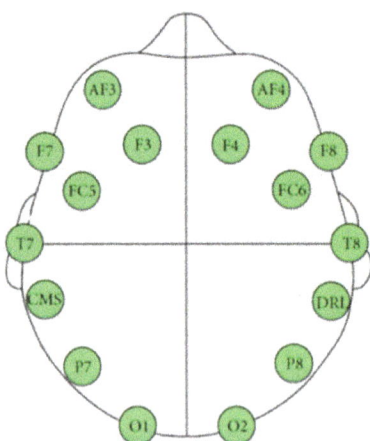

Fig. 2. Emotiv EPOC+ NeuroHeadset electrode placement on scalp using the international 10–20 System.

3 Summary

At present, the data on the use of BCIs to enable communication for patients who are completely paralyzed is limited and finding a way for people affected by locked-in syndrome to communicate is an important endeavour. There currently is no cure for LIS and there probably won't be anytime soon, so finding a way for them to connect with the people around them significantly improves their quality of life. Their complete inability to move is a significant issue, but the inability to communicate poses an even bigger problem as it makes it hard to give them a proper diagnosis as they cannot show that they comprehend what other people are saying and what goes on around them, which results in patients being wrongly diagnosed as being in a coma or a semi-coma. There are several approaches for tackling this problem. In this chapter, only the EEG based P300, SSVEP and a hybrid P300-SSVEP in combination with brain-computer interface systems was presented and analysed. The Emotiv EPOC+ Neuroheadset, which is currently one of the most accessible devices for BCI, was shown as well as its utilization in research combined with a P300 speller. Patients with debilitating neurological disorders, which at this time are extremely difficult or impossible to treat, have benefitted from clinical research into BCIs as it has provided tremendously promising strategies for improving their quality of life. Given the still largely unexplored potential for research in this field, the continued steady growth of BCI and its accessibility, the rising awareness of the syndrome and the challenges it presents, many more fascinating tests, trials and studies could be done. The approaches presented in this chapter provide a solid foundation for future research.

References

1. Wolpaw, J.R., Birbaumer, N., McFarland, D.J., Pfurtscheller, G., Vaughan, T.M.: Brain-computer interfaces for communication and control. Clin. Neurophysiol. **113**(6), 767–791 (2002)
2. Birbaumer, N., Cohen, L.G.: Brain-computer interfaces: communication and restoration of movement in paralysis. J. Physiol. **579**(3), 621–636 (2007)
3. Schnakers, C., Perrin, F., Schabus, M., Majerus, S., Ledoux, D., Damas, P., Boly, M., Vanhau-denhuyse, A., Bruno, M.A., Moonen, G., Laureys, S.: Voluntary brain processing in disorders of consciousness. Neurology **71**(20), 1614–1620 (2008)
4. Sellers, E.W., Donchin, E.: A P300-based brain-computer interface: initial tests by ALS patients. Clin. Neurophysiol. **117**(3), 538–548 (2006)
5. Bauer, G., Gerstenbrand, F., Rumpl, E.: Varieties of the locked-in syndrome. J. Neurol. **221**(2), 77–91 (1979)
6. TVSP, M., Gupta, P.: Locked in syndrome-a case report. Indian J. Anaesth. **49**(2), 143–145 (2005)
7. Smith, E., Delargy, M.: Locked-in syndrome. BMJ **330**(7488), 406–409 (2005)
8. Klein, E., Peters, B., Higger, M.: Ethical considerations in ending exploratory brain-computer interface research studies in locked-in syndrome. Camb. Q. Healthc. Ethics **27**(4), 660–674 (2018)
9. Amiri, S., Rabbi, A., Azinfar, L., Fazel-Rezai, R.: A review of P300, SSVEP, and hybrid P300/SSVEP brain-computer interface systems. Brain-computer interface systems-recent progress and future prospects. InTech **2013**, 195–213 (2013)

10. Sellers, E.W., Krusienski, D.J., McFarland, D.J., Vaughan, T.M., Wolpaw, J.R.: A P300 event-related potential brain-computer interface (BCI): the effects of matrix size and inter stimulus interval on performance. Biol. Psychol. **73**(3), 242–252 (2006)
11. Birbaumer, N., Ghanayim, N., Hinterberger, T., Iversen, I., Kotchoubey, B., Kübler, A., Perelmouter, J., Taub, E., Flor, H.: A spelling device for the paralysed. Nature **398**(6725), 297–298 (1999)
12. Allison, B., Luth, T., Valbuena, D., Teymourian, A., Volosyak, I., Graser, A.: BCI demographics: How many (and what kinds of) people can use an SSVEP BCI? IEEE Trans. Neural Syst. Rehabil. Eng. **18**(2), 107–116 (2010)
13. Guger, C., Daban, S., Sellers, E., Holzner, C., Krausz, G., Carabalona, R., Gramatica, F., Edlinger, G.: How many people are able to control a P300-based brain-computer interface (BCI)? Neurosci. Lett. **462**(1), 94–98 (2009)
14. Paszkiel, S.: Data acquisition methods for human brain activity, analysis and classification of eeg signals for brain-computer interfaces. In: Studies in Computational Intelligence, vol. 852, pp. 3–9 (2020). https://doi.org/10.1007/978-3-030-30581-9_2
15. Ramadhan, M.M., Wijaya, S.K., Prajitno, P.: Development of visual request system by using wireless EEG signal to help communication of patients suffering locked-in syndrome. In: AIP Conference Proceedings, vol. 2168, p. 020018 (2019)
16. Paszkiel, S.: Characteristics of question of blind source separation using Moore-Penrose pseudoinversion for reconstruction of EEG signal. In: Szewczyk, R., Zielinski, C., Kaliczynska, M.: Automation 2017: Innovations in Automation, Robotics and Measurement Techniques. Advances in Intelligent Systems and Computing, vol. 550, pp. 393–400 (2017). https://doi.org/10.1007/978-3-319-54042-9_36
17. Guy, V., Soriani, M.H., Bruno, M., Papadopoulo, T., Desnuelle, C., Clerc, M.: Brain computer interface with the P300 speller: usability for disabled people with amyotrophic lateral sclerosis. Ann. Phys. Rehabil. Med. **61**(1), 5–11 (2018)
18. Lesenfants, D., Habbal, D., Lugo, Z., Lebeau, M., Horki, P., Amico, E., Pokorny, C., Gomez, F., Soddu, A., Müller-Putz, G., Laureys, S.: An independent SSVEP-based brain-computer interface in locked-in syndrome. J. Neural Eng. **11**(3), 035002 (2014)
19. Amiri, S., Fazel-Rezai, R., Asadpour, V.: A review of hybrid brain-computer interface systems. Adv. Hum. Comput. Interact. **2013**, 1–8 (2013)
20. Panicker, R.C., Puthusserypady, S., Sun, Y.: An asynchronous P300 BCI with SSVEP-based control state detection. IEEE Trans. Biomed. Eng. **58**(6), 1781–1788 (2011)
21. Friman, O., Volosyak, I., Graser, A.: Multiple channel detection of steady-state visual evoked potentials for brain-computer interfaces. IEEE Trans. Biomed. Eng. **54**(4), 742–750 (2007)
22. Chaudhary, U., Birbaumer, N., Ramos-Murguialday, A.: Brain-computer interfaces for communication and rehabilitation. Nat. Rev. Neurol. **12**(9), 513 (2016)
23. Alzahrani, S., Anderson, C.W.: EEG P300 wave detection using Emotiv EPOC+: effects of matrix size, flash duration, and colors. PeerJ Preprints **5**, e3474v1 (2017)
24. Paszkiel, S.: Using BCI in IoT implementation. In: Analysis and Classification of EEG Signals for Brain-Computer Interfaces. Studies in Computational Intelligence, vol. 852, pp. 101–110 (2020). https://doi.org/10.1007/978-3-030-30581-9_12

Using Neuralink by Humans: A Process Which Brings Humanity Closer to the Future

Adrian Luckiewicz$^{(\boxtimes)}$ (iD)

Faculty of Electrical Engineering, Automatic Control and Informatics,
Opole University of Technology, Prószkowska 76, 45-758 Opole, Poland

Abstract. In this chapter there are information about brain-computer interface called Neuralink. The future of humanity is closely connected with brain – machine/brain – computer technology. Using this technology have positive effect on humanity overall in many research fields. One of them is treating multiple neurological diseases. Neuralink goal is currently to focus on enhancing the performance of users and give a potentially new approach to treating neurological conditions.

Keywords: BCI · BMI · Neuralink · Human-machine interface

1 Introduction

Brain-machine interfaces have the potential to help people with a wide range of clinical disorders. Current knowledge in this field is limited but every day we expand our knowledge and Neuralink is the perfect interface to help us with multiple problems. There are researchers that have already demonstrated human neuroprosthetic control of computer cursors [1], robotic limbs [2, 3] and speech synthesizers [4] by using older technology that contains no more than 256 electrodes. Although these successes suggest that high-fidelity information transfer between brains and machines is possible, the development of brain-machine interface has been drastically limited by the inability to record signals from a lot of neurons. Approaches that are non-invasive can record the average of millions of neurons, but this signal is disfigured and nonspecific [5, 6]. Electrodes that are invasive, placed on the surface of the cortex can collect functional signals, but they are limited in that they average the activity of thousands of neurons and cannot process signals deep in the brain [7]. Many brain-computer interfaces have used invasive techniques, since the most precise readout of neurons requires recording single action potentials from neurons in distributed, functionally linked groups [8]. Microelectrodes are also called the "gold-standard" technology for recording action potentials, but the fact is that for large-scale recordings there is no clinically translatable microelectrode technology. This technology would require a system with material properties that are biologically compatible, safe, and long useable. What is more, this device also needs a surgical approach which is practical that means using low-powered, high-density electronics to smooth the process of implanted wireless operation [9].

S. Paszkiel (Ed.): ICBCI 2021, AISC 1362, pp. 233–238, 2021.
https://doi.org/10.1007/978-3-030-72254-8_25

Neuralink, unlike other technologies, offers us an alternative approach. It uses polymer probes that are thin, flexible and have multiple electrodes. The increased flexibility of the probes and smaller size in theory should offer greater biocompatibility. The only problem which there is, would be installation process of Neuralink. Construction of the device forces the surgeon robot to perform the operations to make a cut in the skull, a hole with a diameter of about 2 mm. A 2 mm cut would be appropriate to widen the operation to 8 mm [10].

The uncapped part of the skull is sealed with the chipset module after the procedure has ended. To satisfy all of the requirements for a high-bandwidth brain-machine interface, while taking advantage of the properties of thin-film devices. The process is mainly made up from system which has three main components: custom high-density electronics, a neurosurgical robot and polymer probes.

2 Brain-Computer Interface

The so-called brain-computer interfaces are more popular in everyday usage, threating diseases or even helps us achieve more and more ambitious goals like controlling machines with BCI [23, 24]. BCI is giving humanity hope for better future for example it is believed that BCI will restore motor function of humans, also that is believed to be helpful in treatment of neurological disorders. The only problem which is faced was small number of channels in BCI devices which limited their potential [26–28]. The Neuralink is a new, fresh solution to old brain machine interfaces problems. Neuralink offers high-bandwidth brain-machine interface system which has built in arrays of flexible and tiny electrode threads. Although BCI technology has been with us for some time, it is still only solution which we as humans have. Brain-Computer Interface technology is a computer-based system that acquires brain signals, then analyses them carefully, and finally translates them into commands that are relayed to an output device to show result. BCIs do not use the brain's normal output pathways of peripheral nerves and muscles. This definition really limits the term of BCI to systems that measure and use signals produced by the central nervous system (CNS). Voice-activated or muscle-activated communication system is not called a BCI. Furthermore, an electroencephalogram (EEG) machine alone is also not a BCI because it only records brain signals but does not generate an output that has an impact on the user's environment. It is a misconception that BCIs are mind-reading devices. Brain-computer interfaces do not read minds in the sense of extracting information from unwary or unenthusiastic users but enable users to act on the world by using brain signals rather than muscles. The user and the BCI work together. The user, often after a period of training, generates brain signals that encode intention, and the BCI, also after training, decodes the signals and translates them into commands to an output device that accomplishes the user's intention [12]. Neuralink BCI allows people to communicate with external devices.

3 Neuralink Device

The first part of the construction of neuralink device are threads. The Neuralink company developed a custom process of fabricating minimally displace neural probes that employ a collection of thin film materials that are biologically compatible. The main substrate used in these probes is polyimide, which contains a gold thin film trace. Each thin film array is composed of a "thread" area that features electrode contacts and traces and a "sensor" area where the thin film interfaces with custom chips that enable signal amplification and acquisition. Every array has either 48 or 96 threads, each of which contains 32 independent electrodes. Integrated chips are joined to the contacts on the sensor area of the thin film. The process used to join the integrated chips to the contacts is called flip-chip bonding process. This approach aim to maintain a small thread cross-sectional area to minimize tissue displacement in the brain. To achieve this goal, and to keep the channel count high in numbers, microfabrication techniques such as stepper lithography and other are used to form the metal film at submicron resolution.

Neuralink Company invented nearby 20 different thread and electrode types into arrays. The second part of neuralink device construction are electrodes. The third part are chips, the link needs to convert the small electrical signals recorded by each electrode into real-time neural information. Since the neural signals in the brain are small (micro-volts), Link must have high-performance signal amplifiers and digitizers. Also, as the number of electrodes increases, these raw digital signals become too much information to upload with low power devices. So, scaling company devices requires on-chip, real-time identification and characterization of neural spikes. Their own custom chips on the Link meet these goals, while radically reducing power consumption and per-channel chip size. All of that over current technology. The fourth and last part is hermetic packing of neuralink. The Link needs to be protected from the fluid and salts that bathe surrounding tissue. The company made sure that device is well protected and hermetic packed.

4   Implementation and Applications

There are many ways to improve our lives using this device. The enormity of implementation that comes with this small device is big. First of all, it can be used as disease prediction. Just think about sickness-free world where we could know about future mental-health threats like for example Alzheimer Disease. We will have the possibility of learning about electrical signs as well as receiving chemical prompts in the cerebrum and early prevention of diseases as well by using Neuralink.

Solving Mental Diseases. The device has the possibility of placing electrodes into areas of the brain that have been difficult or impossible to reach with existing intra-cortical arrays (e.g., on the medical surfaces of subcortical structures and cortical sluci structures). These new locations could probably provide other types of information for enhancing brain-machine interface performance (e.g., abstract planning of activities, anticipated reward signals, and decision making), as well as better explanation of the interactions between different brain structures (e.g., between different motor areas, integration of sensory-motor activities and processing of sensory information). Such information is very likely to be necessary for lucky performance in more complex brain-machine interface applications, such as multidimensional arm movement which involves complex physical dynamics. As Musk wrote in one of his articles, there is clear value in the advanced brain machine interface systems they describe for therapeutic use. This is a technology that has the potential to provide completely new approaches for addressing a spectrum of neurological conditions, from Parkinson disease to epilepsy or even debilitating migraines. It could also be an extremely improve of users ability to interface with and control prosthetics, including those that replace diminished sensory function [13].

Eliminate Pain. In todays world no one wants to feel pain. There are also many infections in the world which cause pain in many different ways. Using prescribed medication is not always the perfect and long run, way for pain to disappear. Link by partially controlling the brain can help with elimination of receptors in brain which are responsible for feeling pain.

Brain Enhancing. Neuralink could enhance our brain by simply reading our brain waves and alert us about unusual behaviour of our brain or anomalies. In the future there is idea presented by Elon Musk that the Link could automatically correct small brain damage which many diseases do to the brain.

Reading Emotions. By the ability to read the human and animal brain waves the link can determine whether the object which it is examining is happy or sad. The spectrum of emotions is not very big but with the appropriate software this device could work really well. Imagine the situation when you don't know what your partner is felling or close family member which is sick. With special software neuralink can tell you persons emotion.

Controlling Devices. Nowadays there is many old and mentally disabled people. The ability to control their smartphones with their brain, switching the tv channels with their brain, or even opening the door with their brain is just super exciting. Neuralink offers ability to make this everyday problems of disabled people disappear. Using link they could connect by Bluetooth with their smartphones and from that point control almost all compatible devices.

Better control of implants. Neurological control of robotics parts of our body has been with us since nearly 2012. It works well in many cases but some patients who have implants installed complain about not feeling anything they touch in case of robotic limbs. Neuralink thanks to its technology where it records which areas of brain are stimulated can make so called "touch feedback" possible.

Improve brain power. In the near future we can witness something like human-AI hybrid. Although it is very difficult to achieve, humans owning to Neuralink can improve their brain computing power.

5 Summary

Elon Musk described a specific approach [9] which indicates that there are many reasons which could cause worries about current state of the technology and its successful development and use, with many of these worries depends on how companies address potentially serious social issues. Other things worth mentioning are complex issues. The ones about for example around data privacy or user security or autonomy—in particular, fact that the Link is directly in human brain [14].

The risk innovation approach which is described, provides innovative understanding into this worries and is a useful tool for revealing potentially risks that could hit unexpectedly, while there is still time to take rightful action. In the case of advanced brain machine interfaces, it helps to map out opportunities that may otherwise be easy to miss but also challenges that have the ability to ruin progress. The main rule here is to understand that, success depends on the willingness of innovators and others to take ethical and responsible approach to innovation and to draw on the inter and trans-disciplinary expertise that is necessary to translate good intentions into positive outcomes.

Such innovation as neuralink is at current point very young and fresh, but with its potential it will probably dominate the all brain-computer interface field. The more popular it gets, and the more customers will trust it and buy it for own purpose, they will see that it works the better for the humanity and its directed straight towards future.

References

1. Hochberg, L.R., Serruya, M.D., Friehs, G.M., Mukand, J.A., Saleh, M., Caplan, A.H., et al.: Neuronal ensemble control of prosthetic devices by a human with tetraplegia. Nature **442**(7099), 164–171 (2006)
2. Hochberg, L.R., Bacher, D., Jarosiewicz, B., Masse, N.Y., Simeral, J.D., Vogel, J., et al.: Reach and grasp by people with tetraplegia using a neurally controlled robotic arm. Nature **485**(7398), 372–375 (2012)
3. Collinger, J.L., Wodlinger, B., Downey, J.E., Wang, W., Tyler-Kabara, E.C., Weber, D.J., et al.: High-performance neuroprosthetic control by an individual with tetraplegia. Lancet **381**(9866), 557–564 (2013)
4. Anumanchipalli, G.K., Chartier, J., Chang, E.F.: Speech synthesis from neural decoding of spoken sentences. Nature **568**(7753), 493–498 (2019)
5. Buzsáki, G., Anastassiou, C.A., Koch, C.: The origin of extracellular fields and currents–EEG, ECoG, LFP and spikes. Nat. Rev. Neurosci. **13**(6), 407–420 (2012)
6. Pesaran, B., Vinck, M., Einevoll, G.T., Sirota, A., Fries, P., Siegel, M., et al.: Investigating large-scale brain dynamics using field potential recordings: analysis and interpretation. Nat. Neurosci. **21**(7), 903–919 (2018)
7. Kaiju, T., Doi, K., Yokota, M., Watanabe, K., Inoue, M., Ando, H., et al.: High spatiotemporal resolution ECoG recording of somatosensory evoked potentials with flexible micro-electrode arrays. Front. Neural Circ. **11**, 20 (2017)
8. Yuste, R.: From the neuron doctrine to neural networks. Nat. Rev. Neurosci. **16**(8), 487–497 (2015)
9. Musk, E.: An integrated brain-machine interface platform with thousands of channels https://doi.org/10.2196/16194
10. Fadziso, T.: Why Neuralink will change humanity forever? https://doi.org/10.2311/8636
11. Mayo Clin. Proc. **87**(3), 268–279 (2012). https://doi.org/10.1016/j.mayocp.2011.12.008
12. Maynard, A.D., Garbee, E.: Responsible innovation in a culture of entrepreneurship: a US perspective. In: von Schomberg, R., Hankins, J. (eds.) International Handbook on Responsible Innovation: A Global Resource. Edward Elgar, Cheltenham (2019)
13. Maynard, A.D.: The Ethical and Responsible Development and Application of Advanced Brain Machine Interfaces. https://doi.org/10.2196/16321
14. Valle, G.: The Connection Between the Nervous System and Machines: Commentary. https://doi.org/10.2196/16344
15. Stavisky, S.D., Kao, J.C., Nuyujukian, P., Pandarinath, C., Blabe, C., Ryu, S.I., Hochberg, L.R., Henderson, J.M., Shenoy, K.V.: Brain-machine interface cursor position only weakly affects monkey and human motor cortical activity in the absence of arm movements. Sci. Rep. **8**, 16357 (2018)
16. Donepudi, P.K.: AI and machine learning in retail pharmacy: systematic review of related literature. ABC J. Adv. Res. **7**(2), 109–112 (2018)

17. Newman, J.P., Fong, M., Millard, D., Whitmire, C., Stanley, G., Potter, S.: Optogenetic feedback control of neural activity. eLife **4**, e07192 (2015). https://doi.org/10.7554/elife.07192

18. Navarro, X., Krueger, T.B., Lago, N., Micera, S., Stieglitz, T., Dario, P.: A critical review of interfaces with the peripheral nervous system for the control of neuroprostheses and hybrid bionic systems. J. Peripher. Nerv. Syst. **10**(3), 229–258 (2005)

19. Schwemmer, M.A., Skomrock, N.D., Sederberg, P.B., Ting, J.E., Sharma, G., Bockbrader, M.A., et al.: Meeting brain-computer interface user performance expectations using a deep neural network decoding framework. Nat. Med. **24**(11), 1669–1676 (2018)

20. Valle, G., Mazzoni, A., Iberite, F., D'Anna, E., Strauss, I., Granata, G., et al.: Biomimetic intraneural sensory feedback enhances sensation naturalness, tactile sensitivity, and manual dexterity in a bidirectional prosthesis. Neuron **100**(1), 37–45.e7 (2018)

21. da Cruz, L., Dorn, J.D., Humayun, M.S., Dagnelie, G., Handa, J., Barale, P.: Argus II study group. Five-year safety and performance results from the Argus II retinal prosthesis system clinical trial. Ophthalmology **123**(10), 2248–2254 (2016)

22. Dobelle, W.H., Mladejovsky, M.G., Girvin, J.P.: Artifical vision for the blind: electrical stimulation of visual cortex offers hope for a functional prosthesis. Science **183**(4123), 440–444 (1974)

23. Paszkiel, S.: Characteristics of question of blind source separation using Moore-Penrose pseudoinversion for reconstruction of EEG signal. In: Szewczyk, R., Zielinski, C., Kaliczynska, M. (eds.) Automation 2017: Innovations In Automation, Robotics and Measurement Techniques. Advances in Intelligent Systems and Computing, vol. 550, pp. 393–400 (2017). https://doi.org/10.1007/978-3-319-54042-9_36

24. Paszkiel, S.: Augmented reality of technological environment in correlation with brain computer interfaces for control processes, recent advances. In: Szewczyk, R., Zielinski, C., Kaliczynska, M. (eds.) Automation, Robotics and Measuring Techniques. Advances in Intelligent Systems and Computing, vol. 267, pp. 197–203 (2014). https://doi.org/10.1007/978-3-319-05353-0_20

25. ApyNews. https://apynews.pl/. 05 January 2021

26. Paszkiel, S.: Using BCI in IoT implementation, analysis and classification of EEG signals for brain-computer interfaces. In: Studies in Computational Intelligence, vol. 852, pp: 101–110 (2020). https://doi.org/10.1007/978-3-030-30581-9_12

27. Paszkiel, S.: The use of facial expressions identified from the level of the EEG signal for controlling a mobile vehicle based on a state machine. In: Szewczyk, R., Zielinski, C., Kaliczynska, M. (eds.) Automation 2020: Towards Industry Of The Future. Advances in Intelligent Systems and Computing, vol. 1140, pp. 227–238 (2020). https://doi.org/10.1007/978-3-030-40971-5_21

28. Paszkiel, S.: Using neural networks for classification of the changes in the EEG signal based on facial expressions, analysis and classification of EEG signals for brain-computer interfaces. In: Studies in Computational Intelligence, vol. 852, pp. 41–69 (2020). https://doi.org/10.1007/978-3-030-30581-9_7

29. Spectrum. https://spectrum.ieee.org/. 05 January 2021

The Developments of Text Entry and Recognition Using Brain-Computer Interfaces

Szymon Kaszura$^{(\boxtimes)}$ (iD)

Faculty of Electrical Engineering, Automatic Control and Informatics,
Opole University of Technology, Prószkowska 76, 45-758 Opole, Poland

Abstract. As civilization progresses the percentage of people with disabilities in society increases. A part of this group has trouble using conventional methods of communication. For this reason, many new systems that allow physically disabled people to interact with the outside world have been getting developed. Brain-Computer Interfaces are one of them and although, the performance they provide is significantly lower than other solutions it has the lowest access requirement with voluntary eye movement being one of the only constraints. This paper discusses some of the techniques and paradigms used in state-of-the-art BCI Spellers as well as considers potential threats to security of the system.

Keywords: Brain-Computer Interface (BCI) · Speller · Text entry · User Interface (UI)

1 Introduction

The main goal of all BCI-based (Brain-Computer Interface) spellers is to make communication easier or even possible at all for people with certain physical disabilities disallowing them from using other means of communication. There are many different text entry systems in development as of right now, improving rapidly with every passing day. However, none are yet ready for commercial use.

The most important issue right now with all state-of-the-art BCI spellers is the difficulty of using such a device as an inexperienced user [1]. This is an ongoing discussion among the researchers, who have presented many promising solutions [2–7]. There has been a lot of studies done on the importance of implementing a sophisticated language model [4] into the speller, as the speed of manual typing is below industry standards compared to other text entry systems.

User Interface (UI) is also an important aspect in creation of a reliable, easy to use text entry system [8]. The general consensus is that simplicity should be at the forefront of design ideology [1, 13] with the user being able to understand the controls without any issues and having to perform as little unnecessary operations as possible. The majority of work should be done by the back end, using language prediction models and automation algorithms to streamline the process.

S. Paszkiel (Ed.): ICBCI 2021, AISC 1362, pp. 239–245, 2021.
https://doi.org/10.1007/978-3-030-72254-8_26

Evaluation of BCI-based text entry systems has been questioned [7, 11, 12], as copy-spelling, the current method of measuring the quality of the system leaves many questions unanswered and is not task-specific enough [11].

There have also been efforts at implementing BCI spellers in Chinese writing system hanzi, consisting of over 7000 characters, a vast difference compared to the 26 characters in the English alphabet. Two methods have achieved similar amounts of success; First-Last (FLAST) [9] and Row-Column [10] both based on the P300 wave, utilized in an oddball paradigm.

Security of EEG-based systems has mostly been overlooked in favor of performance gain. Yet, recent study [15] shows that there exist vulnerabilities in the system ready to be exploited and presents ways to counteract it. The focus of this chapter is highlighting the developments in text entry using Brain-Computer Interfaces and considering what obstacles may come up for future iterations of the system [16, 17].

2 Evaluation of BCI Speller Systems

Before discussing different approaches to BCI-based Spellers the methodology of evaluating them should be considered. Thus far, these systems were tested using a pre-written text given to the user for them to copy. However, it may be the case that while this type of test gives a measurable outcome, it is not as optimal as manually created tests that fit the task. In one study, researchers found that when the test subject was asked to create a new text instead of copying one, the precision decreased significantly [11].

One of the most common way of measuring the effectiveness of the system is the speed at which text is entered by the user. Despite that, users in a study unanimously decided to choose the slowest of given methods as their preferred method of text entry (Dasher-BCI being the slowest method and keyboard being the fastest) [12]. The physical ability of the target audience has to be considered in this field. Although Dasher-BCI was performing the worst according to standard evaluation, it was the most comfortable to use for wheelchair-bound subjects, usually only able to perform eye movements.

In the RSVP paradigm, one experiment [7] replaced rapidly flashing letters with familiar to the user pictures that would act as a stimulus. The results were inconclusive; user performance did indeed increase, however, the author speculates as to the cause of that. The improvement could have been caused by the physical properties of the pictures (color, shape, size) or the way the brain processes images. More testing needs to be done to conclude the reason for the increase in performance.

3 Language Model

A language model assesses text that has been written so far and based on it determines the probability distribution on all symbols, which can be used to form a dynamic decision tree from which the more probable symbols are more easily reachable [13].

The interaction between the user and the BCI Speller is quite similar to that of the user and mobile phone's touchscreen keyboard. There is a disparity between what the user is able to process and what they are able to input. The error correction is also slow and irritating [13].

Proper implementation of a sophisticated language model would expedite the input process by predicting the next letter based on contextual information from the already entered text.

Thus far, Brain-Computer Interface text entry systems required the user to select the same character multiple times in a row to confirm that it had not been a mistake. However, with a high-order language model the speller would be able to instead rely on the model and switch to a single-trial mode. Such a method would only work for subsequent letters of the word, as the language model is not able to establish the probability distribution for the first letter of the word.

The main drawback of this solution is that it is challenging for users to learn the layout of the system. The characters are constantly adapting its' position based on previous input and the person using the speller has to focus and read through the suggestions instead of memorizing the layout as is the case with a physical keyboard and touch-typing.

Nonetheless, implementing language models has proven successful so far and regardless of sophistication, accuracy increase has been reported [3, 8]. However, it is worth mentioning that most of the subjects that have been tested were unfamiliar with the technology before the experiments and were they to gain more experience with the equipment the results may change [4].

4   Design Paradigms for BCI Spellers

One common characteristic all Spellers are opting for is the simplicity of the design of User Interface [13]. Since the intended audience for such systems are people with limited movement ability it is crucial that majority of tasks is done behind the scenes without user's input. For this reason, language models have been getting implemented as well as many frontend paradigms aiming to speed up the typing process without actual performance increase from the user.

Hex-O-Spell is a User Interface design coupled with Berlin Brain-Computer Interface. During a live demonstration, in a highly unfavorable environment, users were able to reach typing speed of up to 7.6 characters/minute. The author of the paper attributes this to the simplicity of the User Interface coupled with a powerful language model and layout reordering utilizing a "minimal reorganization algorithm", which reduces the search time done by the user before each selection. The organized, compact layout of the interface also allows to scale it for smaller displays, such as phones for example or as dynamically appearing context menu for editing text files [13].

Dasher-BCI, though initially developed for use with a mouse pointer, has proven operational coupled with Brain-Computer Interface. Dasher uses adaptive target resizing, which calculates the size of a letter based on the probability of likelihood of that character being chosen. The user then uses a vector to point at the direction Dasher should "zoom in" and show next letters. Users were able to reach an average of 3.9 characters/minute, which is considerably lower compared to previously mentioned Hex-O-Spell. However, the author remarks that "the concept is indeed viable and worthy of further development". To add a further comparison, the UI of Dasher is much less compact and user-friendly compared to the Berlin alternative [12].

Rapid Serial Visual Presentation or RSVP is a paradigm, which relies on presenting one stimulus (character) at a time at the focal point of the screen. The method uses

Evoked Response Potential created by the P300 wave, which occurs when the user sees a "sought-after target and allows to make a binary decision about the user's intent" [4]. In the experiment the time each character was on screen was set to 400 ms and two subjects were tested: one expert, healthy subject and one non-expert subject with Locked-In Syndrome (LIS). The results showed that while the healthy user needed only 1.4 sequences on average to select the desired character, the subject with LIS needed 14.4 sequences per character, going over the set limit of 6 tries per character multiple times. The justification given by authors saying that it could have been caused by the lack of inexperience with the system seems reasonable as the paradigm is reflex based.

Hierarchical keyboard is an interface design similar to Hex-O-Spell. It also groups characters into clusters of 5 from which the user then selects the desired letter. However, this model also includes two other components: a control layer at the bottom and a prediction layer at the top, which employs a language model to deduce what the user is trying to type. Navigation is achieved by collecting the EEG data from thinking about arm movement and nodding, depending what direction the user wants to go. Experiment subjects using the hierarchical keyboard were able to reach 88% accuracy, but when compared to other solutions in the field it was described as convoluted. The author of the experiment remarked that a better guidance system is required to help users understand how the UI works [3].

Speller with adaptive queries is akin to the hierarchical keyboard design when selecting the first letter; characters are split into equal groups and the Steady State Visually Evoked Potentials are used to detect which group the user concentrates on. However, after the first selection language prediction model starts calculating probabilities to which the letter groups (called "range queries") adapt. The text speller is capable of 2925 range queries and 20475 character queries, a total of 23400 queries, which is a lot bigger than in other similar systems. A pool like this allows the interface "freedom to pick the right question" [14]. The experiment conducted by authors of [14] yielded performance of 11.93 characters/minute in multiple-word texts scenario, making it one of the fastest BCI Spellers.

User-made errors are one of the biggest contributors to reducing the average typing speed across all spellers. To reduce the time it takes to deal with these mistakes researchers introduced Error-related Potentials (ErrPs) [1], which generate a signal 50–100 ms after user detects they have made an error and use it to automatically detect and remove the mistyped character.

5   Brain Computer Interface Spellers for Chinese Text Input

Even though it is the most spoken language, written Chinese is one of the most complicated languages in the world. With over 7000 characters it is impossible to implement it in the same way English is implemented on Brain-Computer Interface Spellers. So far, two methods have emerged amongst researchers, with similar effectiveness.

Row/Column system presents a 6 × 6 matrix of Chinese character components and an error removal symbol. In the related study, different paradigms were studied to determine the one with highest accuracy.

Paradigms I and II were highlighting Single Characters (SC) foreground and background, respectively. Paradigms III and IV highlighted a row/column's foreground color and paradigms V, VI highlighted row/column's background color.

Although the experiment has been purely theoretical, the study did indicate that communication rate of 1.1 characters per minute is possible while also maintaining high accuracy [10].

First-Last or FLAST employs a similar technique to the Row/Column system, where the user selects components of a hanzi sinogram. However, in this method the user makes only 3 selections. The first component of the character, the last component and the desired sinogram. During the input mode an 8×8 matrix is displayed with the first 7 rows displaying sinogram components in stages 1 and 2, then entire sinograms. The last row contains controls, such as: delete component, delete sinogram, copy, paste.

FLAST users achieved a typing speed of 0.56 sinograms per minute with accuracy of 82.8% [9]. The author of the experiment remarks that selection speed drops significantly when selecting the entire sinogram (12.93 bits/minute when selecting components to 4.23 bits/minute selecting sinograms). The time taken looking for desired symbols and the pauses between selections further lowers the typing speed. The proposed solution is to reduce the amount of possible sinograms, thus making the matrix smaller and easier to navigate [9].

6   Hybrid Solutions for Text Entry System

A gaze-based text entry system combining SSVEP-based BCI Speller with an eye-tracking module in a VRHMD (Virtual Reality Head-Mounted Display) has been implemented in Virtual Reality on an HTC Vive platform. A 40-element virtual keyboard is displayed (26 letters, 10 numbers and 4 other symbols) on the HMD and a "Hybrid Data Fusion" method is employed to process and blend the signals from both modules [2].

The hybrid text entry system achieved spelling speed of 10 words/minute (1.1 character/second), which places it magnitudes above traditional BCI Spellers [2].

7   Security of BCI Spellers

The performance of Brain-Computer Interface text entry systems has always been a priority for researchers. However, in 2020 an experiment [15] raised valid concerns about the security of such systems. The issue is that small, unnoticeable adversarial perturbations can be added to the EEG signal, allowing the attacker to use the system to spell anything they want. The author remarks, that it could lead to critical errors such as misdiagnosis in a medical application using BCI.

The experiment confirms that an attack like this is possible, even though there are certain limitations that if solved could make the issue even more alarming. The attacker has to create an adversarial perturbation template, however that template could then be used on other models even with different architecture. The second limitation is that the attack must be synchronized with the EEG signal to maximize attack performance. While this may not seem critical for BCI Speller, since the worst-case scenario is misspelling, for certain people with disabilities this could be the only way of communication with the outside world and an attack like that would be effectively silencing them.

8 Note on BCI Text Readers

While this paper focuses on text entry systems it is worth mentioning a recent implementation of a Brain-Computer Interface within a Text Reader. The application contains a dictionary with problematic words that the user might not understand, and it can be accessed by focusing and double blinking when the cursor hovers over the desired word. Such a system provides a novel facilitation for physically disabled users on the other side of the spectrum, during reading instead of writing.

9 Conclusion

With the developments of both the sophistication of the backend with language prediction models and algorithms automating a lot of the work previously required to be done by the user the Brain-Computer Interface Text Spellers are looking promising. However, as mentioned in [1, 18] other than a much wider userbase potential - due to virtually no voluntary muscle movement required to operate it - the system faces many setbacks compared to other solutions (e.g. eye tracking, chin joystick, tongue controls): needing to set up electrodes at fixed positions, applying gel to electrodes and low performance in comparison. The BCI-based Spellers are also mostly difficult to use for inexperienced users with each of them requiring additional instructions before being able to use them and the need for continuous focus throughout usage as any unrelated thoughts could generate EEG signals that will be incorrectly interpreted.

Out of the presented designs, Hex-O-Spell [13] coupled with the BBCI and the speller using adaptive queries paradigm [14] hold the most promise with the former having more useful application due to its' compactness and the latter being a faster typing solution. Both utilize powerful language prediction models, and this area should be further researched by experimenters as it shows the most promise for achieving viable typing speeds. Outside of pure BCI-based Spellers, hybrid solution presented in [2] without a doubt shows superiority of utilizing multiple systems at once, yet almost no studies have been done in this regard.

Security of Brain-Computer Interfaces concerns have been marginalized so far, yet study [15] shows that it can be a real issue for future consumers and has to be taken seriously into consideration developing future iterations of the system.

References

1. Rezeika, A., Benda, M., Stawicki, P., Gembler, F., Saboor, A., Volosyak, I.: Brain-computer interface spellers: a review. Brain Sci. **8**, 57 (2018)
2. Ma, X., Yao, Z., Wang, Y., Pei, W., Chen, H.: Combining brain-computer interface and eye tracking for high-speed text entry in virtual reality. In: 23rd International Conference on Intelligent User Interfaces (IUI 2018), pp. 263–267. Association for Computing Machinery, New York (2018)
3. Hayet, I., Haq, T.F., Mahmud, H., Hasan, M.K.: Designing a hierarchical keyboard layout for brain computer interface based text entry. In: 2019 International Conference on Electrical, Computer and Communication Engineering (ECCE), Cox'sBazar, Bangladesh, pp. 1–6 (2019)

4. Kenneth, II., Orhan, U., Erdogmus, D., Roark, B., Oken, B., Purwar, S., Nezamfar, H., Fried-Oken, M.: An ERP-based brain-computer interface for text entry using rapid serial visual presentation and language modeling, pp. 38–43 (2011)
5. George, K., Iniguez, A., Donze, H.: Automated sensing, interpretation and conversion of facial and mental expressions into text acronyms using brain-computer interface technology. In: 2014 IEEE International Instrumentation and Measurement Technology Conference (I2MTC) Proceedings, Montevideo, pp. 1247–1250 (2014)
6. Barsim, K.S., Zheng, W., Yang, B.: Ensemble learning to EEG-based brain computer interfaces with applications on P300-spellers. In: 2018 IEEE International Conference on Systems, Man, and Cybernetics (SMC), Miyazaki, Japan, pp. 631–638 (2018)
7. Fernández-Rodríguez, Á., Velasco-Álvarez, F., Medina-Juliá, M.T., Ron-Angevin, R.: Evaluation of emotional and neutral pictures as flashing stimuli using a P300 brain-computer interface speller. J. Neural Eng. **16**, (2019)
8. Paszkiel, S., Dobrakowski, P., Lysiak, A.: The impact of different sounds on stress level in the context of EEG, cardiac measures and subjective stress level: a pilot study. Brain Sci. **10**(10), 728 (2020). https://doi.org/10.3390/brainsci10100728
9. Minett, J., Zheng, H.-Y., Fong, M., Zhou, L., Peng, G., Wang, W.: A Chinese text input brain–computer interface based on the P300 speller. Int. J. Hum. Comput. Interact. – IJHCI **28**, 472–483 (2012)
10. Minett, J., Peng, G., Zhou, L., Zheng, H.-Y., Wang, W.: An assistive communication brain-computer interface for Chinese text input. In: 4th International Conference on Bioinformatics and Biomedical Engineering, iCBBE 2010 (2010)
11. Huggins, J.E., Alcaide-Aguirre, R.E., Hill, K.: Effects of text generation on P300 brain-computer interface performance. Brain Comput. Inter. (Abingdon) **3**, 112–120 (2016)
12. Welton, T., Brown, D., Evett, L., Sherkat, N.: A brain–computer interface for the Dasher alternative text entry system. Univers. Access Inf. Soc. **15**, 77–83 (2016)
13. Blankertz, B., Krauledat, M., Dornhege, G., Williamson, J., Murray-Smith, R., Müller, K.-R.: A note on brain actuated spelling with the Berlin brain-computer interface (2007)
14. Akce, A., Norton, J., Bretl, T.: An SSVEP-based brain-computer interface for text spelling with adaptive queries that maximize information gain rates. IEEE Trans. Neural Syst. Rehabil. Eng. **23**, 857–866 (2014)
15. Zhang, X., Wu, D., Ding, L., Luo, H., Lin, C., Jung, T.-P., Chavarriaga, R.: Tiny noise, big mistakes: adversarial perturbations induce errors in brain-computer interface spellers. Natl. Sci. Rev. (2020)
16. Paszkiel S.: Data acquisition methods for human brain activity, analysis and classification of EEG signals for brain-computer interfaces. In: Studies in Computational Intelligence, vol. 852, pp. 3–9 (2020). https://doi.org/10.1007/978-3-030-30581-9_2
17. Paszkiel, S., Szpulak, P.: Methods of acquisition, archiving and biomedical data analysis of brain functioning, biomedical engineering and neuroscience. In: Hunek, W.P., Paszkiel, S. (eds.) Advances in Intelligent Systems and Computing, vol. 720, pp. 158–171 (2018). https://doi.org/10.1007/978-3-319-75025-5_15
18. Paszkiel, S.: The use of facial expressions identified from the level of the EEG signal for controlling a mobile vehicle based on a state machine. In: Szewczyk, R., Zielinski, C., Kaliczynska, M. (eds.) Automation 2020: Towards Industry of the Future. Advances in Intelligent Systems and Computing, vol. 1140, pp: 227–238 (2020). https://doi.org/10.1007/978-3-030-40971-5_21

Selected Aspects of the New Recommendation on Subjective Methods of Assessing Video Quality in Recognition Tasks

Mikołaj Leszczuk(✉) and Lucjan Janowski

AGH University of Science and Technology, 30059 Kraków, Poland
vq@kt.agh.edu.pl
http://vq.kt.agh.edu.pl

Abstract. It was once thought that high QoS (Quality of Service) performance solves recurrent problems of low-quality multimedia services. Since then, solutions have been proposed to ensure a high level of QoE (Quality of Experience). In this document, the authors attempt to outline his understanding of an accurate meaning of quality of multimedia services. Starting from QoS and passing through generalised QoE, the authors focus on aspects of subjective and objective quality modelling and optimisation of visual performance for TRV (Target Recognition Video) applications (such as video surveillance), outlining the path of ITU-T standardisation in this area. The authors revised the ITU-T Recommendation P.912 to reflect improved subjective test techniques developed since this Recommendation was approved. The authors also attempt to predict at least some existing errors of reasoning, which are likely to become evident for the industry in the next decade.

Keywords: QoS · QoE · ITU-T · TRV · P.912 · CCTV

1 Introduction

Decades ago, the telecommunications industry believed that high-performance Quality of Service (QoS) techniques resolve any recurrent problems of low-quality multimedia services. For many applications, buffering multimedia data streams can alleviate significant delays and jitter. Since discovering that QoS is not a useful metric of network quality, most proposals suggest that quality should be measured on the user level. These structures are increasingly being filled with solutions that attempt to model the overall quality, operating at the intersection of QoS and Quality of Experience (QoE) [11] or only in QoE. However, it has become apparent that such a general approach does not work for many visual applications such as target recognition applications [8,13].

2 Target Recognition Video

In many visual applications, the motion picture's quality is not as important as the ability of the visual system to perform specific tasks for which it is created, given the processed video sequences. These tests' basic premise is to find Target Recognition Video (TRV) quality limits for which the task can be performed with the desired probability or accuracy. Given the use of TRV, qualitative tests do not focus on the subject's satisfaction with the video sequence's quality, but instead, they measure how the subject uses TRV to accomplish specific tasks.

3 Methods for Subjective Evaluation of TRV

Recommendation ITU-T P.912 [2] "Subjective Video Quality Assessment Methods for Recognition Tasks" addresses questions formulated in the previous section. Recommendation P.912 organises terminology related to subjective TRV testing, introducing appropriate definitions for testing methods. The initial version of Recommendation P.912 was only the first step in standardising subjective TRV testing methods. In the authors' opinion, based on research results and observations conducted during numerous experiments with TRV, many claims of P.912. In this situation, the authors have taken steps to introduce significant modifications to P.912.

4 Source Signal

In Clause 5, P.912 (08/08) stated that test sequences should follow the general principles stated in [5]. Unlike other subjective assessment methods developed for quality evaluations, this Method is directed at the usefulness of the video material to complete a task and not the video's quality. Another example concerns research on the impact of CCTV recordings on the accuracy of licence plate recognition [10]. For this study, a unique video database was created [7]. The recordings have been created using fixed CCTV cameras, recording cars entering the car park at the AGH University of Science and Technology in Kraków, Małopolska, Poland. Again, it is clear that due to the conditions mentioned above of acquisition, recordings represent a particular CCTV camera, its specific location and direction, a specific distance from the object, and specific lighting conditions.

What is more, since the recordings were made in Kraków, most of the licence plates have the letter "K" in the first position on the plate and "R" in the second position. As shown, contrary to P.912 (08/08), it was tough to ensure complete coverage of the potential applications of the recordings. Any expansion of the record database was laborious, time-consuming, or even impossible.

The authors revised Clause 5 of P.912 with the following amendments: test sequences should follow the general principles stated in [5] and [3], which specify that scenes should be consistent with the transmission service under test, and should span the full range of spatial and temporal information. For example, the results should not be generalised. Unlike other subjective assessment methods developed for quality evaluations, this Method is directed at the usefulness of the video material to complete a task and not the video's quality.

5   Testing Methods and Experimental Design

To assess the quality level of TRV, methods that reduce subjective factors and measure a participant's ability to perform a task are useful in that they avoid ambiguity and personal preference.

In Clause 6, P.912 stated that the application of TRV is directly related to the user's ability to recognise targets at increasing levels of detail. When determining the DC for particular scenarios, it needs to be considered that, for a set distance from the camera to the object of interest, the DC directly correlates to decreasing video resolution of the target. Fewer CPD of the resolution also means that the object subtends less of the video's information content, making identification of the target more difficult.

Consequently, the authors revised P.912 with the following contribution: CPD, the critical parameter, is affected by the resolution of the object and the distance between the camera and the object [9]. Consequently, it relates to achievable DC. Experimental methods should consist of responding to questions related to the content in the image or video.

5.1   Multiple Choice Method

In Clause 6.1, P.912 (08/08) stated that "Unsure" might be one of the listed choices. This problem has been observed when applying a Comparison Category Rating, as defined in Recommendation ITU-T P.800 [4], in which subjects tend to abuse the response "0" ("About the Same").

The authors revised that entry in P.912 (08/08) amending it as follows: **The use of** "Unsure" **as** one of the listed choices **is discouraged but allowed.**

5.2   Single Answer Method

In Clause 6.2, P.912 (08/08) stated that if there is a non-ambiguous answer to an identification question, the single answer method may be used. This Method is appropriate for alphanumeric character recognition scenarios. This is because even in the event of a plate being recognised incorrectly, by correlating it with a vehicle database containing the vehicle's make and colour, we substantially reduce the risk of the vehicle being identified incorrectly.

The authors revised the single choice method description, expanding it as follows: if there is a non-ambiguous answer to an identification question, the single answer method may be used. **Alternatively, fuzzy logic may be used (e.g. Hamming distance or Levenshtein distance), as seen in** [10].

6   Evaluation Procedures

In Clause 7, a laboratory test is described. Description of a crowdsourcing environment is described in Appendix I (see Sect. 8).

6.1   Subjects

In Clause 7.3, P.912 (08/08) stated that The number of subjects should follow the recommendations of [5]. In the first experiment, the subjects were experts – law enforcement officers [14,15]. When the experiment was repeated with non-experts, very similar results were obtained, as long as the non-experts were compensated for their time [12].

The authors introduced an entry which allows the use of non-expert subjects providing they are correctly motivated. Subjects who are experts in the application field of the TRV should be used. **For certain areas of application testing, where neither specific experience nor expertise is required, non-expert subjects may also be used. In [12] shows the validity of this approach.** The number of subjects should follow the recommendations of [5].

6.2   Instructions to Subjects and Training Session

In Clause 7.4, P.912 (08/08) stated that the subject should be given the context of the task before the video clip is played, and told what they are looking for or trying to accomplish. If questions are to be answered about the video's content, the questions should be posted before the video is shown so that the viewer knows what the task is.

However, the topic was not exhausted. Therefore, the authors extended this clause, adding it means the instructions must clearly state what subjects must-do. In this case, the running of pretest on a small group before running any more massive experiment is strongly recommended. A typical number of subjects for a pretest is approximately 20% of the total. A pretest group can consist of a single person. Feedback from the pretest is used to improve the experiment before running it with actual subjects.

7   Statistical Analysis and Reporting of Results

The first step of the analysis is subject screening to eliminate those who did not pay attention or who did not understand the task. Further statistical analyses vary slightly depending on the scoring method.

7.1   Subject Screening

P.912 did not contain information about subject screening at all. Therefore, the authors added a new subsection stating that in order to detect abnormal subjects, it is not enough to compare the results obtained by one subject to the average obtained in the experiment, since in a typical experiment different subjects perform different tasks. An algorithm for solving the problem of different task difficulty performed by different subjects is proposed in [6]. The algorithm proposed assumes that tasks can be partially ordered. The Processed Video Sequences obtained for the same source, and lower bit-rate are likely to have less information, and likely the detection is not easier than for a higher bit-rate.

7.2   Further Statistical Analysis

The statistical analysis for each Method varies slightly. For all conditions, a correlation and understanding of the number of CPD or area subtended of the target are considered to determine the correlation between success and CPD. For cases where there are multiple answers, a statistical validity indicator is required.

Multiple Choice. P.912 stated only that for multiple-choice answers, the probability of an incorrect answer needs to be balanced against the ability to answer the questions correctly. Therefore, the authors extended this clause, adding information about recognition probability as a function and comparing different conditions.

Single Answer. P.912 (08/08) stated only that for single answer conditions, where answers are either correct or incorrect, a statistical metric to determine whether the subject performs above the random chance of answering correctly should be implemented. "Unsure" answers should be pooled with incorrect answers. However, the topic was not exhausted. Therefore, the authors extended this clause, adding that the correctness of the answer can be analysed on a different scale for a single answer. The correctness threshold can be different depending on the specific analysis. Since the final results are of the 0–1 type, the results obtained are similar to those for the multiple-choice case and the same analytical tools must be used. It is not easy to describe all the options since the answer can differ depending on the answer type.

Timed Task. P.912 (08/08) stated that for the correctness factor, the same statistical analysis as for single answer conditions is also applied. However, the topic was not exhausted. Therefore, the authors extended this clause, adding that the statistical analysis must incorporate time as an explanatory variable for timed tasks. Time can be a numerical value "how long it took to finish the task, in seconds" or it could be "several replays of the movie before a decision was made." The analysis must indicate the influence of time on the result obtained.

8   Crowd-Sourcing Environment

P.912 did not contain information about the crowdsourcing environment at all. This appendix does not form an integral part of this Recommendation.

8.1   Introduction

One of the main problems of recognition tasks is the apparent limitation of source sequences reuse described in Clause 6.5. Nevertheless, such a solution has an obvious drawback: it requires a much larger number of subjects. A natural solution is crowdsourcing, which gives access to thousands of potential subjects at the same time.

8.2   Definitions

NOTE – 8.2 follows terminology presented in [1].

- **Crowd-sourcing**: obtaining the needed service by a large group of people, most probably an on-line community.
- **Test**: subjective assessments in a crowd-sourcing environment.
- **Worker**: person participating in a crowdsourcing test.
- **task**: set of actions that a worker needs to perform to complete a subscribed part of the test.
- **Question**: a single event that requires an answer for a worker.
- **Campaign**: a group of similar tasks; it also contains a more detailed description of the part of the test under investigation, like workers' payment, and indicates subjective assessments in a crowdsourcing environment.

8.3   Software

In order to be able to run a crowdsourced test, a worker has to have access to the test environment. Of course, other solutions, like software code or an application, can also be used, but the number of workers willing to install additional software, compared to those willing to access a specific web page, is much smaller.

8.4   Designing a Task

Consultation with non-native speakers to ask their opinion is a good idea, since it is probable that some workers do not speak English well. For the same reason, use simple English in all descriptions, questions and messages presented to workers.

8.5　Distribution of the Campaign

After creating the test platform, distribute it among subjects. There are two main ways to advertise a specific campaign.

1 Using social media and mailing lists
   a Advantages: (i) it is possible to get to a specific group, e.g., policemen; (ii) quite often it does not include additional costs; (iii) workers willing to make a task for free are most of the time honest.
   b Disadvantages: (i) the mailing list or social media generate(s) a very specific (probably biased) group of workers; (ii) since no payment is made for the task, a large number of tests will not be completed, unless the test is concise or involves gamification; (iii) the speed of collecting the data is, most of the time, very rapid just after announcement, but falls away rapidly, meaning that the web server can be overloaded; (iv) it is difficult to predict how many answers will be collected; (v) checking whether an individual ran the task once only is problematic.
2 Using specific services (called crowdsource platforms) gathering people willing to make micro tasks
   a Advantages: (i) the speed of collecting the data can be adjusted; (ii) the task is advertised constantly by the service; (iii) a large number of data can be collected in a short time.
   b Disadvantages: (i) some workers will use the test to get money, and their answers are random; (ii) every answer, even those given by workers answering randomly, costs some money; (iii) workers are pooled from a specific group of people willing to make money by doing microtasks.

8.6　Data Analysis

Even with careful subject validation, assume that subjects are different. Since a diverse subgroup of subjects validates each sequence, the difference in recognition probability can be characterised only by a subgroup of subjects, not by a difference in conditions. Nevertheless, results show a high correlation between results obtained in a laboratory environment and those from crowdsourcing.

9　Conclusions

The discussion of statements in ITU-T Recommendation P. 912 showed that some of the findings and observations required the verification of specific provisions of the Recommendation. The authors revised Recommendation P. 912 to reflect improved subjective test techniques developed since this Recommendation was approved.

References

1. Hoßfeld, T., Hirth, M., Redi, J., Mazza, F., Korshunov, P., Naderi, B., Seufert, M., Gardlo, B., Egger, S., Keimel, C.: Best practices and recommendations for crowdsourced QoE - lessons learned from the qualinet task force "Crowdsourcing", October 2014. https://hal.archives-ouvertes.fr/hal-01078761, lessons learned from the Qualinet Task Force "Crowdsourcing" COST Action IC1003 European Network on Quality of Experience in Multimedia Systems and Services (QUALINET)

2. ITU-T: ITU-T P.912, Subjective video quality assessment methods for recognition tasks. https://www.itu.int/rec/T-REC-P.912

3. ITU-T: ANSI T1.801.01, Digital Transport of Video Teleconferencing/Video Telephony Signals - Video Test Scenes for Subjective and Objective Performance Assessment (1995). http://engineers.ihs.com/document/abstract/POWPBBAAAAAAAAAA

4. ITU-T: ITU-T P.800, Methods for subjective determination of transmission quality (1996). http://www.itu.int/rec/T-REC-P.800-199608-I

5. ITU-T: ITU-T P.910, Subjective video quality assessment methods for multimedia applications (1999). http://www.itu.int/rec/T-REC-P.910-200804-I

6. Janowski, L.: Task-based subject validation: reliability metrics. In: 2012 Fourth International Workshop on Quality of Multimedia Experience, pp. 182–187, July 2012. https://doi.org/10.1109/QoMEX.2012.6263863

7. Leszczuk, M., Janowski, L.: Database for video quality assessment in license plate recognition. In: Signal Processing: Algorithms, Architectures, Arrangements, and Applications (SPA), vol. 2013, pp. 51–55, September 2013

8. Leszczuk, M., Stange, I., Ford, C.: Determining image quality requirements for recognition tasks in generalized public safety video applications: definitions, testing, standardization, and current trends. In: 2011 IEEE International Symposium on Broadband Multimedia Systems and Broadcasting (BMSB), pp. 1–5, June 2011. https://doi.org/10.1109/BMSB.2011.5954938

9. Leszczuk, M., Janowski, L., Romaniak, P., Głowacz, A., Mirek, R.: Quality assessment for a licence plate recognition task based on a video streamed in limited networking conditions. In: Dziech, A., Czyżewski, A. (eds.) Multimedia Communications, Services and Security, pp. 10–18. Springer, Berlin Heidelberg (2011)

10. Leszczuk, M.: Optimising task-based video quality. Multimedia Tools Appl. 68(1), 41–58 (2014). https://doi.org/10.1007/s11042-012-1161-6, http://dx.doi.org/10.1007/s11042-012-1161-6

11. Leszczuk, M., Janowski, L., Romaniak, P., Papir, Z.: Assessing quality of experience for high definition video streaming under diverse packet loss patterns. Signal Process. Image Commun. 28(8), 903–916 (2013). http://www.sciencedirect.com/science/article/pii/S0923596512001804

12. Leszczuk, M., Kon, A., Dumke, J., Janowski, L.: Redefining ITU-T p.912 recommendation requirements for subjects of quality assessments in recognition tasks. In: Dziech, A., Czyzewski, A. (eds.) Multimedia Communications, Services and Security, Communications in Computer and Information Science, vol. 287, pp. 188–199. Springer Berlin Heidelberg (2012). https://doi.org/10.1007/978-3-642-30721-8_19, http://dx.doi.org/10.1007/978-3-642-30721-8_19

13. Möller, S., Raake, A. (eds.): Quality of Experience: Advanced Concepts, Applications and Methods. Springer, Cham (2014). https://doi.org/10.1007/978-3-319-02681-7

14. VQiPS: Video quality tests for object recognition applications. Public Safety Communications DHS-TR-PSC-10-09, U.S. Department of Homeland Security's Office for Interoperability and Compatibility, June 2010
15. VQiPS: Recorded-video quality tests for object recognition tasks. Public Safety Communications DHS-TR-PSC-11-01, U.S. Department of Homeland Security's Office for Interoperability and Compatibility, June 2011

Object Description Based on Local Features Repeatability

Michal Tomaszewski, Pawel Michalski, and Jakub Osuchowski[✉]

Institute of Computer Science, Faculty of Electrical Engineering, Automatic Control
and Informatics, Opole University of Technology,
Prószkowska 76, 45-758 Opole, Poland
{m.tomaszewski,p.michalski}@po.opole.pl,
j.osuchowski@doktorant.po.edu.pl
https://www.po.opole.pl/

Abstract. This chapter describes a methodology for selecting characteristic local features, to facilitate the detection of specified objects on
power supply lines in diversified scenes. The objective of this works was
to automate a process intended to search for characteristic objects in
digital images, utilizing a limited training set. Description of the insulator in the test was based on a set of distinctive local features, which in
the individual images of the test set, demonstrated the highest degree of
repeatability. Such set allowed the location of insulators in other samples
obtained from a diagnostic flight. To verify the method, the verification
sets were proposed. To search and describe local features the SURF algorithm was used. The elaborated training set and adopted methodology of
the test provided promising results within the analyzed scenario (flight
along the power supply line). Analyses that were carried out, aimed at
the development of methods supporting the power network maintenance,
basing on the advanced techniques of digital image processing.

Keywords: Distinctive local feature · Insulator · Local feature
repeatability · Bag of features · SURF algorithm

1 Introduction

Computer Vision is used for many purposes and in a variety of applications,
such as object detection in images, regardless of the composed scene. Detection
can also be used during the inspection of technical objects in processes that
support their operation. The authors are involved in addressing issues with power
network maintenance and supporting the use of advanced techniques in digital
image processing.

Overhead lines insulators fulfill two primary functions; insulates conductors
from the ground and the tower structure and provides mechanical support. It
is estimated that about 50% of line maintenance costs are connected with the
replacement, repair, and diagnostics of insulators. Insulator failure is the cause

© The Author(s), under exclusive license to Springer Nature Switzerland AG 2021
S. Paszkiel (Ed.): ICBCI 2021, AISC 1362, pp. 255–267, 2021.
https://doi.org/10.1007/978-3-030-72254-8_28

of approximately 70% of line operation standstills [2, 22]. These statistics demonstrate the importance of diagnosing and detecting insulator failures early.

Powerline inspections are performed in two ways. The first one is sending diagnostic groups on foot to assess the line condition and collect photographic documentation. This approach is highly time-consuming due to the size of the area being diagnosed. The second approach uses an aircraft, such as a helicopter, light aircraft, or drone. In this case, using the installed detectors video material is obtained, documenting the overhead powerline and its important elements. Diagnostic assessment of individual elements is completed manually during data postprocessing. The volume of collected material is substantial; however, only a small portion of this material is valuable from the diagnostics point of view. The duty of the operator is to assess the degree of corrosion in the supporting structures by first manually identifying such structures in the entire set – a time-consuming process.

Similar problems related to insulators, bridges, and other elements are included in the assessment. Thus the research objectives of these authors to automatize the searching process; especially as it concerns characterizing elements on digital images when utilizing limited training sets. Such an approach is justified in this case due to the specificity of the aerial data acquisition. The parameters of flight distance within the line surroundings and the size of individual elements on the digital image remain on a similar level. In addition, parameters such as lens optics setting, defects of an image, and the number of frames obtained per second are subject to small changes during the flight, which positively influences the process of detection of the interesting elements. Images selected this way can be assessed quickly by the operator, as they can select only those images of insulators that require replacement or renovation. An additional benefit of this approach is that samples, indicated by the operator, requiring intervention could be used for building diagnostic standards and support network operators on whether an element requires or does not require replacement.

2   Materials and Methods

Local features describe the characteristic of a digital image and assume the form of points, regions, edges or straight lines [21] because in those places characteristic signal changes occur most often. Those signal changes can be detected using dedicated detectors of local features and then described by descriptors, taking into consideration neighborhood of the point. For each image several points can be found and defined by descriptors, obtaining a set of vectors that represent the contents of the presented scene. Specified local features can be localized in different images. This repeatability of occurrence can be used within the range of the computer vision for many objectives, such as classification, localization or detection of objects in digital images.

The number of available algorithms responsible for the detection stages of key features and the building of descriptors is large. The algorithms most frequently used to detect key features are MSER, HESSIAN AFFINE, FAST, GFTT, and KAZE. The algorithms responsible solely for the creation of a descriptor, which also features a significant group, includes BRIEF, FREAK, RIFF, BinBoost, and LATCH. The last group of algorithms is those that combine the two functionalities; which consists of the detection of key points and creation of the descriptor. Algorithms that fall into this group are SIFT, SURF, BRISK, and ORB [13,18]. In this paper the SURF algorithm was used, which is a modification of the SIFT (Scale Invariant Feature Transform) algorithm and recommended by D. Lowe [15]. Operation of the SIFT algorithm can be divided into several key stages. The first stage is the creation of the scale space of images, using the gauss broadening in different scales. To detect differences in particular images the pyramid DoG (Difference of Gaussian) method is used, which is an acceptable approximation of the LoG (Laplacian of Gaussian) function. The next stage consists of searching for local extrema in the multi-scale space and selecting key points. Then, each of the points orientation is assigned on the basis of scale and value of surrounding gradients, in respective directions. For each of the obtained points, a descriptor of its surrounding is created in the dimensions 16×16. This region is further divided into 16 smaller elements with 4×4 dimensions, and for each of those dimensions, an 8-bin histogram is created. This results in a 128-element vector that describes one key point. The algorithm that solves the problem of the SIFT algorithm speed is the SURF (Speeded up robust features) algorithm [3]. The main difference between the operation of both algorithms consists in replacing the scale space stage by DoG in the SIFT algorithm, to construct a box filter relying on integral images which is much faster. Due to the above, SURF is less resistant and less accurate, an estimated 10% loss, but is three times faster [4,9,24].

In the initial works [11,12], attempts were made to use descriptors of local features for the texture classification problem. In the work of Matas et al. [17] the method for the description of IPSS (Invariant Pixel Set Signature) objects was presented. The objects are represented with a probability density on the space of invariants local features inside convex hulls of n-tuples of interest points. Description of the object was performed in the following steps: detection of local features (using a Harris corner detector) for several different model images of the object, generation of convex sets, and computation of invariants. In many subsequent works [1,7,8,10,14,16,23,25,26], object identification methods were developed, using local features that treated an image as a set of patches, which are the descriptors of transmitted information. The collected sets of descriptor vectors were statistically analyzed to quantify the images. Patches can be chosen in one or several scales, densely or at random, according to the selected criterion.

The critical stage of the described methods development was presented by Csurka et al. [5], - the currently popular "bag of keypoints" method. The method was applied during the visual categorization of images. A "bag of keypoints" corresponds to a histogram representing the number of occurrences of particular local features in a given image. Main steps of the presented method include:

- Detection and description of a set of key features for the images from the training set.
- Assignment of the local features descriptors to a set of the predefined classes (creation of words).
- The building of a "bag of keypoints," which counts the number of local features assigned to each cluster.
- Application of a classifier that compares the set of local features (conventional "word") of an image from the test set with the "bag of keypoints," and on this basis, assigning the image to a specified category....

The method was tested on our dataset and provided promising results. Research into the methods of local feature identification and object detection in images is ongoing, to allow for the identification and detection of a class of objects from the images natural scenes as in the article of Dork and Schmid [6]. The first step of the method consists of grouping the local scale-invariable descriptors for characterization of an object class appearance. Then the class classifiers are trained together with the isolation and selection of the local features, specifying the most discriminating elements. The training was conducted without the need for marking of the image elements or separation of the objects from the background. The method was also tested for highly complex backgrounds and partly covered objects.

Researchers often discuss the problems in acquiring and determining an appropriately sized training set. The current deep learning techniques developed here were possible due to a large dataset and the ability to process the data quickly. Nonlinear algorithms are typically more flexible and non-parametric (they can calculate how many parameters are required for modeling of a problem, apart from the value of those parameters), they are also highly diversified meaning that the forecasts differ depending on the concrete data used to train them; size of the training data sample is very important. Two factors influence sample size, the complexity of a problem, which best binds the input data with the variable output, and complexity of the learning algorithm used for inducing the unknown basic mapping function, on the basis of concrete examples. As described in the first part of this chapter, the image data obtained in the process of acquisition are to a large extent homogeneous, therefore at the stage of post-processing, they can be segmented into the training, testing, and validation data. After localizing essential elements in the training samples, on their basis, we can train classifier by modifying its parameters based on results obtained from the test data. The conducted research was completed to obtain the smallest possible set of local features, that describes an insulator based on the training set.

Such a set will enable the location of insulators in other samples obtained from a diagnostic flight. Film sequences, for this research, were made using the BSL with the high-resolution camera installed. The flight was performed along the line, maintaining a distance from the line on the level of ca. 10 m. The images obtained in the sequences were similar to those from the diagnostic flight and captured the supporting structures, conductors, and insulators.

Set of data on the basis of different variants of key features set were built, was called the Dataset D1 (training set) and contained single frames from the sequences obtained. For each of the frames, ROI (Region of Interest) was assigned manually with the location of the insulator in the scene. Frame examples included in the Dataset D1 are presented in Fig. 1a.

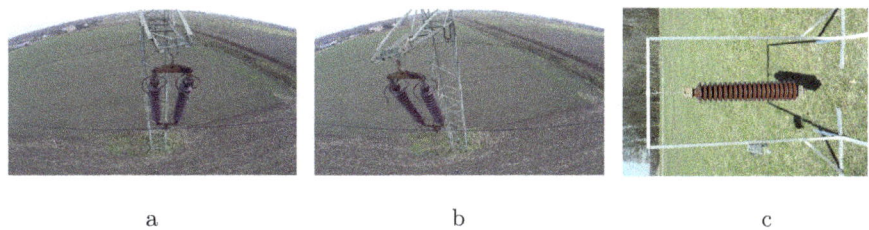

a   b   c

Fig. 1. Example samples from: a = Dataset D1 (training dataset), b = Dataset D2 (test dataset), c = Dataset D3 (test dataset).

To verify the method two sets of test data were elaborated. Pictures included in the D2 set also include frames from the same flights as the D1 set but in different shots. Such an approach to assess whether the algorithm used can localize the same object when varying scale and perspective. Picture examples included in the D2 set are presented in Fig. 1b.

In the last set, named D3, the authors collected highly diversified digital images of ceramic insulators of different types. The authors succeeded in assembling a collection of over 100 photographs across a range of resolutions, illumination, and background elements that show the insulators. In the D3 set no frames were included during the flights and used to build a set of repeatable features. Digital picture examples included in the D3 set are presented in Fig. 1c.

The purpose of the experiment was to check the feasibility of building a set of repeatable local features describing the object in the digital images. In the test, the object for which the set of repeatable features was developed is the long-rod ceramic power insulator. One of the test stages attempted to select features that best described the object in different scenes and illumination conditions. The Ratio parameter, which defines a key point's range of filtering and repeatability, which indicates number of repetitions of a specific feature in the remaining scenes. The purpose of the test was to assess and select the optimal values for both parameters.

The next aim of the research attempted to create the set of features that describe the object based on the reduced training set. Both the detector and the descriptor of key points in the test used the SURF (Speeded Up Robust Features) algorithm. The SURF algorithm was chosen for its superior performance in retrieving characteristic invariable features from images. It is invariable in the case of affine transformations and changes in light intensity. Moreover, thanks to the use of the Fast-Hessian detector, SURF is much faster in comparison to algorithms that use DoG (e.g., SIFT).

Key parameters from the conducted research are the Local Point Repeatability (LPR) and Ratio parameters. LPR is a feature commonly used as the metrics of assessment for feature point detectors [20] operation efficiency. This property carries information on the number of repetitions of a chosen feature in all images of the training dataset, imaging the power insulator in different scenes at the variable illumination conditions. In the current test, the LPR property was used not as a measure for the detector assessment, but as a measure of feature suitability when building the set of the key features to describe the analyzed object. Detecting distinctive features in test samples allows to state that the scene contains a tested object with high probability In this work, LPR is expressed as a percentage shared matchings of a specified feature in the images from the training dataset.

The Ratio parameter is used for limiting of points matching as redundancy. It assumes the value within the range 0–1 (describe precisely our repeatability and provide the range of changes of this parameter).

The simplified test algorithm is presented in Fig. 2.

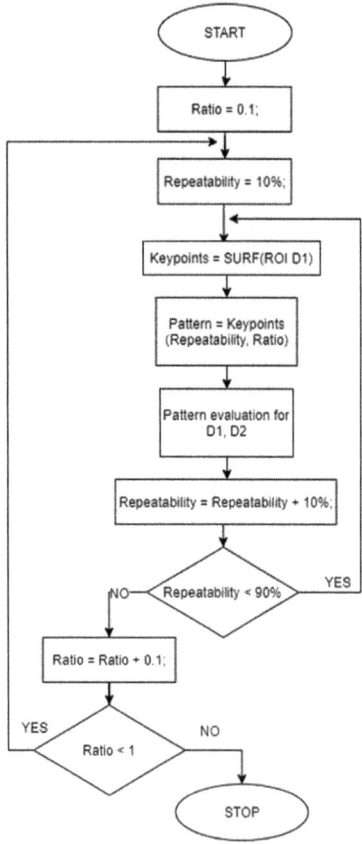

Fig. 2. Diagram of the test variants creation process for the repeatable local features set at the change of the ratio and repeatability parameters.

The set of distinctive features is built in the following four steps:

1. For all images in the test dataset key features were for searched and vectorized, using the SURF algorithm to accomplish both stages.
2. Each detected key feature was fit to all features from the collected images.
3. Features that were correctly matched (depending on the established Ratio level) were added to the temporary set of features.
4. In the last step, the set is filtered and only distinctive features remained, with a repeatability greater than or equal to the assumed LPR level.

The above process is repeated for different values of Ratio (0.1 to 1) and Repeatability (10% to 90%).

A set of repeatable features was created on the basis of Region of Interest (ROI) and determined the position of the insulators in the D1 dataset. Ratio parameter was the primary influence on the size of the created set; its task was to filter out a portion of the key features found at the stage of detection in scenes. The number of features obtained for the test set at changing of the ratio parameter is presented in Fig. 2 and has characteristics similar to a high-pass filter. Consequently, the Ratio parameter, which is set at the level of 0.1–0.5 resulted in too many high number points. When the Ratio parameter is between 0.7 and 1, numerous key points were generated, which could become an obstacle in regards to the processing time for the method to be effective.

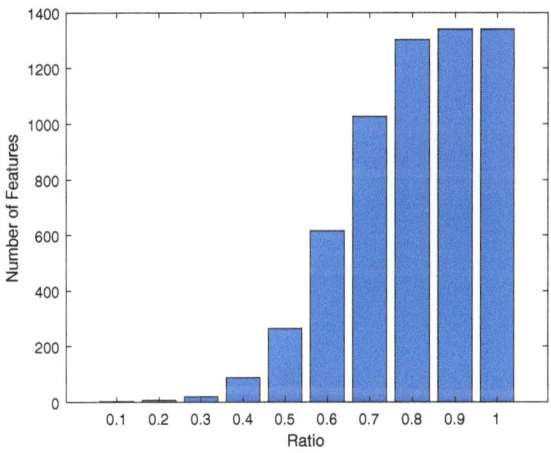

Fig. 3. The number of matching features in ROI for the D1 dataset for the change the ratio parameter within the range of 0.1–1 (X - ratio Y - the number of matching features)

To limit the calculation complexity of the feature matching process the size of the repeatable features set should also be limited. To those features whose repeatability exceeds a certain threshold. As presented in Fig. 3, increasing the

Ratio parameter also increases the number of bag features, which in turn influences the repeatability parameter of individual features. During the analysis, a ratio parameter of 0.6 was adopted as the level of compromise between the size of the distinctive features set and repeatability of the specified features in the set.

3   Results

For the test dataset, created on the basis of ROI, the insulators specified in the dataset D1 for the 0.6 ratio parameter, the following repeatable feature sets were obtained for the specified repeatability: 615 features for the ratio equal to 10%, 157 features for the ratio equal to 20%, 34 features for the ratio equal to 30%, 4 features for the ratio equal to 40%, 1 feature for the ratio equal to 50% and zero features for the other cases.

As expected, increasing the cut-off level (specified level of LPR parameter) decreased the number of repeating features in the obtained set. The purpose of this research was to extract the appropriate number of features that describe a power line insulator. Figure 4 presents the average number of repeatable features matching in all D1 and D2 dataset scenes.

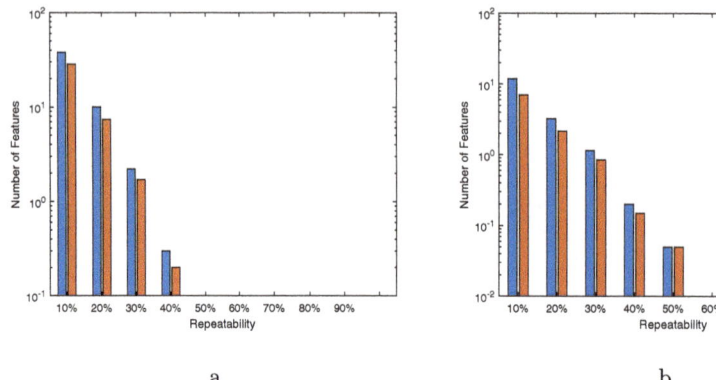

a   b

Fig. 4. The average number of repeatable features matching in all D1 and D2 scenes (blue) to the features matching on the insulator (red): a) for A = dataset D1, b) for B = dataset D2 (x - Repeatability, y - Average number of matching repeatable features)

The graph (Fig. 4) indicates that only a small number of repeatable features in set D1 are on the insulators; therefore, to limit the calculation complexity of the whole process, we should limit the size of the set, matching the LPR parameter respectively.

Figure 5 presents the results of matching the distinctive features, extracted in test sets D2 and D3 when changing the LPR parameter. The blue bar presents the number of images, in which at least one feature was matched, while the other bars present, respectively, the number of images matching at least two features were found (red bar) and the number of images matching at least three features were found (orange bar).

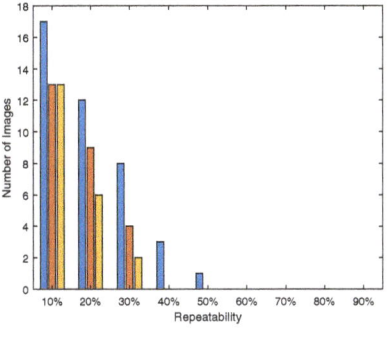 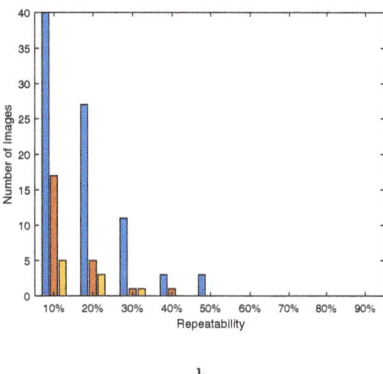

a   b

Fig. 5. Number of images with: a) dataset D2, b) dataset D3, in which respectively at least 1,2, or 3 features were matched depending on the value of LPR parameter (X - LPR, Y1 - Number of pictures, on which 1 feature on the insulator was matched, Y2 - number of images on which 2 features on the insulator were matched, Y3 - Number of images, on which 3 features on the insulator were matched)

As the LPR parameter is increased, the number of matching features decreases for each of the analyzed cases. The analyses aimed to detect a specified object and appropriately match it as many key features as possible in the scene. As we can see, matching of one feature was possible within repeatability range of 10 to 50%, but this substantially restricts the possibility to use different techniques for detection of the objects.

Analysis of the data represented in Fig. 5 suggested that the sets of features used for object detection in a scene should be as large as possible to obtain a high level of matching features. However, increasing the size of the distinctive features sets leads to an increased detection time for those features in the test scenes.

Figure 6 presents the dependence of the detection time analysis for scenes from datasets D2 and D3 on the size of the distinctive features set. Scene analysis time includes reading the scene, detecting features in the scene, using the SURF algorithm, and matching of individual features in walls and additional calculations, establishing the positions of the local features found in relation to ROI.

The test was carried out using the MathWorks MATLAB environment, and a computer equipped with an Intel Core i7-3517U processor and an 8 GB RAM. Figure 6 shows that increasing the size of the distinctive features set significantly influences the time needed to analyze the scenes and increases the calculation cost of the matching process.

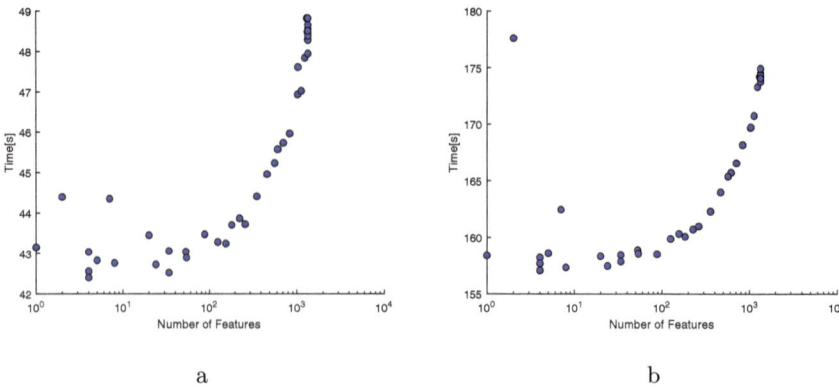

a   b

Fig. 6. Comparison of processing times of: a) dataset D2 and b) dataset D3 for different sizes of sets for selected local features (x - size of the set of selected local features, y - processing time [s])

Feature set, obtained on the basis of the training dataset was verified using dataset D3; results of the analysis are presented in Figs. 7 and 8.

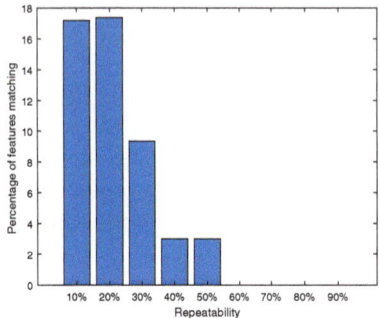

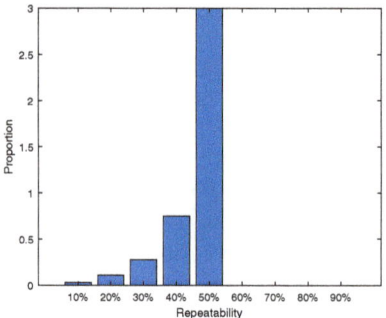

Fig. 7. The proportion of features matching on the insulator to all features matching in the entire scene for different values of LPR (dataset D3) (x - repeatability, y - the percentage of features matching on the insulator to all features matching in the image)

Fig. 8. Proportion of a number of features matching on the insulator to all features matching in the entire scene in relation to the size of repeatable features set for different values of LPR (dataset D3) (x - repeatability, y - percentage of features matching on the insulator to all features matching in the image in relation to the size of the repeatable features set)

The purpose of the test was to answer the question, what percentage of the matching features was included in the insulator ROI and what percentage was outside. Figure 8 presents the statistics in relation to the size of the distinctive features set.

The analyses, shown in Figs. 7 and 8, concluded that decreasing the LPR parameter impacts the growth of the distinctive features set but does not translate linearly into an increasing proportion of features matching on the insulator to all features matching in the dataset D3.

To verify the results of this research a visualization of matching was performed for selected distinctive features that describe the testing object. Figure 9 presents example features contained in the composition of the distinctive features set and the features matching in the datasets D2 and D3.

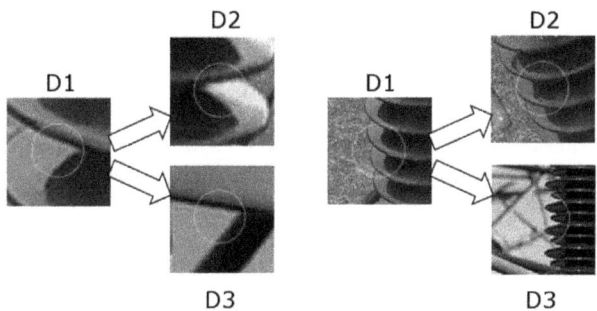

Fig. 9. Example of features from the distinctive features set created on the basis of the training dataset and matching them in the datasets D2 and D3 Example of features from the distinctive features set created on the basis of the training dataset and matching them in the datasets D2 and D3

This comparison was elaborated to determine whether the matching distinctive features accurately describe characteristic fragments of the analyzed object. As we can see in Fig. 9 the example distinctive features were correctly matched, the figure also shows local features describing a fragment of the analyzed object - in this case, it is the area between the insulator disks.

4 Summary

According to other publications, the effectiveness of machine learning depends on the size of the training dataset. Using a limited training dataset, as in the research described here, yielded promising results within the analysed scenario (imaging performed during the flight along the power supply line). Thus far we have concentrated on building a limited training dataset (D1) of the distinctive features, which would allow for matching in the images contained in the test datasets (D2 and D3) of at least one feature that describes the searched object – a starting point for the object detection methods. As shown, increasing the LPR parameter limits the number of matching features in the training set.

Such limitations benefit the features location time (Fig. 6). At the same time, the number of local features that describe the object decreases and may adversely impact the object detection process. Continuation of the research concerning the methods that allow to describe the object in the digital image, basing on the limited number of key features. As presented in Fig. 5, it appears appropriate to develop the object detection method based on one matching distinctive feature. However, this method would require highly complex calculations, and decision concerning the detection of an object burdened with high uncertainty. Other directions of further works include:

1. Application of other methods for selection of local features to create object patterns.
2. Application of other detectors and descriptors of local features (e.g., Harris, SIFT, MSER).
3. Extension of the dataset, among the others, through:Image acquisition from ground level and using UAV vehicles, Acquisition of image material from distribution and transmission network operators and service companies involved in repairs of power supply lines, Utilization of the satellite imaging.
4. Application of the convolution neural networks for detection of insulators, comparison of results obtained using different methods.

References

1. Agarwal, A., Triggs, B.: Hyperfeatures - multilevel local coding for visual recognition. In: European Conference on Computer Vision (2006)
2. Anjum, S.: A study of the detection of defects in ceramic insulators based on radio frequency, University of Waterloo (2014)
3. Bay, S., Ess, A., Tuytelaars, T., Van Gool, L.: Speeded-up robust features (SURF). Comput. Vis. Image Underst. **110**, 346–359 (2008)
4. Cheng, C., Wang, X., Li X.: UAV image matching based on surf feature and Harris corner algorithm. In: ICSSC 4th International Conference on Smart and Sustainable City, Shanghai, pp. 1–6 (2017)
5. Csurka, G., Dance, C.R., Fan, L., Willamowski, J., Bray, C.: Visual categorization with bags of keypoints. In: ECCV International Workshop on Statistical Learning in Computer Vision (2004)
6. Dork, G., Schmid, C.: Object class recognition using discriminative local features. Technical report RR-5497 (2005)
7. Fei-Fei, L., Perona, P.: A Bayesian hierarchical model for learning natural scene categories. In: International Conference on Computer Vision and Pattern Recognition (2005)
8. Geest, R.D., Tuytelaars. T.: Dense interest features for video processing. In: Image Processing (ICIP), pp. 5771–5775 (2014)
9. Geng, Z.X., Qiao, Y.Q.: An improved illumination invariant surf image feature descriptor. In: 2017 International Conference on Virtual Reality and Visualization (ICVRV), Zhengzhou, China, pp. 389–390 (2017)
10. Iscen, A., Tolias, G., Gosselin, P.H., Jégou, H.: A comparison of dense region detectors for image search and fine-grained classification. IEEE Trans. Image Process. **24**, 2369–2381 (2015)

11. Lazebnik, S., Schmid, C., Ponce, J.: Sparse texture representation using affine-invariant neighborhoods. In: CVPR (2003)
12. Leung, T.H., Malik, J.: Representing and recognizing the visual appearance of materials using three-dimensional textons. In: ICCV (1999)
13. Li, Z., Chen, C.: an improved adaptive threshold BRISK feature matching algorithm based on SURF. In: CAC Chinese Automation Congress, Xi'an, China, pp. 2928–2932 (2018)
14. Liao, H., Xiang, J., Sun, W., YuL, S.: Adaptive aggregating multiresolution feature coding for image classification. Math. Probl. Eng. **2014**, 1–10 (2014)
15. Lowe, D.: Distinctive Image Features from Scale-invariant Keypoints. Int. J. Comput. Vis. **60**(2), 91–110 (2004)
16. Luo, H.L., Wei, H., Lai, L.L.: Creating efficient visual codebook ensembles for object categorization. IEEE Trans. Syst. Man Cybern. **41**, 238–253 (2011)
17. Matas, J., Burianek, J., Kittler, J.: Object recognition using the invariant pixel-set signature. In: BMVC (2000)
18. Muhathir Rizal, R.A., Sihotang, J.S., Gultom, R.: Comparison of SURF and HOG extraction in classifying the blood image of malaria parasites using SVM. In: ICoS-NIKOM International Conference of Computer Science and Information Technology, Medan, Indonesia, pp. 1-6 (2019)
19. Ordóñez, Á., Heras, D.B., Argüello, F.: Surf-based registration for hyperspectral images. In: IGARSS 2019–2019 IEEE International Geoscience and Remote Sensing Symposium, Yokohama, Japan, pp. 63–66 (2019)
20. Schmid, C., Mohr, R., Bauckhage, C.: Evaluation of interest point detectors. Int. J. Comput. Vis. **37**, 151 (2000)
21. Szeliski, R.: Computer vision: algorithms and applications. Computer **5**, 832 (2010)
22. Tomaszewski, M., Osuchowski, J., Debita, L.: Effect of spatial filtering on object detection with the SURF algorithm. In: Advances in Intelligent Systems and Computing, vol. 720, pp. 141–149 (2018)
23. Tuytelaars, T.: Dense interest points. In: IEEE Computer Society Conference on Computer Vision and Pattern Recognition, pp. 2281–228 (2010)
24. Wu, S., Oerlemans, A., Bakker, E.M., Lew, M.S.: A comprehensive evaluation of local detectors and descriptors. Signal Process. Image Commun. **59**, 150–167 (2017). https://doi.org/10.1016/j.image.2017.06.010
25. Xie, L., Tian, Q., Zhang, B.: Simple techniques make sense: feature pooling and normalization for image classification. Circuits Syst. Video Technol. **26**, 1251–1264 (2016)
26. Zhang, J., Marszalek, M., Lazebnik, S., Schmid, C.: Local features and kernels for classification of texture and object categories: a comprehensive study. Int. J. Comput. Vis. **73**, 213–238 (2007)

Monitoring Vegetation Changes Using Satellite Imaging – NDVI and RVI4S1 Indicators

Michał Tomaszewski[1]([✉]), Rafał Gasz[1], and Krzysztof Smykała[1,2]

[1] Faculty of Electrical Engineering, Automatic Control and Computer Science, Opole University of Technology, ul. Prószkowska 76, 45-758 Opole, Poland
m.tomaszewski@po.edu.pl
[2] QZ Solutions Sp. z o.o., ul. Ozimska 72A, 45-310 Opole, Poland

Abstract. Modern technologies are often used in agriculture, which allow increasing yields while minimising production costs. Many tools are created to support farmers' activities in the so-called precision agriculture consisting of adapting agrotechnical operations to changing conditions in different areas of the cultivated field. For field yield variability mapping and other precision agricultural applications is used high-resolution satellite imagery. This paper presents the possibility of monitoring the areas of crops specified by the user and informing about the hazards that appear thereby detecting areas showing vegetation degradation based on a series of satellite spectral and radar data recorded at specific points in time. This paper presents an example of the application of imaging performed by the Sentinel-1 and Sentinel-2 satellites as well as advanced imaging techniques and digital image processing for precision farming. The possibility of using satellite images to calculate indices NDVI and RVI4S1 that determine changes in vegetation in the study area was presented. Combined analysis of changes in these indicators in the analysed period may be performed to support the decision-making process and perform actions adequate to the identified situation. The research presented in this paper is conducted to determine the possibility of using combined, satellite spectral and radar data (so-called hybrid set) to detect areas of vegetative degradation in crops in the absence of an element of the spectral data series.

Keywords: Satellite images · Synthetic aperture radar · NDVI · RVI4S1 · Precision agriculture · Vegetation monitoring

1 Introduction

Optimised and efficient crop management requires good knowledge of crop development. Continuous information on the state of a crop can be provided through crop growth modelling. Numerous crop biomass and yield models have been developed since the late 1970s when the science of agroecosystem modelling was initiated [1]. Crop growth models were developed and applied for three primary purposes [2]. First, as a research tool to understand and analyse processes and crop growth. Second, for diagnostic evaluation and analysis within short time frames to manage crop systems with water and fertiliser application assistance, the models can provide in-season decision aid for

S. Paszkiel (Ed.): ICBCI 2021, AISC 1362, pp. 268–278, 2021.
https://doi.org/10.1007/978-3-030-72254-8_29

farmers or precision farming. Third, models can give long-term prognostic information for policy assistance by developing best management practices or evaluating climate change effects on agriculture.

Significant changes occur in agricultural business management, the decision-making process and the automation of activities. Technologies such as automatic expert systems, autonomous tractors with agricultural machines, innovative agricultural robots based on machine learning algorithms, modelling and simulation tools developed using current knowledge and experience in agriculture are developed [3, 4].

Currently, for the precision farming (computer-aided agriculture, based mainly on the collection of data on spatial variation of yields within a field [5]), precise information on spatial variability within fields is essential. This variability is influenced by various factors- including yield, soil properties and nutrients, crop plant nutrients, plant volume, water content, and pest conditions (weeds and insects). These factors can be measured using various methods and instruments, such as electronic sensors, vision systems, including satellite imagery. The most advanced are systems for detecting weeds, soil properties and nutrients.

Satellite remote sensing is one of the main sources of information on the environment and natural processes occurring on the Earth's surface. Based on satellite images with a different spatial, spectral and radiometric resolution, spatial information is obtained, which can be interpreted directly or used in mathematical modelling processes. On their basis, it is possible to determine, i.a. the classes of land cover and land use as well as the condition of vegetation caused by natural and anthropogenic factors. Due to different characteristics, satellite images can be used both in overview scales and in local, multi-time scales with various repetition periods.

2 Agricultural Vegetation Monitoring

Key roles in precision farming implementations are played by both crop state determination and change analysis of the field. Currently, visual diagnostics methods are used for this purpose, particularly with the use of various types of aerial platforms. Images taken from the bird's eye perspective acquired during the vegetation season can be used to plan appropriate field treatments.

Besides standard imaginary of visual wavelength range, multispectral imagery taken by appropriate cameras are useful in visual crop diagnostics. A crucial role in this area plays very near-infrared (VNIR) and near-infrared imagery (NIR), because of green plants high reflectance in this wavelength due to chlorophyll activity [6]. Multispectral imagery systems with narrow spectral bands provide images with very high spatial resolution and real-time crop monitoring. Extension of mentioned technology is hyperspectral imagery (HSI). HSI sensors can simultaneously register dozens or even hundreds of spectral bands, allowing for more accurate measurements and analyses.

Remote sensing data allows determining plants biophysical properties [7–11]. Depending on registered radiation wavelength these properties may be, inter alia, chlorophyll and other pigments content, active photosynthesis radiation (VIS, VNIR), internal leaf structure (VNIR), water content (SWIR, short-wave infrared), leaf surface temperature (TIR, thermal infrared) and spatial vegetation structure (microwaves). The unique biophysical feature is usually defined on vegetation indices, developed based on radiation registered in the visible and near-infrared spectral range.

The use of vision systems in agriculture tasks requires advanced image processing techniques to identify specific spectral signatures. Determining vegetation indices' level requires isolating certain spectral features from the pictures by combining two or more spectral bands of plant surface reflectance [12]. Standard 3-bands RGB images have limited agricultural applicability compared to multispectral and hyperspectral techniques. However, even this basic type of imaging enables to identify two essential crop parameters. One of them is detecting green plants and their parts, such as weeds areas, plants in rows, leaves, and other vegetation plant parts by green band enhancement. The other applications use the intensified red band for a soil analysis to determine i.a. soil type, moisture, etc.

Noticeable computer vision and machine learning progress created an opportunity to extend and improve precision plant protection practices and implement these practices in precision farming [e.g. 13–15]. It is currently possible to identify crop areas affected by disease and classify the pathogen based on shape and colour characteristics of leaf, plant stem, or fruit within image processing methods or machine learning techniques. [16] describes a wide variety of studies on land use, plant species and soil type classification, weed identification, fruit counting, animal growth, weather, product yield index, and moisture assessment to develop solutions for agricultural and food production challenges using deep learning techniques. In these studies, deep learning techniques performed better and were more accurate than popular image processing methods.

3   Satellite Imaging for Crop Vegetation Monitoring

Earth surface is monitoring by various instruments installed onboard many satellites. These measurement devices register electromagnetic radiation in varied spectral ranges such as visible wavelengths, near-infrared, short-wave infrared, far-infrared, including thermal infrared, and microwaves. The most important parameters of the satellite imaging are spatial resolution, spectral resolution, and temporal resolution. The observation level depends on spatial resolutions, often presented as ground sampling distance (GSD). The number of registering bands and their bandwidths (spectral resulution) determines the ability to distinguish and recognise objects on the Earth's surface. The parameter of increasing importance is the frequency of acquiring data (temporal resolution). Satellite orbit altitude, registering data not only in nadir, but also in off-nadir angles, and availability of data acquired by satellite constellations make observations of the same place possible every day, or even several times a day [17].

Besides observing wide areas, the main purpose of remote sensing data usage in agriculture is information about crops and its vegetation condition [18]. These parameters are described by empirical models limited to local areas, or by physical models, known as Radiative Transfer Models (RTM). The physical models simulate the transmission of radiation through the plant and take into account the physical processes of light scattering and absorption by chemical plant elements such as chlorophyll, water, carotene, etc. RTMs are generally considered to be a robust application for combining remote sensing data with crop growth parameters [19]. This kind of models is widely used in plenty of agriculture utilisations for crops, e.g., corn, potato, wheat and grassland. The studies' results showed the validity and usefulness of RTMs, especially in homogeneous agricultural land use periods [20–22].

In many fields increasingly, radar imaging (especially SAR imaging) is used, which differs from the methods using optical instruments. Optical scanners rely on light produced by external sources. Light reflected from the object hits the lens and allows the image to be obtained. Active radiometric instruments send radio waves towards the subject and then measure how they reflect. The synthetic aperture radar (SAR) technology operates based on "virtual" increase in the antenna's size by transmitting and receiving probing signals using a sensor placed on a mobile platform – a satellite or an aeroplane. The registered signals are assembled as if they came from one large antenna.

The theoretical resolution of the SAR in the direction of a moving platform (aircraft or satellite) is half the antenna length (in the direction of movement). The SAR resolution can be increased by reducing the antenna size. The radar's spatial resolution is determined by the probe pulse width or its wavelength if using a composite signal. To obtain high-quality SAR images, very precise information about the platform trajectory, lower than 0.1 wavelengths, is required. The platform movement's trajectory is usually not so accurately known, and it is necessary to use automatic focusing techniques [23].

High-resolution radar images of small areas can appear similar like familiar optical images. Due to the nature of the registered data, consisting of amplitude and phase of the signal reflected from the Earth's surface, advanced image processing methods are required to change the measurements to image. However, mentioned methods provide a possibility to observe the planet even during unfavourable weather condition, such as overcast clouds, that prevent imaging with optical instruments.

Thanks to satellite imaging of the planet's surface, the information gained nowadays in various areas and branches of the economy, including geodesy, cartography, energy [24], forestry, and water management, for instance, to change analysis of riverbeds. Satellites simplify ensuring the security, tightness of borders and crisis management, they turn out to be an indispensable tool for rescue and neutralising the effects of natural disasters.

The main advantage of radar technology is the possibility of conducting it in a continuous mode. Contrary to optical instruments, SAR can execute remote sensing observation despite disadvantageous weather or light condition over a given patch of the planet. Observations obtained using SAR equipment can be successfully carried out either at night, in heavily clouded areas, or even during precipitation. The critical aspect is the continuity of monitoring, ensured by the ability to perform it regardless of the circumstances mentioned above or weather conditions. Therefore, radar satellite instruments are perfect for monitoring land movement changes, building landslides, rising water levels in rivers or flooded areas, or changes in crop yield growth in farmland [25–27]. Another advantage of SAR is the ability to imaging large areas in a short time. For example, it takes about 30 s to acquire a 100×50 km rectangle.

Satellite images analysis can be performed in an object-oriented manner, automatically classifying the images according to the area of occurrence of a given band [9]. Several parameters are taken into account in the recognition process, e.g., the objects' spectral features, relations between them, statistical parameters calculated based on the object pixels and parameters describing the objects' shape. Currently, the use of vision methods, especially those based on Artificial Neural Networks (ANN) [28], is increasingly gaining popularity in many areas [29, 30], including agriculture. Both pixel-based and object-oriented ANN analysis are utilising, especially for hyperspectral data, and the classification is performed based on a comparative study of spectral libraries. Regardless of the algorithms used, an important issue is the optimal spectral band selection. Classification processes are also performed using processed data (inter alia texture functions, indicators, canonical transformations). Furthermore, research related to the vegetation condition assessment is being developed. For this purpose, vegetation indicators (of varying complexity) and mathematical modelling are used, in which remote sensing data is not the only source of input information. The high frequency of acquiring satellite images enables analysis of time series made of data representing the same area. Based on the time series analysis, it is possible to determine changes in plants' condition, which are correlated, among others, with the availability of water in the soil, fertilisation, and conducted agrotechnical treatments.

4 Normalised Difference Vegetation Index (NDVI)

The primary indicator determining plant vegetation are among other Normalised Difference Vegetation Index (NDVI), Red Edge Inflection Point (REIP), Relative Reflectance Index (RRI), Simple Ratio (SR), Photochemical Reflectance Index (PRI), etc.

The Normalised Difference Vegetation Index (NDVI) is applied in remote sensing measurements to determine vegetation development status and condition. NDVI is based on the contrast between reflectance in the red and near-infrared spectral ranges. Green and healthy plants absorb solar radiation from 400 to 700 nm, known as photosynthetically active radiation (PAR), and use it as an energy source for photosynthesis. Moreover, photon energy for very near-infrared (VNIR) and near-infrared (NIR) spectral ranges (longer than 700 nm) is too small to be used for the synthesis of organic molecules, and approximately half of the solar energy is reflected from leaves. Plants reflect VNIR and NIR radiation to prevent itself from overheating and tissues damage. Therefore, live, green plants appear dark in PAR and relatively bright in VNIR and NIR. On the other hand, clouds and snow appear bright in the red band and dark in NIR. The index value increases in direct proportion to the biomass of a given area. Researchers studied relationships between gathered data and the imaged area because of the data type availability [31].

The NDVI indicator is used to collect information about the intensity of photosynthesis and the ecosystem's biomass. This index was presented for the first time by J.W. Rouse in 1973 [32] and is calculated according to the Eq. 1

$$NDVI = (NIR - VIS)/(NIR + VIS) \tag{1}$$

where VIS and NIR are spectral reflectance measurements from red and near-infrared spectral range, respectively.

Values of NDVI fall within a range from -1.0 to $+1.0$ and is the non-linear equivalent of the simple VIS-NIR ratio indices. As a general rule, green, healthy plants' spectral signature is characterised by a high NIR response and high absorption in the red band.

The spectral signature of brown dried up and unhealthy plants is similar in both VIS and NIR is not so significant. The ratio of these spectral regions' values, presented by Eq. 4.1, shows that the higher the NDVI value, the healthier the plant is. A low NDVI value indicates invalid plants vegetation or lack of it. Values equal to 0.0 represents bare soil when a negative result – a body of water, clouds, and snow.

5 Radar Vegetation Index for Sentinel-1 SAR Data (RVI4S1)

With the use of SAR radar images, research is conducted to determine the state of plant vegetation [7, 9, 33–38] and cloud cover over the analysed agricultural area. On their basis, several indicators have been developed, e.g. Quad-pol Radar Vegetation Index (RVI), Radar Forest Degradation Index (RFDI), Dual Polarisation SAR Vegetation Index (DPSVI), Dual Polarimetric Radar Vegetation Index (DpRVI), Radar Vegetation Index for Sentinel-1 SAR data (RVI4S1), Compact Polarimetric Radar Vegetation index (CpRVI) etc.

The Sentinel-1 satellite has a SAR synthetic aperture radar operating in the C band, i.e. with a frequency of 5.405 GHz. It is used for continuous imaging of land and sea surfaces with a 5 to 25 m resolution per pixel [39]. The data recorded by the Sentinel-1 satellite provided an opportunity to monitor crops thanks to the high return frequency and extensive spatial coverage. Data from the dual-pol (VV-VH bands) Sentinel-1 SAR are used for the European Common Agricultural Policy (CAP) and other projects that provide information to support the crop monitoring network.

In [37] the RVI4S1 indicator was proposed to assess crops' vegetation state using satellite radar data. This index uses images from the Sentinel-1 GRD product –a combination of VV and VH polarities. Soil or other non-vegetated areas show a very low RVI4S1, close to 0.0, and conversely, fully developed cultivated areas increase the index towards 1.0.

This indicator was defined as:

$$RVI4S1 = sqrt(DOP)(4VH/(VV + VH)) \tag{2}$$

DOP is equivalent to the degree of polarisation calculated as:

$$DOP = VV/SPAN = VV/(VV + VH) \tag{3}$$

where SPAN is total power received at both the channel and can be treated as VV + VH.

Digital images from SAR systems are characterised by a specific spotting effect (speckle effect), which is a significant limitation when performing various analyses using radar imaging. The spotting effect manifests itself as the spatial variation of adjacent pixels' brightness (so-called salt and pepper). It is typical for coherent systems, where the value of the recorded signal is the vector sum of reflections from objects inside the resolution cell [40]. The spotting effect can be reduced in two ways (despeckling):

- at the stage of creating a radar image, as a result of incoherent averaging of several signal samples (the so-called looks); this process, the so-called multi looking improves the image's radiometric resolution, but at the expense of spatial resolution.

- as part of preprocessing recorded radar images with the use of spatial filters; by using adaptive filters that reduce noise and at the same time preserve the edges and features of the image (e.g. Lee filter, extended Lee filter, Frost filter, Gamma filter, Kuan filter, bilateral filter) or by using advanced algorithms to remove noise from digital images (e.g. BM3D, PPB or other solutions based on convolutional neural networks).

For radar satellite images shown below was used BM3D algorithm [41].

6   Analysis of Changes in Crops Status Based on NDVI and RVI4S1 Indices

One of the basic directions of satellite image analysis is the detection of changes occurring in fields. It consists of determining the differences in the land cover depicted in the photos recorded on two dates. The simplest method used is visual interpretation, making it possible to find differences relatively quickly in a small area. In large areas depicted in high-resolution images, we use automatic tools to detect changes.

NDVI - 2018.05.13 - 2018.05.23   RVI4S1 -2018.05.14 - 2018.05.23

Fig. 1. Change in crop status determined based on NDVI and RVI4S1 indices - field "corn"

The attached figures present selected examples of changes in the vegetation state, illustrated by comparing the images illustrating the changes in the values of NDVI and RVI4S1 indices in a specific period. Sample images are provided for three different types of crops, i.e. corn (Fig.1), wheat (Fig.2) and potatoes (Fig. 3).

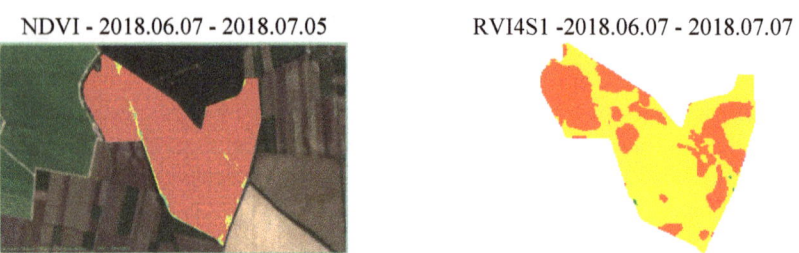

NDVI - 2018.06.07 - 2018.07.05   RVI4S1 -2018.06.07 - 2018.07.07

Fig. 2. Change in crop status determined based on NDVI and RVI4S1 indices - field "wheat"

The yellow colour means no change in the value of a given indicator within the specified tolerance range (e.g. for NDVI the change is less than 0.1). Red indicates a deterioration in vegetation, and green indicates its improvement. The analyses carried out for different periods, and different arable fields showed that both indicators correctly indicate specific changes. It was approved that the NDVI index is more accurate than RVI4S1 and more precisely illustrates the areas of changes in the vegetation state. Although the RVI4S1 index allows determining changes regardless of the cloud cover, a much obstacle in interpreting radar images is the necessity to perform despeckling.

NDVI - 2018.06.07 - 2018.07.05   RVI4S1 -2018.06.07 - 2018.07.07

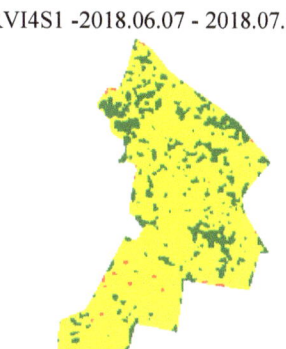

Fig. 3. Change in crop status determined based on NDVI and RVI4S1 indices - field "potatoes"

7   Summary

According to the cited literature and the performed analyses, the Normalised Difference Vegetation Index (NDVI) and Radar Vegetation Index for Sentinel-1 SAR data (RVI4S1) can be jointly used to assess the state of vegetation in the observed area. During the conducted research, changes in both tested vegetation indices were observed in the designated time intervals. These changes correspond to the state of crops in the presented area to a varying extent. During the comparative analysis of NDVI and RVI4S1 indices, in some cases, similar results regarding vegetation changes were obtained, and in others, the obtained images differed significantly. These results may be due to the following factors:

- both satellites (Sentinel-1 and Sentinel-2) fly over the area non-synchronously, taking satellite images on other days. Thus, it is impossible to perform the differential analysis for precisely the same period in every case. The offset is usually from 1 to 3 days, which may result in unequal indications in vegetation changes when comparing the differential indices calculated based on the NDVI and RVI4S1,
- significant is the time interval of satellite imaging and the corresponding differential analysis when the longer time interval between observations was considered, the obtained results were more similar for both indicators,
- the parameters and the selected method of de-noise of radar images used for the despeckling impact the obtained results' accuracy. Averaging filters and other noise

reduction algorithms remove extreme values from the digital image, but in some cases, these values may also occur due to changes in the vegetation state in the studied area. It is appropriate to carry out further work to optimise this preprocessing process of radar satellite images.

In this study, the obtained RVI4S1 values were compared to the calculated NDVI values in a given area. It is advisable to perform analyses using in situ measurements of other indicators describing the state of vegetation in the observed area. Then it will be possible to more accurately evaluate the changes in vegetation with the help of indicators based on radar satellite images. Further work aims to develop a hybrid index based on both presented methods of imaging the state of vegetation.

Acknowledgement. The presented results were made as a part of the project "Development of an application detecting the vegetative degradation of crops based on the time series of satellite data – RPOP.01.01.00-16-0001/20". The project is co-financed from The European Regional Development Fund under the Opole Voivodeship Regional Operational Programme (RPO WO) for the years 2014-2020, Action 1.1 Innovation in Enterprises.

References

1. Passioura, J.B.: Simulation models: science, snake oil, education, or engineering? Agron. J. **88**, 690–694 (1996)
2. Boote, K.J., Jones, J.W., Pickering, N.B.: Potential uses and limitations of crop models. Agron. J. **88**, 704–716 (1996)
3. Ozguven, M.M.: The newest agricultural technologies. Curr. Investig. Agric. Curr. Res. **5**(1), 573–580 (2018). https://doi.org/10.32474/ciacr.2018.05.000201
4. Singh, V., Misra, A.K.: Detection of plant leaf diseases using image segmentation and soft computing techniques. Inf. Process. Agric. **4**, 41–49 (2017). https://doi.org/10.1016/j.inpa.2016.10.005
5. Faber, A.: System rolnictwa precyzyjnego. I. Mapy plonów, Fragmenta Agronomica **57**, 4–15 (1998)
6. Ollinger, S.V.: Sources of variability in canopy reflectance and the convergent properties of plants. New Phytol. **19**, 375–394 (2011)
7. El Hajj, M., Baghdadi, N., Wigneron, J., Zribi, M., Albergel, C., Calvet, J., Fayad, I.: First vegetation optical depth mapping from sentinel-1 C-band SAR data over crop fields. Remote Sens. **11**(23), 2769 (2019). https://doi.org/10.3390/rs11232769
8. Mantovani, E., Althoff, D.: Crop NDVI monitoring based on sentinel 1. Remote Sens. **11**(12), 1441 (2019). https://doi.org/10.3390/rs11121441
9. Yamada, Y.: Preliminary study on the radar vegetation index (RVI) application to actual paddy fields by Alos/Palsar full-polarimetry SAR data, Conference. ISRSE36At, Berlin, Germany (2015)
10. Vuolo, F., Atzberger, C., Richter, K., D'Urso, G., Dash, J.: Retrieval of biophysical vegetation products from rapideye imagery. In: ISPRS TC VII Symposium – 100 Years ISPRS, XXXVIII, pp. 281–286 (2010)
11. Rouse, J.W., Haas, R.H., Scheel, J.A., Deering, D.W.: Monitoring vegetation systems in the great plains with ERTS. In: Proceedings, 3rd Earth Resource Technology Satellite (ERTS) Symposium, vol. 1, pp. 48–62 (1974) https://ntrs.nasa.gov/archive/nasa/casi.ntrs.nasa.gov/19740022592.pdf

12. Meyer, G.E., Camargo-Neto, J.: Verification of color vegetation indices for automated crop imaging applications. Comput. Electron. Agri. **63**(2), 282–293 (2008)
13. Ruszczak, B., Smykała, K., Dziubański, K.: The detection of Alternaria solani infection on tomatoes using ensemble learning. J. Ambient Intell. Smart Environ. **12**(5), 407–418 (2020). https://doi.org/10.3233/AIS-200573
14. Smykała, K., Ruszczak, B., Dziubański, K.: Application of ensemble learning to detect Alternaria solani infection on tomatoes cultivated under foil tunnels. In: Iglesias, C.A. (ed.) Intelligent Environments 2020. Workshop Proceedings of the 16th International Conference on Intelligent Environments, Ambient Intelligence and Smart Environments, Amsterdam, vol. 28, pp. 127–132. IOS Press (2020). ISBN 978-1-64368-090-3, https://doi.org/10.3233/aise20 0033
15. Słapek, M., Smykała, K., Ruszczak, B.: Brassica napus florescence modeling based on modified vegetation index using sentinel-2 imagery. In: Rutkowski, L. (ed) Artificial Artificial Intelligence and Soft Computing: 18th International Conference, ICAISC 2019, Zakopane, Poland, 16–20 June 2019, Proceedings, Part II, Lecture Notes In Computer Science, Springer, vol. 11509, pp. 80–90 (2019). ISBN 978-3-030-20914-8, https://doi.org/10.1007/978-3-030-20915-5_8
16. Sladojevic, S., Arsenovic, M., Anderla, A., Culibrk, D., Stefanovic, D.: Deep neural networks based recognition of plant diseases by leaf image classification. Comput. Intell. Neurosci. **2016**, 11 (2016)
17. Banaszkiewcz, M., Lewiński, S.: Zastosowanie technik satelitarnych w rolnictwie zrównoważonym - wybrane przykłady zastosowań. Prob. Agric. Eng. PIR 2012 (VII–IX). **3**(77), 109–122 (2012). ISSN 1231-0093
18. Moulin, S., Bondeau, A., Delecolle, R.: Review article: combining agricultural crop models and satellite observations: from field to regional scales Int. J. Remote Sens. **19**, 1021–1036 (1998)
19. Dorigo, W.A., Zurita-Milla, R., de Wit, J.W., Brazile, J., Singh, R., Schaepman, M.E.: A review on reflective remote sensing and data assimilation techniques for enhanced agroecosystem modeling. Int. J. Appl. Earth Obs. Geoinf. **9**, 165–193 (2007)
20. Koetz, B., Baret, F., Poilvé, H., Hill, J.: Use of coupled canopy structure dynamic and radiative transfer models to estimate biophysical canopy characteristics. Remote Sens. Environ. **95**, 115–124 (2005)
21. Vuolo, F., Atzberger, C., Richter, K., D'Urso, G., Dash J.: Retrieval of biophysical vegetation products from rapideye imagery ISPRS TC VII Symp. – 100 Years ISPRS, XXXVIII, pp. 281–286 (2010)
22. Wang, P., Sun, R., Zhang, J., Zhou, Y., Xie, D., Zhu, Q.: Yield estimation of winter wheat in the North China Plain using the remote-sensing–photosynthesis–yield estimation for crops (RS–P–YEC) model. Int. J. Remote Sens. **32**, 6335–6348 (2011)
23. Crockett, M.T.: An introduction to synthetic aperture radar: a high-resolution alternative to optical imaging. https://digitalcommons.usu.edu/cgi/viewcontent.cgi?article=1012&con text=spacegrant

24. Michalski, P., Ruszczak, B., Lorente, P.J.N.: The implementation of a convolutional neural network for the detection of the transmission towers using satellite imagery. In: Świątek, J., Borzemski, L., Wilimowska, Z. (ed.) Information Systems Architecture and Technology: Proceedings of 40th Anniversary International Conference on Information Systems Architecture and Technology – ISAT 2019. Part II. Advances in Intelligent Systems and Computing, vol. 1051, pp. 287–299. Springer, Cham (2020). ISBN 978-3-030-30603-8. https://doi.org/10.1007/978-3-030-30604-5_26

25. Wang, R., Deng, Y.: Bistatic SAR System and Signal Processing Technology. Springer Nature (2018), ISBN 978–981-10-3078-9. https://doi.org/10.1007/2F978-981-10-3078-9

26. Carvalho, E.A., Ushizima, D.M., Medeirs, F.N.S.: SAR imagery segmentation by statistical region growing and hierarchical merging. Digit. Signal Process. **20**(5), 1365–1378 (2010)

27. Singhroy, V., Moloch, K.: Characterising and monitoring rockslides from SAR techniques. Adv. Space Res. **33**(3), 290–295 (2004)

28. Michalski, P., Ruszczak, B., Tomaszewski, M.: Convolutional neural networks implementations for computer vision, w: biomedical engineering and neuroscience. In: Hunek, W.P., Paszkiel, S. (ed.) Proceedings of the 3rd International Scientific Conference on Brain-Computer Interfaces, BCI 2018, 13–14 March, Opole, Poland. Advances in Intelligent Systems and Computing, vol. 720, pp. 98-110. Springer, Cham (2018), ISBN 978-3-319-75024-8. https://doi.org/10.1007/978-3-319-75025-5_10

29. Tomaszewski, M., Michalski, P., Ruszczak, B.: Detection of power line insulators on digital images with the use of laser spots. IET Image Process. **13**(12), 2358–2366 (2019). https://doi.org/10.1049/iet-ipr.2018.6284

30. Tomaszewski, M., Michalski, P., Osuchowski, J.: Evaluation of power insulator detection efficiency with the use of limited training dataset. Appl. Sci. Basel **10**(6), 1–12 (2020)

31. Measuring Vegetation. https://earthobservatory.nasa.gov/features/MeasuringVegetation

32. Rouse, J.W., Haas, R.H., Scheel, J.A., Deering, D.W.: Monitoring vegetation systems in the great plains with ERTS. In: Proceedings, 3rd Earth Resource Technology Satellite (ERTS) Symposium, vol. 1, pp. 48–62 (1974). https://ntrs.nasa.gov/archive/nasa/casi.ntrs.nasa.gov/19740022592.pdf

33. Mandal, D., et al.: A Radar vegetation index for crop monitoring using compact polarimetric SAR data. IEEE Trans. Geosci. Remote Sens. **58**(9), 6321–6335 (2020). https://doi.org/10.1109/TGRS.2020.2976661

34. Filgueiras, R., Mantovani, E.C., Althoff, D., Fernandes Filho, E.I.: Crop NDVI monitoring based on sentinel 1. Remote Sens. **11**, 1441 (2019)

35. Juan, M., Heather, M., Yalamanchili, S.: Dual polarimetric radar vegetation index for crop growth monitoring using Sentinel-1 SAR data. Remote Sens. Environ. **247**, (2020)

36. Bai, Z., Fang, S., Gao, J., Zhang, Y., Jin, G., Wang, S., Zhu, Y., Xu, J.: Could vegetation index be derive from synthetic aperture radar - the linear relationship between interferometric coherence and NDVI. Sci. Rep. **10**, 6749 (2020)

37. Mandal, D.: Radar Vegetation Index for Sentinel-1 SAR data - RVI4S1 Script. https://custom-scripts.sentinel-hub.com/custom-scripts/sentinel-1/radar_vegetation_index/

38. Mandal, D., Kumar, V., Ratha, D., Dey, S., Bhattacharya, A., Lopez-Sanchez, J.M., McNairn, H., Rao, Y.S.: Dual polarimetric radar vegetation index for crop growth monitoring using Sentinel-1 SAR data. Remote Sens. Environ. **247**, (2020). https://doi.org/10.1016/j.rse.2020.111954

39. Sentinel-1. https://sentinel.esa.int/web/sentinel/missions/sentinel-1

40. Yuan, J., Lv, X., Li, R.: A speckle filtering method based on hypothesis testing for time-series sar images. Remote Sens. **10**(9), 1383 (2018). https://doi.org/10.3390/rs10091383

41. Lebrum, M.: An analysis and implementation of the BM3D image denoising method. IPOL J. Image Process. Line (2021), ISSN 2105-1232

Subannual, Seasonal and Interannual Variability of Data on Residential Construction Market - Case Study for Poland

Lukasz Mach[1]([✉]), Szczepan Paszkiel[2] [ID], and Michał Grubiak[3]

[1] Faculty of Economics and Management, Opole University of Technology,
Luboszycka 7, 45-036 Opole, Poland
l.mach@po.edu.pl
[2] Faculty of Electrical Engineering, Automatic Control and Informatics,
Opole University of Technology, Prószkowska 76, 45-758 Opole, Poland
[3] Regional Branch in Opole, Narodowy Bank Polski, Opole, Poland

Abstract. The chapter reports the results of research whose objective involved the analysis of the time-frequency waveforms representing variables affecting the residential construction market. The investigation of the time and frequency characteristics of the time series provides the identification of subannual, seasonal and interannual components in them with the purpose of defining variables describing the stages of the construction process, i.e. variables expressing the number of building permits issued, the number of dwellings under construction and the number of new dwellings completed. The adequate determination of periodic characteristics offers the simultaneous parameterization of subannual, seasonal and long-term variabilities of the resulting waveforms. The proposed algorithm uses combination of FFT, inverse FFT method and subannual spectra peak model. The identification provides adequate analysis of the residential construction market, and at the same time contributes to reducing the information gap in the decision-making process in the field of the housing market for the subjects that operate on the demand and supply sides of this market.

Keywords: Residential construction market · Housing market · Time-frequency analysis · Time series components · Subannual · Seasonal · Interannual

1 Introduction

The dwelling market forms an important component with a direct effect on the condition of the economy in many of its dimensions and the economy as a whole (Chan et al. 2016; He and Xia 2019; Lee and Lee 2019; Mach 2019; Mach et al. 2020a; Rokita-Poskart and Mach 2019; Silva et al. 2019; Tupenaite et al. 2017). The adequate analysis of diagnostic variables that have a significant impact on the residential construction market plays a crucial role in the parameterization of this market. Parametrization enables the determination of the actual potential characterized by this market and identification of the risk of investment activities performed on it (Chan et al. 2016; He and Xia 2019;

© The Author(s), under exclusive license to Springer Nature Switzerland AG 2021
S. Paszkiel (Ed.): ICBCI 2021, AISC 1362, pp. 279–294, 2021.
https://doi.org/10.1007/978-3-030-72254-8_30

Lee and Lee 2019; Silva et al. 2019; Tupenaite et al. 2017). The potential of the residential construction market can be directly identified in terms of its components, which in economic analysis and in this study were defined by means of the number of building permits issued, the number of dwellings under construction and the number of completed dwellings. The cascade effect and the specific order of implementation of these diagnostic variables means that they should be analyzed as a system of interdependent variables. The simultaneous analysis of all components of the residential construction market can provide a comprehensive assessment of its potential at every stage of this process, and any deviations from the anticipated market values, if any, are likely to offer the grounds to introduce adequate decisions at every stage of the investment process. The potential of the housing market and the degree of its predictability can affect the price of dwellings to a significant extent, as demonstrated in various studies (Deng et al. 2019; Gong et al. 2018; Hofman and Aalbers 2019; Mach 2020; Martins et al. 2019), as well as on speculative bubble development (Escobari and Jafarinejad 2016; Fernández Muñoz and Collado Cueto 2017; Wang et al. 2019; Zhao et al. 2017) and on the investment activity in this market (Hartzell et al. 2014; Hei-Ling Lam and Chi-Man Hui 2018; Phasuphan et al. 2019; Seok et al. 2019; Wright and Yanotti 2019). Given the above, the ability to analyze phenomena that occur on the housing market takes on a key role. The analysis involving the adequate decomposition of the investigated diagnostic variables takes into account specific cases in terms of variable components that in economic perspective are characterized as periodic frequencies over various time windows. Identification of periodic components in the time series describing the real estate market can be found, among others in works (Bengtsson et al. 2018; Devaney and Xiao 2017; Fan et al. 2019; Gabauer and Gupta 2020; Liow and Newell 2016; Mach et al. 2021, 2020a, b, c; Yépez 2018). In this paper, the identification of the diversity and complexity of time windows forms tools is applied for the purposes of examining the characteristics of the time series of the investigated time series in terms of subannual, seasonal and interannual variability. Research was carried out with the purpose of identifying the frequency components for various time windows that take the form of diagnostic variables that affect the potential of residential construction market. The feasible decomposition of the diagnostic variables can offer a comprehensive picture that captures the potential of residential construction market and thus identify the relations that occur on it. This can reduce the information gap in the process of making decisions in this market with regard to investments and thus provide investors with a competitive advantage. The element of novelty and proprietary approach applied in the implementation of the described research is associated with the use of the classification of decomposed time series components into periodic characteristics that apply to subannual, seasonal and interannual periodicity as well as the with the application and demonstration of utilitarian characteritics of the application of signal processing methods in the analysis of the phenomena affecting the potential of the housing market.

2 Implementation of Research and Methodology of Study

The research is carried out in four main stages. In the first of them, untransformed data is presented and discussed. This stage utilizes data on the number of building permits issued, the number of dwellings under construction and the number of new dwellings completed. In the second stage, the procedure for identifying subannual fluctuations is carried out. For this purpose, the FFT procedure and the peak search algorithm for harmonic frequencies representing subsequent subannual frequencies is applied (findpeaks).

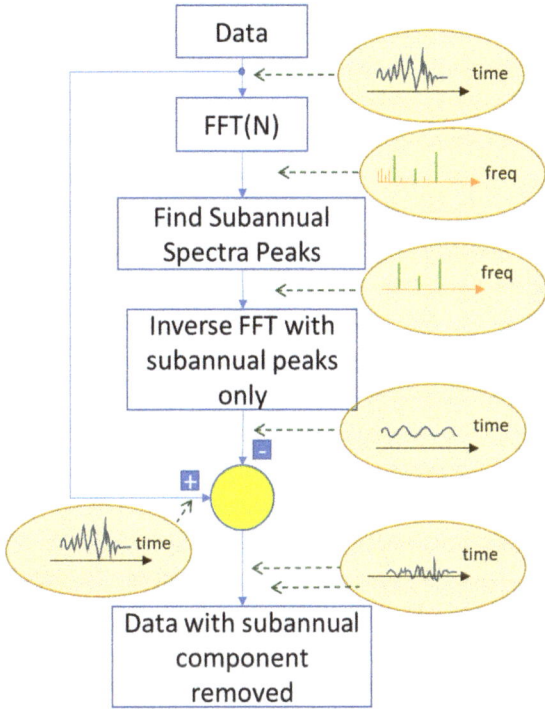

Fig. 1. Block scheme of removing subannual component in analyzed data

In the third stage, the so-called sub-seasonal periodicity is identified by means of inverse Fourier transform. It is possible to identify sub-seasonal periodicity since it is feasible to transform the frequency domain into the time domain as a result of employing FFT inverse algorithm. On the other hand, in the fourth stage, using long term spectra components, long-term variations are identified, which are commonly identified in business cycles in the area of business studies. The same algorithm of the described stages of research is applied for three diagnostic variables, i.e. building permits, flats under construction and new flats commissioned (see Fig. 1). Data with a monthly measuring window for 2005–2018 was used (156 data sets for each variable studied were recorded).

3   Analysis of Up-to-Date Data in Residential Construction Market

Throughout the study of the data on the residential construction market, the analysis of waveform characteristics for three variables describing the components of the stages of the construction process was performed in the first stage. The first component is formed by the variable that defines the number of building permits issued, the second represents the number of apartments under construction, while and the third represents the number of new dwellings completed. This analysis involved the form of graphic presentation of waveforms representing each of the investigated variables combined with a synthetic note to describe or account for the implementation of the examined time series.

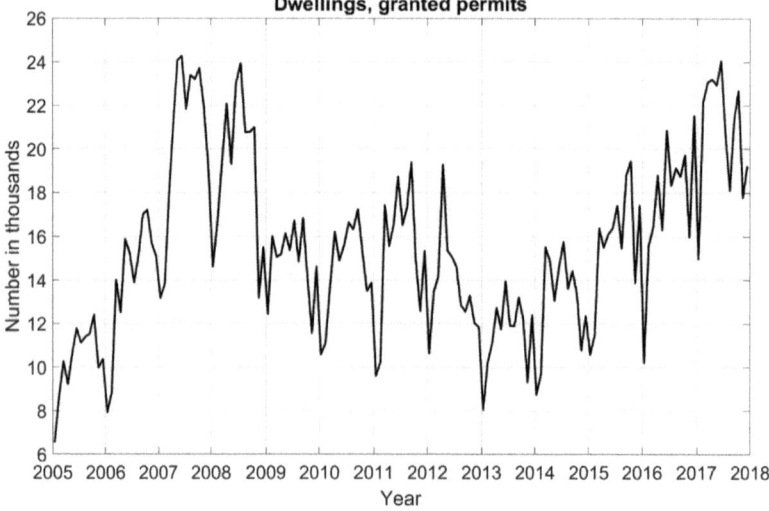

Fig. 2.   Granted building permits in Poland.

Figure 2 was developed for the description carried out for the first variable, i.e. number of granted building permits. In terms of macroeconomic development, three time periods can be identified on the basis of the data in Fig. 2. The first period involved years from 2005 to 2009, when the dwellings market in Poland was developing very dynamically. Such dynamic characteristic (i.e. increase in the number of building permits issued) resulted for a number of reasons, among which two were found to play a key role in economic terms, i.e. the relative ease of gaining funding the purchase of dwellings through mortgage and the stability and even continuous improvement of the economic condition.

The second period, characterized by a decrease in the number of issues building permits occurred between 2009 and 2013. The change in the development trend of the number of building permits issued in 2009 was caused by the global economic crisis that began on the US housing market in 2008 (Coombs and Laufer 2018; Degl'Innocenti et al. 2018; Gallegati and Delli Gatti 2018; Makin 2019; Manjunath et al. 2019). Since 2013, we can notice the third period of time in which the dynamic characteristics of the phenomenon increased again. Assuming that the number of building permits issued reflects the mood of investors on the dwellings market, we can conclude that in the opinions of the investors, the period of the economic crisis ended in 2013.

However, from the point of view of the phenomenon of seasonal variability, we can note that in each of the investigated annual periods, Q2 and Q3 are characterized by a greater number of building permits issued compared to Q1 and Q4.

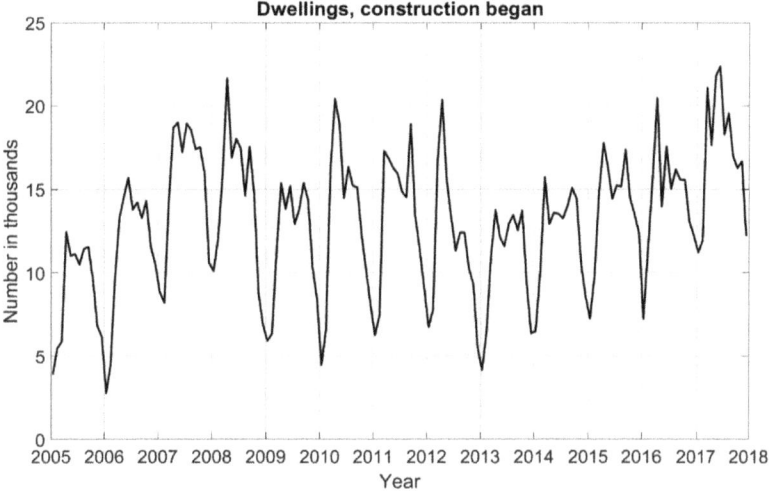

Fig. 3. Number of started constructions in Poland.

Figure 3 contains data regarding the number of apartments under construction in Poland in the period from 2005 to 2018. Just as in the case of the first variable analyzed here, i.e. the number of building permits issued, we can observe three characteristics of a growth trend in the construction of new dwellings. The periods from 2005 to 2009 and 2013 to 2018 were characterized by an increase in this area, whereas a decline was recorded the period from 2009 to 2012. The period of negative dynamics of change was caused by the effects of the global economic crisis.

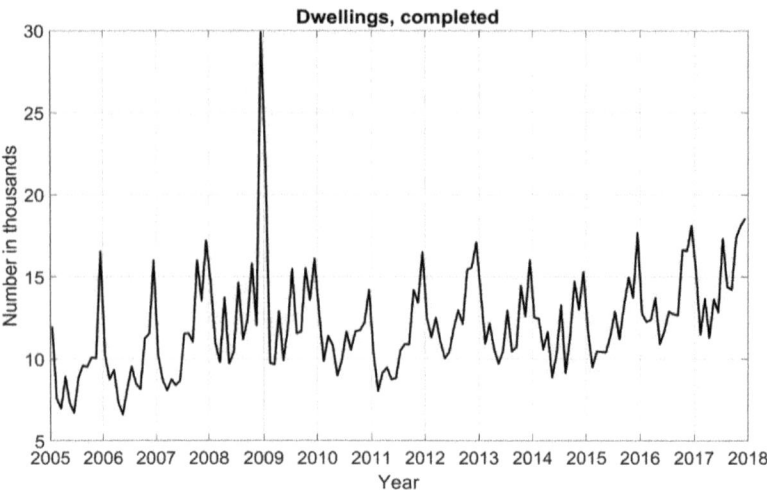

Fig. 4. Completed construction of dwellings in Poland.

Figure 4 contains data for completed dwellings, and we can note that it has one outlier, which was recorded in December 2008. We can note that the value that significantly deviated from the economic conditions represents a value which actually took place on the residential construction market. The significant increase in the number of dwellings completed is due to two co-existing factors, i.e. legal changes that took place in the period and the forms of the dwellings built[1]. From January 2019, the law imposed the need to draw up a document called the energy performance certificate for commissioning newly built dwellings. In the consideration of the fact that in Poland, in 2008, the vast majority of flats that were put into use were commissioned by individual investors, the new legal provision generated two aspects for them, one including cost and the second – uncertainty of investment. These aspects led to a significant increase in the number of dwellings completed prior to the law changes.

The above general characteristics of the variables affecting the condition of the housing construction market in Poland demonstrated the their time-frequency variability. This offered the basis for preliminary identification of subannual, seasonal and long-term fluctuations. The first stage of the descriptive research was applied to determine the directions of changes in the considered variables and for preliminary and general indications of the occurrence of fluctuations in the nature of subannual, seasonal and cyclic changes in terms of interannual characteristics. In the following stage of the research, using the fast Fourier transformation procedure, the identification of subannual variability of the examined time series was performed. The parameterization of sub-seasonal characteristics of periodicity in the examined time series will allow examining the significance of the impact of periodic fluctuations with a time window of one month, with their maximum values of 1 year (12 months).

---

[1] Forms: individual, cooperative ownership, built for sale or rent, intended for rent, communal.

4   Identification of Subannual Time-Frequency Variability

The process of identifying time-frequency variability was aimed at identifying subannual changes with a particular emphasis on dynamic frequency characteristics. The analysis of these variations illustrates the evolution of the investigated relations in terms of its implementation within a period of up to one year. Fast Fourier Transform (FFT) and the

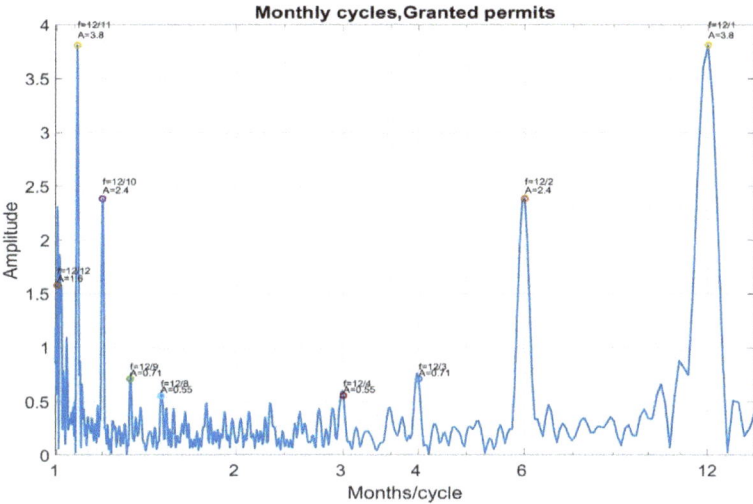

Fig. 5. Amplitude spectrum of granted permits with estimated subannual frequency components

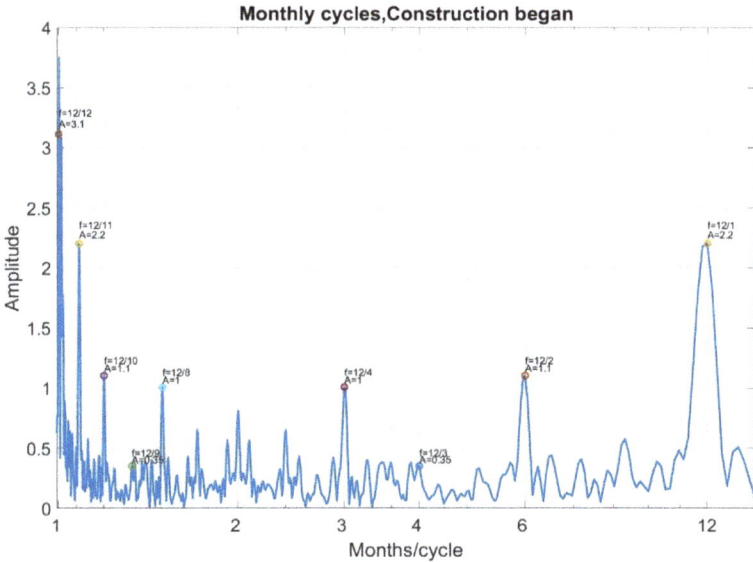

Fig. 6. Amplitude spectrum of construction began data with estimated subannual frequency components.

findpeaks procedure were applied for this purpose. The results of the transformations of the examined time series are presented in Figs. 5, 6 and 7. Figure 5 presents the annual implementation of the spectrogram representing the number of building permits issued, Fig. 6 contains the spectrogram for a variable housing under construction, whereas Fig. 7 contains the spectrogram with the number of new dwellings completed.

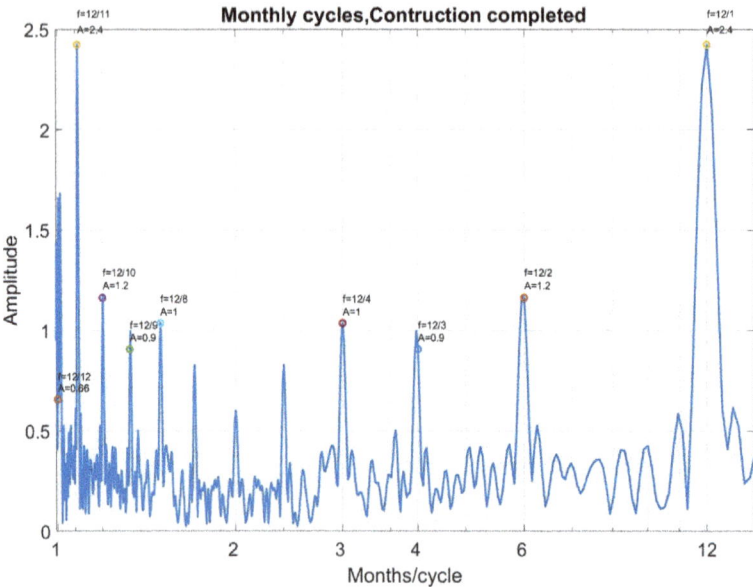

Fig. 7. Amplitude spectrum of construction completed data with estimated subannual frequency components.

We can note that in Figs. 5, 6 and 7, on the x-axis, the data representing a single year is divided into 12 months (logarithmic values), whereas the y-axis contains the amplitude of the fluctuations.

When an interpretation of the results is sought, we can see that in each of the examined cases we can identify an annual sub-period of 12 months (i.e. 1 year), an annual sub-period of 6 months (half a year) and an annual sub-period of 3 months (quarter). For variables that define the number of building permits and dwellings completed, we can conclude about the existence of a significant sub-period lasting four months. The parameters of the results obtained are given in Table 1 to demonstrate significant sub-seasonality within an annual period. This table presents the significant subannual frequency and the corresponding amplitude and observed frequency.

Table 1. Parameters of subseasonality for investigated variables.

Variable	Subannual seasonality	Amplitude	Frequency
Building permits issued	12	A = 3.8	A = 12/1
	6	A = 2.4	A = 12/2
	4	A = 0.71	A = 12/3
	3	A = 0.55	A = 12/4
Dwellings under construction	12	A = 2.2	A = 12/1
	6	A = 1.1	A = 12/2
	3	A = 1.0	A = 12/4
New dwellings completed	12	A = 2.4	A = 12/1
	6	A = 1.2	A = 12/2
	4	A = 0.9	A = 12/3
	3	A = 1.0	A = 12/4

5 Identification of Seasonal Subannual Variability

Following the identification of the variable component that describes subannual variability, an attempt is made in the latter part of present study to perform a decomposition of the tested time-frequency waveforms, which can provide the isolation of clear seasonal fluctuations (subannual seasonality). The identification of subannual seasonality can provide the basis for an viable analysis of the number of building permits issued, the number of flats under construction and the number of new flats commissioned as a result from the occurrence of season-related phenomena. Seasonal fluctuations play an important role in the housing market for at least two reasons, i.e. due to technological processes and their links to weather conditions, and because of the time period of cash flows that appear on this market over the year (balance year).

For the purposes of decomposition of seasonal fluctuations, a procedure was applied, which involved the use of an inverse Fourier transform. Figures 8, 9 and 10 present the characteristics of decomposed seasonal fluctuations (sub-seasonal component) combined with initial data. We can notice note that with regard to each examined variable there is a very clear subannual variable component. The variable applied to define the number of building permits issued (see Fig. 8) as well as the variable with concerning construction of new dwellings (see Fig. 9) are characterized by almost identical seasonal characteristics. These variables have very similar annual seasonality profiles. The nature of the annual seasonal profile for the described variables is reflected in the process of housing construction technology, which means that fewer apartments are commissioned in winter, followed by a greater number in summer. We can also note that the variable describing the number of completed dwellings has different characteristics of seasonal subannual fluctuations, characterized by the fact that the majority of dwellings are completed every year just before the winter off-peak season (see Fig. 10).

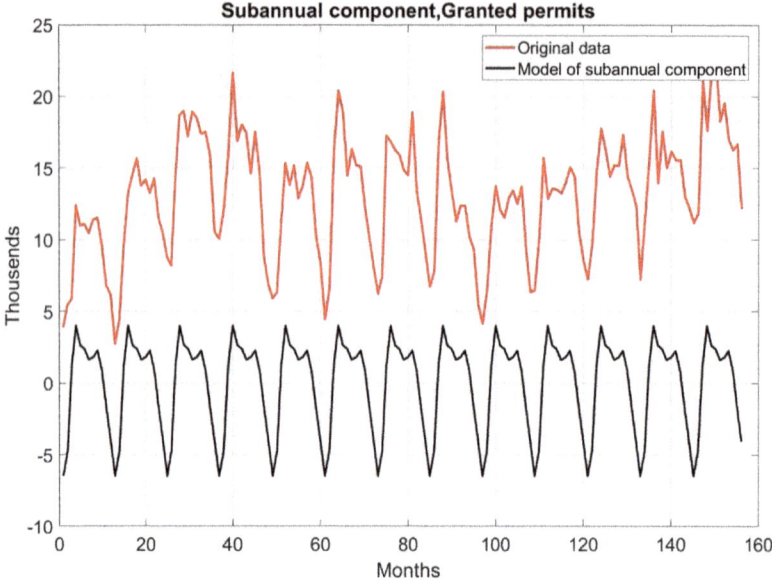

Fig. 8. Model of subannual component signal compared with original granted permissions data.

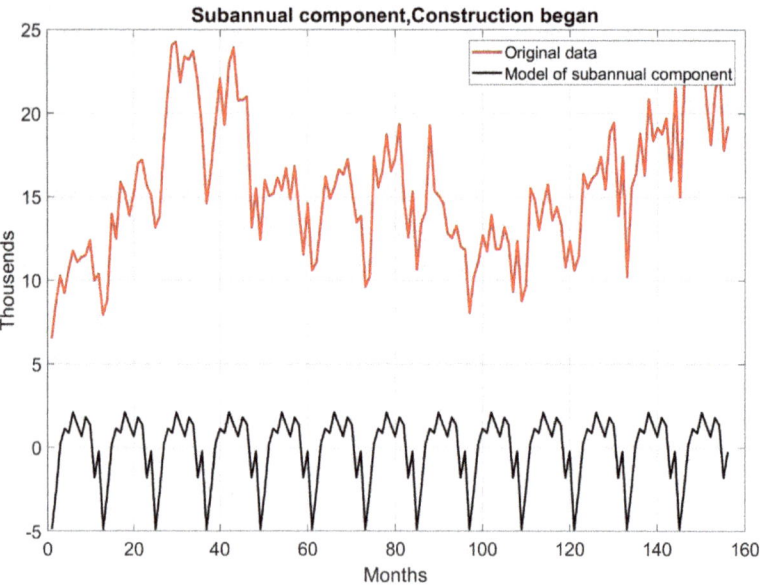

Fig. 9. Model of subannual component signal compared with original construction started data.

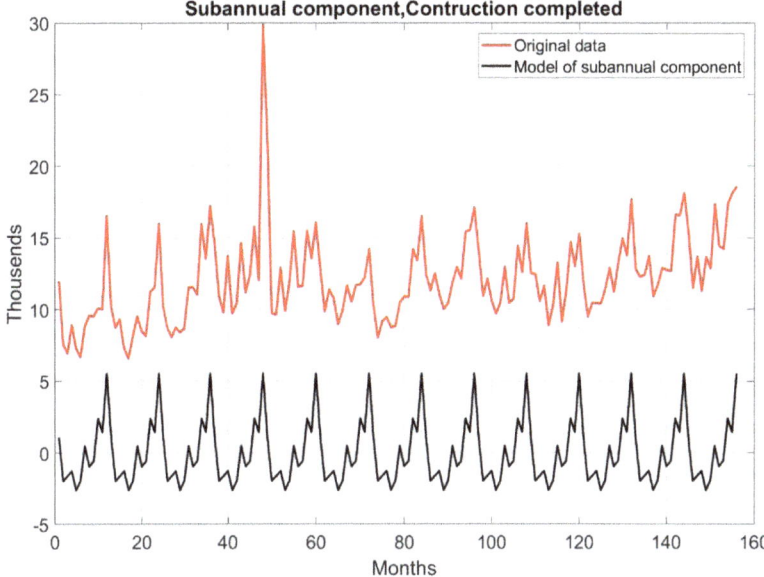

Fig. 10. Model of subannual component signal compared with original construction completed data.

6   Identification of Time-Frequency Variability in the Interannual Context

The final stage of decomposing the frequency components in the examined diagnostic variables involved the extraction cyclical fluctuations that occur over many years from their characteristics. This identification was carried our for the variable with subannual component removed by application of the Fourier transform over a selected window width. In the described research stage, the zero padding method was used. Figures 11, 12 and 13 present spectrograms with a slow-frequency component, where long-term spectra components of granted permits data, long spectra of construction granted data and long-term spectra components of construction completed data are listed, respectively. The analysis of individual charts could be applied to identify the occurrence of cycles and sub-cycles in the housing construction market. When an analysis was performed of the initial variables hat describe the number of permits issued for the construction of new dwellings, we can see that periodicity characteristics for cycles lasting 55, 41, 31 and 22 months is significant (see Fig. 11). For the variable describing the number of apartments under construction, the leading periodic cyclicality is primarily 55 months and 39 months (see Fig. 12).

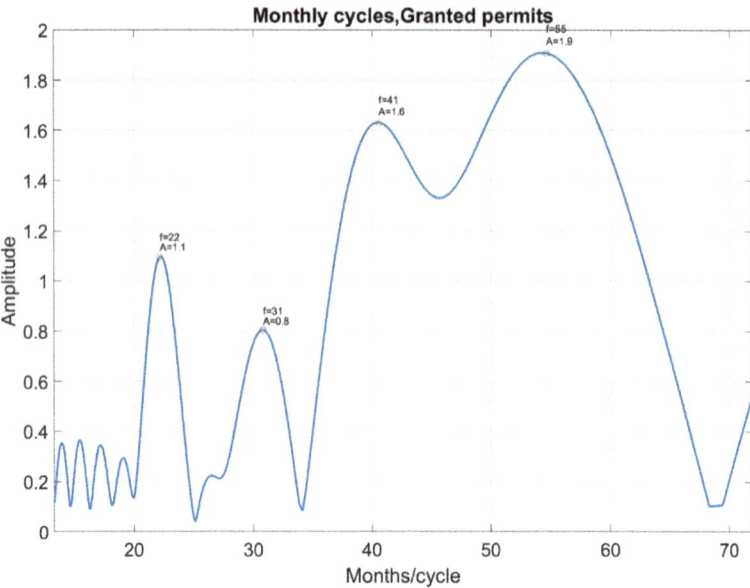

Fig. 11. Long-term spectra components of granted permits data with subannual component removed (low frequency component).

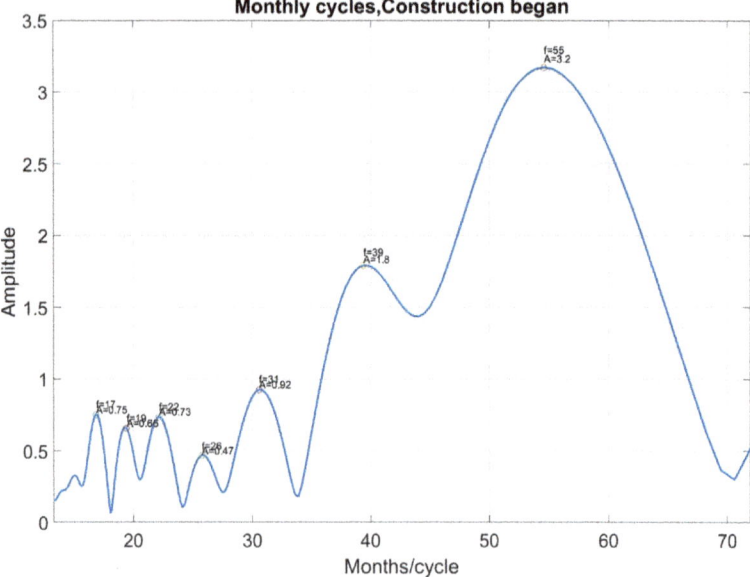

Fig. 12. Long-term spectra components of construction started data with subannual component removed (low frequency component).

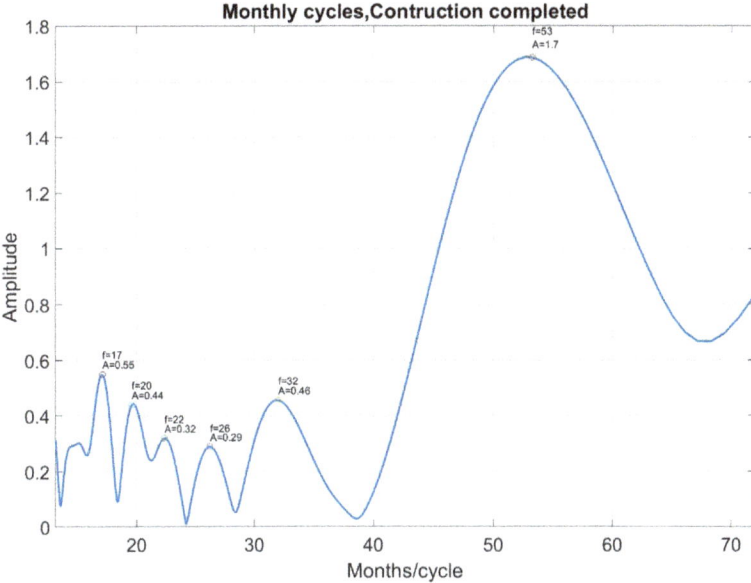

Fig. 13. Long-term spectra components of construction completed data with subannual component removed (low frequency component).

For the third investigated diagnostic variable that defines the number of new dwellings completed, periodicity occurs over intervals lasting 53 months (see Fig. 13).

7 Conclusions

The research involved the decomposition of frequency components for three variables that define the condition of the housing construction market. The first was associated with the use of the variable that describes the number of building permits issued, the second was related to the number of dwellings under construction, whereas the third was concerned with the number of new dwellings completed. The decomposition of these variables provided an adequate identification of frequency variations in terms of subannual waveforms, subannual seasonality and interannual cyclicality. As a result of performing a synthetic analysis of the results derived for subannual periodicity, we can see that significant level of periodicity is characterized for periods lasting 12, 6 and 3 months.

These values represent the significance of annual, semi-annual and quarterly regularities in the examined time series. However, as a consequence of the analysis of the results derived in the process of identifying seasonal fluctuations, we can apply them to create an annual seasonality profile for each of the three investigated variable components. We should note here that there is a significant similarity in seasonal characteristics between the variables describing building permits and flats under construction. When we take the last stage of the research into account, whose aimed was to identify cyclical variations, we can see that a significant period lasts 53 to 55 months for each examined variable, which is characterized by the highest amplitude waveform. Therefore, we can hypothesize that the cyclicality lasting four and a half years forms is the leading cyclic characteristic of the housing construction market.

The results of the present study, whose effect took the form of the adequate identification of subannual, seasonal and interannual variation, can be employed as premises for the support in the decision-making process taking place on the housing market. We can also note that the conducted research confirmed the applicability of the computational engineering for the decomposition of time series with waveforms representing the variability of the residential construction market.

Acknowledgement. The paper presents the personal opinions of the authors and does not necessarily reflect the official position of the Narodowy Bank Polski.

References

Bengtsson, E., Grothe, M., Lepers, E.: Home, safe home: cross-country monitoring framework for vulnerabilities in the residential real estate sector. J. Bank. Financ. **112**, 105268 (2018). https://doi.org/10.1016/j.jbankfin.2017.12.006

Chan, S., Han, G., Zhang, W.: How strong are the linkages between real estate and other sectors in China? Res. Int. Bus. Financ. **36**, 52–72 (2016). https://doi.org/10.1016/j.ribaf.2015.09.018

Coombs, W.T., Laufer, D.: Global crisis management – current research and future directions. J. Int. Manag. **24**, 199–203 (2018). https://doi.org/10.1016/j.intman.2017.12.003

Degl'Innocenti, M., Grant, K., Šević, A., Tzeremes, N.G.: Financial stability, competitiveness and banks' innovation capacity: evidence from the Global Financial Crisis. Int. Rev. Financ. Anal. **59**, 35–46 (2018). https://doi.org/10.1016/j.irfa.2018.07.009

Deng, Y., Zeng, Y., Li, Z.: Real estate prices and systemic banking crises. Econ. Model. **80**, 111–120 (2019). https://doi.org/10.1016/j.econmod.2018.09.032

Devaney, S., Xiao, Q.: Cyclical co-movements of private real estate, public real estate and equity markets: a cross-continental spectrum. J. Multinatl. Financ. Manag. **42–43**, 132–151 (2017). https://doi.org/10.1016/j.mulfin.2017.10.002

Escobari, D., Jafarinejad, M.: Date stamping bubbles in real estate investment trusts. Q. Rev. Econ. Financ. **60**, 224–230 (2016). https://doi.org/10.1016/j.qref.2015.10.003

Fan, Y., Yang, Z., Yavas, A.: Understanding real estate price dynamics: the case of housing prices in five major cities of China☆. J. Hous. Econ. **43**, 37–55 (2019). https://doi.org/10.1016/j.jhe.2018.09.003

Fernández Muñoz, S., Collado Cueto, L.: What has happened in Spain? The real estate bubble, corruption and housing development: a view from the local level. Geoforum **85**, 206–213 (2017). https://doi.org/10.1016/j.geoforum.2017.08.002

Gabauer, D., Gupta, R.: Spillovers across macroeconomic, financial and real estate uncertainties: a time-varying approach. Struct. Chang. Econ. Dyn. **52**, 167–173 (2020). https://doi.org/10.1016/j.strueco.2019.09.009

Gallegati, M., Delli Gatti, D.: Macrofinancial imbalances in historical perspective: a global crisis index. J. Econ. Dyn. Control **91**, 190–205 (2018). https://doi.org/10.1016/j.jedc.2018.01.026

Gong, P., Zou, D., Wang, J.: Pricing and simulation for real estate index options: radial basis point interpolation. Phys. A Stat. Mech. Appl. **500**, 177–188 (2018). https://doi.org/10.1016/j.physa.2018.02.135

Hartzell, J.C., Sun, L., Titman, S.: Institutional investors as monitors of corporate diversification decisions: evidence from real estate investment trusts. J. Corp. Financ. **25**, 61–72 (2014). https://doi.org/10.1016/j.jcorpfin.2013.10.006

He, Y., Xia, F.: Heterogeneous traders, house prices and healthy urban housing market: a DSGE model based on behavioral economics. Habitat Int. **96**, 102085 (2019). https://doi.org/10.1016/j.habitatint.2019.102085

Hei-Ling Lam, C., Chi-Man Hui, E.: How does investor sentiment predict the future real estate returns of residential property in Hong Kong? Habitat Int. **75**, 1–1 (2018). https://doi.org/10.1016/j.habitatint.2018.02.009

Hofman, A., Aalbers, M.B.: A finance-and real estate-driven regime in the United Kingdom. Geoforum **100**, 89–100 (2019). https://doi.org/10.1016/j.geoforum.2019.02.014

Lee, H.S., Lee, W.S.: Cross-regional connectedness in the Korean housing market. J. Hous. Econ. **46**, 101654 (2019). https://doi.org/10.1016/j.jhe.2019.101654

Liow, K.H., Newell, G.: Real estate global beta and spillovers: an international study. Econ. Model. **59**, 297–313 (2016). https://doi.org/10.1016/j.econmod.2016.08.001

Mach, Ł.: Prices of accommodation rental as functioning on the basis of a sharing economy in the capitals of cee states. Argumenta Oeconomica **45**, 141–162 (2020). https://doi.org/10.15611/aoe.2020.2.06

Mach, Ł.: Measuring and assessing the impact of the global economic crisis on European real property market. J. Bus. Econ. Manag. **20**, 1189–1209 (2019). https://doi.org/10.3846/jbem.2019.11234

Mach, Ł, Zmarzły, D., Dąbrowski, I., Frącz, P.: Identification and parametrization of polycyclicity in the primary housing market. J. Hous. Built Environ. (2021). https://doi.org/10.1007/s10901-020-09817-6

Mach, Ł, Zmarzły, D., Dąbrowski, I., Frącz, P.: comparison on subannual seasonality of building construction in European countries. Eur. Res. Stud. J. **23**, 241–257 (2020a)

Mach, Ł., Zmarzły, D., Dąbrowski, I., Frącz, P.: A time-frequency analysis of the housing construction time as the basis for making decisions (the case study of poland). Pr. Nauk. Uniw. Ekon. we Wrocławiu **64**, 54–71 (2020b). https://doi.org/10.15611/pn.2020.8.05

Mach, Ł., Zmarzły, D., Dąbrowski, I., Frącz, P.: Identification of harmonic models based on fourier transforms as a tool for parameterization of the dwelling construction market. In: Khalid, S. (ed.) Education Excellence and Innovation Management: A 2025 Vision to Sustain Economic Development during Global Challenges. Proceedings of the 35th International Business Information Management Association Conference (IBIMA), pp. 11295–11316. International Business Information Management Association (2020c)

Makin, A.J.: Lessons for macroeconomic policy from the Global Financial Crisis. Econ. Anal. Policy **64**, 13–25 (2019). https://doi.org/10.1016/j.eap.2019.07.008

Martins, A.M., Serra, A.P., Stevenson, S.: Determinants of real estate bank profitability. Res. Int. Bus. Financ. **49**, 282–300 (2019). https://doi.org/10.1016/j.ribaf.2019.04.004

Phasuphan, W., Praphairaksit, N., Imyim, A.: Removal of ibuprofen, diclofenac, and naproxen from water using chitosan-modified waste tire crumb rubber. J. Mol. Liq. **294**, 111554 (2019). https://doi.org/10.1016/j.molliq.2019.111554

Rokita-Poskart, D., Mach, Ł: Selected meso-economic consequences of the changing number of students in academic towns and cities (a case study of Poland). Sustainability **11**(7), 1901 (2019). https://doi.org/10.3390/su11071901

Manjunath, S.V., Baghel, R.S., Kumar, M.: Performance evaluation of cement–carbon composite for adsorptive removal of acidic and basic dyes from single and multi-component systems. Environ. Technol. Innov. **16**, 100478 (2019). https://doi.org/10.1016/j.eti.2019.100478

Seok, S.I., Cho, H., Ryu, D.: The information content of funds from operations and net income in real estate investment trusts. North Am. J. Econ. Financ. **51**, 101063 (2019). https://doi.org/10.1016/j.najef.2019.101063

Silva, M.E.A., da Nóbrega Besarria, C., Baerlocher, D.: Aggregate shocks and the Brazilian housing market dynamics. Economia **20**, 121–137 (2019). https://doi.org/10.1016/j.econ.2019.08.001

Tupenaite, L., Kanapeckiene, L., Naimaviciene, J.: Determinants of housing market fluctuations: case study of Lithuania. Procedia Eng. **172**, 1169–1175 (2017). https://doi.org/10.1016/j.proeng.2017.02.136

Wang, L., Li, S., Wang, J., Meng, Y.: Real estate bubbles in a bank-real estate loan network model integrating economic cycle and macro-prudential stress testing. Phys. A Stat. Mech. Appl. **542**, 122576 (2019). https://doi.org/10.1016/j.physa.2019.122576

Wright, D., Yanotti, M.B.: Home advantage: the preference for local residential real estate investment. Pacific Basin Financ. J. **57**, 101167 (2019). https://doi.org/10.1016/j.pacfin.2019.06.014

Yépez, C.A.: Financial intermediation and real estate prices impact on business cycles: a Bayesian analysis. North Am. J. Econ. Financ. **45**, 138–160 (2018). https://doi.org/10.1016/j.najef.2018.02.006

Zhao, S.X.B., Zhan, H., Jiang, Y., Pan, W.: How big is China's real estate bubble and why hasn't it burst yet? Land Use Policy **64**, 153–162 (2017). https://doi.org/10.1016/j.landusepol.2017.02.024

Automation of Administrative Processes in Referees' Committee of the Polish Football Association

Patryk Mauer[1]([✉]) and Damian Picz[2]

[1] Opole University of Technology, Prószkowska 76, 45-758 Opole, Poland
`patryk.mauer@student.po.edu.pl`
[2] Polish Football Association, Bitwy Warszawskiej 1920r. 7, 02-366 Warsaw, Poland

Abstract. This chapter is devoted to the automation of administrative processes of the Referees' Committee of the Polish Football Association with the use of applications written in the C# programming language and describes the localization of these processes, application development, their implementation and its results. The first chapters describe the institutions which processes were being automated. The main part is the processes location and automation with the description of development of three desktop applications which reduce the workload of the administrative employee of the Referees' Committee. The last sections present the implementation of the application into a business process and present the result of measurements that indicate the profitability of automation. This work emphasizes the importance of automation in the professional area and thus encourages the continuation of works towards automation.

Keywords: Automation · Business process automation · Console application · C#

1 Introduction

Nowadays automation is present in almost every area of one's live. There is a high need of saving time where it is possible with use of technology. The world is heading towards more automated systems where human plays a role more of a supervising unit than executing one. This chapter is focused on one of the branches of business process automations which is automation of administrative processes.

Polish Football Association is a football federation which is a member of UEFA and FIFA. It is the largest and one of the oldest Polish sports associations Polish Football Association was founded on 20 and 21 December 1919 in Warsaw. The Statute of the Association was created by Józef Lustgarten, Jan Polakiewicz and Jan Weyssenhoff. Its main objectives are, among others, education and development of grassroots football and refereeing, development of football, leagues, clubs, competitions, referees' association on each level across Poland. There are sixteen Regional Football Associations which are members of Polish Football Association, however, they are independent legal bodies. Polish Football Association is responsible for organization of competitions, i.e. football matches in League 2, League 1, Ekstraklasa, Women's Ekstraklasa, Women's League 1, Central Youth League, polish Cup and all international matches of the Polish national team [1].

1.1 Referees' Committee

The Referees' Committee of the Polish Football Association is one of the bodies of the Polish FA which is legally completely dependent on the Polish Football Association. It has around 8000 members who are referees and referee observers (former referees).

The Referees Committee of the Polish FA was the pioneer in regard to the system of assessments which results in promotions and demotions of referees and assistant referees.

The system is called "triple-checking" and its most important assumption is that referees are subjected to a three-dimensional assessment, i.e. the Referees' Committee gathers three documents after each match day during a round and at a meeting all three documents are juxtaposed and the Referees' Committee agrees on the final mark of the referee (this final mark and points are the basis for further decisions concerning promotions and demotions). The three above-mentioned documents are: referee's self-assessment report; referee observer's report and TV referee observer's report.

A self-assessment report – is a report (document) written by a referee after his match where the referee includes the most important situations from his match performance, both negative and positive. It is a way of a self-reflection analysis aimed at rising self-awareness of the referees and prompting their developing by strengthening their abilities to notice, analyze and find a way to eliminate any mistakes they make.

A referee – is a person who has obtained a special license confirming that appropriate theoretical and fitness tests have been passed and that the holder of this license is entitled to officiate matches in an appropriate league (based on promotions and demotions).

A referee observer – is a former referee (usually retired) who is appointed to matches to assess performances of referees and assistant referees. Based on his expertise he writes a report emphasizing positives and negatives, areas which require development and he marks all member of a refereeing team.

A TV referee observer - is a former referee (usually retired) who is appointed to matches to assess performances of referees and assistant referees. Based on his expertise he writes a report emphasizing positives and negatives, areas which require development and he marks all member of a refereeing team. However, his assessment is based on a live TV broadcast, i.e. such an observer is watching a match live on TV. What is important to mention is the fact that in Ekstraklasa TV observers are exclusively former referees, in League 1 and League 2 TV observers are also Ekstraklasa referees and assistant referees.

The above-mentioned triple-checking system requires considerable workload. Therefore, two full-time employees of the Refereeing Unit are dedicated to run this system technical and administration wise.

Until now the only automated process introduced in the Referees' Committee was the system of referees and referee observers appointments called Extranet.

2 Process Location

The first thing which is needed to be done to automate any process is to simply identify it. The crucial part is to gain the trust of an organization to have access to its interior processes. Finding the one which requires huge amounts of time or engages many people to execute tasks and/or has significant impact on other processes and systems should be analyzed in order to check the possibility to automation [2].

The Referees' Association of the Polish Football Association utilizes processes with factors indicating the need for automation. To locate these processes, the interview with administrative employee has been conducted, resulting with three processes which were the most susceptible to automation and the most important for business goals. These processes are as follows: 1. Process of folder hierarchy creation on Google Drive and managing folders' permissions. 2. Process of folder checking and notification sending. 3. Process of record creation.

All these processes require huge amount of repetitive task which are time-sensitive, what matches the criteria of key factors indicating the need for automation which are: high-volume of tasks, time-sensitive nature, significant impact on other processes and systems. The first three Figures depict these processes which for the need of this chapter are being referred to as "manual" (Figs. 1, 2 and 3).

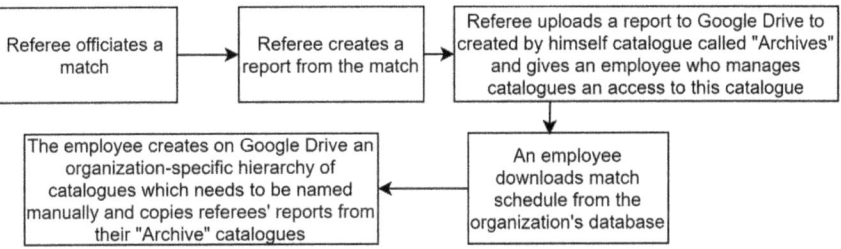

Fig. 1. The manual process of folder hierarchy creation on Google Drive.

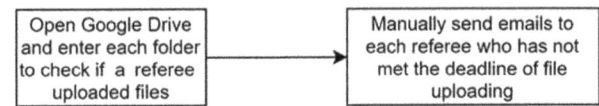

Fig. 2. The manual process of folder checking and notification sending.

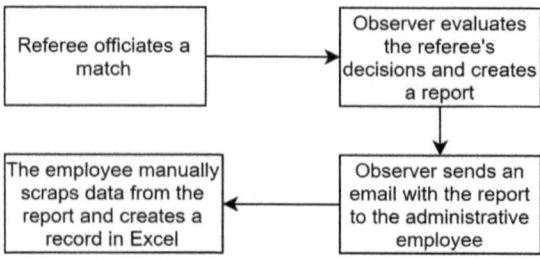

Fig. 3. The manual process of record creation.

3 Process Automation

To automate a process, it is necessary to start with a clear understanding of what tasks are involved, who is responsible for it and when each task needs to be executed. The goal of automation needs to be clearly defined and profitability of automation should be possible to estimate. The most important factor indicating worthiness of automation is an estimated ratio between amount of work being taken when utilizing a manual process and the cost of its automation. If the cost is manageable to be taken and a specified process takes a significant amount of resources, it is highly recommended to perform automation.

This section presents in detail the automated processes' flows which were constructed minding the lowest possible cost of automation.

3.1 Folder Hierarchy Creation

The new process replacing the manual folder hierarchy creation, which utilizes a software looks like the Fig. 4 presents.

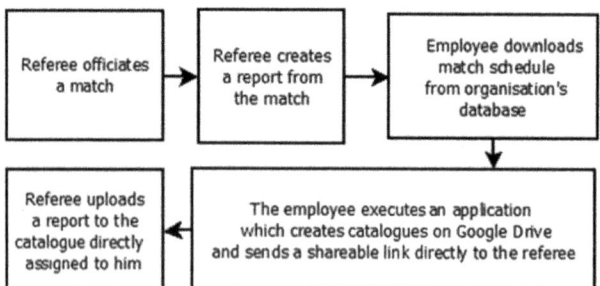

Fig. 4. The automated process of folder hierarchy creation on Google Drive with use of a software.

To be more detailed, the Fig. 5 shows a part of match schedule file.

			Referees' appointments First round, 1. matchday				
No.	League	Home	Away	Date	Time	Place	Referee
1	II league	Stal Rzeszów S.A.	MKP POGO Ń Siedlce	2019-07-27	17:00	STADION MIEJSKI W RZESZOWIE (Rzeszów, Hetma ńska 69)	Karol Iwanowicz (Kra ś nik)
2	II league	KS Legionovia	KS SKRA CZ Ę STOCHOWA S.A.	2019-07-27	18:00	Stadion Miejski im Ks. P łk. Jana Mrugacza w Legionowie (Legionowo, Parkowa 27)	Patryk Gryckiewicz (Toru ń)

Fig. 5. A part of match schedule file.

After downloading the file, the employee executes the application which scraps data this data to name folder to be created on Google Drive. The Fig. 6 presents the part of result of the application execution.

Fig. 6. The result of the application execution.

The application created a directory named 1. kolejka in a folder II liga 2019/2020. Inside the created folder the data scrapped from the match fixtures were used and folders were created. Subsequently, emails with upload permission were sent to the referees responsible for file uploading.

3.2 The Automated Folder Checking

The new process replacing the manual folder checking, which utilizes the software looks as in the Fig. 7 presents.

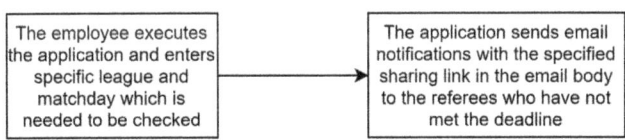

Fig. 7. The automated process of folder checking on Google Drive with use of the software.

The employee does not need to manually send emails to the referees who have not met the deadline. As the result of an execution of the application, only after navigation to the specified matchday, the emails are sent. The result is shown in the Fig. 8.

```
Choose League:
0 - Ekstraklasa
1 - I Liga
2 - II Liga
2
II liga 2019/2020
II liga 2019/2020
Choose Matchday:
Type number of matchday
1
GKS Giek Katowice   - Widzew Lódz   Szymon Lizak contains 14 files
Stal Stalowa Wola   - Elana Torun Robert Marciniak contains 17 files
Resovia Rzeszów - Górnik Polkowice Marcin Szrek contains 5 files
MKS Znicz Pruszków - KKS Lech II Poznan Jacek Lis contains 5 files
Gryf Wejherowo WKS - Garbarnia Kraków Filip Kaliszewski contains 15 files
Bytovia Bytów - KS Legionovia Lukasz Ostrowski contains 23 files
Blekitni Stargard - Stal Rzeszów  Maciej Pelka  EMPTY
ZKS Olimpia Elblag - KS Skra Czestochowa  Konrad Gasiorowski contains 13 files
```

Fig. 8. The result of application execution.

3.3   The Automated Record Creation

The new process replacing the manual folder checking, which utilizes the software looks as in the Fig. 9.

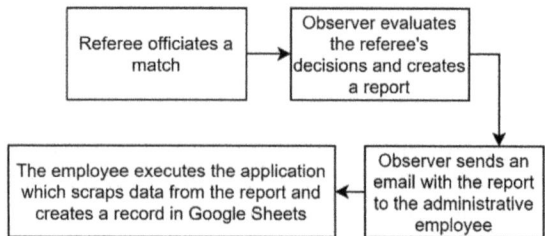

Fig. 9. The automated process of record creation with use of a software.

The employee receives reports from observers and executes the application. As these reports are in a .docx format, it is difficult to scrap the data from it. Therefore, the application uploads these reports to the Google Drive, where .docx is converted to a Google Document, to which it is possible to make API calls. This format is perceived as a structural tree of data from which the application scraps necessary data. In the next step, the application needs to replicate the table below which is provided to the employee in a Google Excel format. With use of Google API it is possible to have an access to each cell in the table and copy its data. The Fig. 10 shows the part of the table from which the data is copied.

II league - 13. matchday					
Match		Date	Time	Referee	Observer
TS PODBESKIDZIE S.A.	GKS PIAST GLIWICE S.A.	2020-12-11	18:00	Łukasz Szczech - Kobyłka	Marcin` Szulc
ZAGŁĘBIE LUBIN	ŚLĄSK WROCŁAW	2020-12-11	20:30	Zbigniew Dobrynin - Łódź	Katarzyna Wierzbowska

Fig. 10. Part of the table with data provided to the employee.

When the employee has received all data needed to create the record, he puts it into the folder and executes the application. To fill the rest of the table, the application uploads .docx files to the Google Drive in order to use the automatic data conversion to the Google Document format from which data is being read. The Figs. 11 and 12 show crucial (for the application flow) parts of the observer report in Google Document format.

POLISH FOOTBALL ASSOCIATION
Observer's report
Referee: ZBIGNIEW DOBRYNIN

Referee assistant 1: KONRAD SAPELA
Referee assistant 2: KAMIL WÓJCIK
Fourth official: MACIEJ PELKA

Observer: Katarzyna WIERZBOWSKA

Match: ZAGŁĘBIE LUBIN– ŚLĄSK WROCŁAW   Date: 11.12.2020

Fig. 11. The header of an observer's report.

The application scraps data from the header of an observer's report and using names of teams participating in a match localizes the row to which to upload the data. The process is performed with use of Google Docs API [3]. Using the data from the header, the application fills cells under the "referee" and "observer" columns. Moreover, it remembers the row to which to upload the data in the following steps. To fill cells under the "mark" column the evaluation part of the report needs to be scrapped. The Fig. 12 presents this part of the report.

	Referee	A1	A2
Mark > 8,4			
Mark 8,3 – 8,4	X	X	X

Fig. 12. A part of the evaluation table of an observer's report.

The application looks for the cell to which the data has been put by the observer and takes its value and fills the cell under the "mark" column. Other columns are filled with use of the second, similar report created by the second observer of the referee. As this process is very similar, the explanation of this is skipped.

As the result of the application execution, the record is generated. A part of the record is depicted in the Fig. 13.

Date	Match		Referee	Observer	Mark	Observer TV	Mark
2020-12-11	TS PODBESKIDZIE S.A.	GKS PIAST GLIWICE S.A.	Łukasz Szczech - Kobyłka	Julian Pasek	8,3	Marcin Szulc	8,3
2020-12-11	ZAGŁĘBIE LUBIN	ŚLĄSK WROCŁAW	Zbigniew Dobrynin - Łódź	Robert Małek	7,9	Katarzyna Wierzbowska	7,8

Fig. 13. The result of the application execution.

4 Third Party Technology

The crucial libraries for correct functioning of software which automates the Referee's Association's process are Google.Apis which are the client libraries designed to make requests to Google API which operate on Google Service. This communication for Google Drive is depicted on Fig. 14.

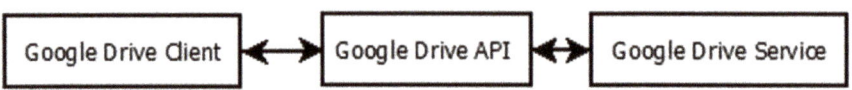

Fig. 14. Google Drive communication model.

The application Google API [4] uses OAuth flow to grant the application access to user account, what is shown in the Fig. 15.

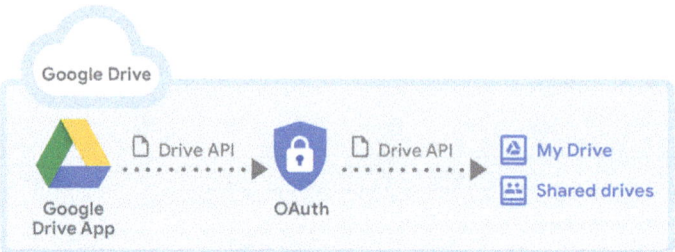

Fig. 15. Relationship diagram between Google Drive Application, Google Drive and Google Drive API.

- Google Drive App is the application which uses storage solution provided by Google Drive,
- OAuth is the authorization protocol that is required by Google Drive API to authenticate Google Drive App and handles OAuth 2.0 flow and application access tokens,
- Drive API is the REST application programming interface created by Google that allows to use Google Drive storage from within the application,
- My Drive/Shared drives represent Google Drive cloud storage.

In order to leverage the flow presented in the Fig. 15, there is a need to create a Cloud Platform project and enable the Google Drive API on it. This can be achieved in the Developers Console web page [5] (Fig. 16).

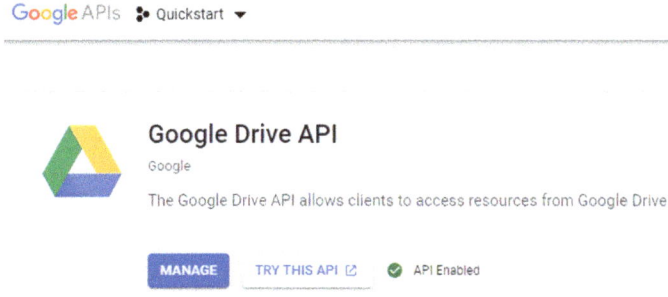

Fig. 16. The view after enabling the Google Drive API in the project.

Methods provided by Google API libraires were used in the automating applications development. The most useful ones were: Files.List(), Data.File(), Files.Create(), Permissions.Create(), Spreadsheets.Values.Update(), Spreadsheets.Values.Insert(), Spreadsheets.Values.Get(). The explanation and exemplary use cases can be found in Google APIs Docs. To leverage Google Drive API it is needed to install a NuGet package which consists of Http Client and tools facilitating communication with Google Drive API.

5   Development of the First Application

The software requires an html file which is generated by the employee from the organizations' database. This file contains data about teams participating in particular matchdays and referees who officiate in these matches. Another input is a configuration file which determines the actual year of the season. When these files are in the same folder as an application the business logic runs and uses the database to save and get referee emails and create catalogues on Google Drive and sends emails with access links.

Crucial for this application are Google.Apis.Drive.v3 NuGet package which provides ways to call the Google Drive API and HtmlAgilityPack which helps to scrap data from .html files [6]. This application uses Autofac as its IoC container. The Strategy pattern implementation [7] in this application allows to scrap data from the.html file in the way that the Strategy object indicates the part which is to be scrabbed. These specified strategy objects can be passed to the context to steer the flow of the method which returns needed lists of data. With the data scrapped, the folders can be created by creating metadata of Google Drive file and execution of Files.Create() method.

6   Development of Second Application

The software requires txt configuration files which determine the actual year of a season. When these files are in the same folder as an application, the business logic can run. The user is asked to provide input depending on what kind of a result is wanted to be shown.

The Google policy allows to download only a thousand metadata files per request, hence multiple call results must be stored in-memory. Useful for managing the stored metadata can be the Repository pattern implementation in the application which would facilitates retrieving data from memory [7].

When the specific league and matchday where chosen by the employee, the application starts to look for the folders containing league name, then inside its children finds specific matchday folder. Inside the matchday folder, there are usually 9 folders in which some files should be placed by the referees after each match. The key activity of checking if these files are placed is usage of File.Parents property.

7   Development of Third Application

The software requires the observer reports from which it scraps data and a table with initial pieces of information provided to the employee. As a result of the application execution, the table filled with data is created.

The application starts with requesting input from the user, who types the needed number of matchday from which the report is need to be made. Then the Google Sheet document storing the record is found in Google Drive.

Then the initial data provided to the employee is downloaded and placed to the final sheet. Subsequently, observer reports are being uploaded to the Google Drive and its IDs are listed in memory. After scrapping the from the observers reports, the imperative part of filling cells with evaluations is performed in a LINQ operation which looks for correct rows in which the destination evaluation cells are located. Searching for rows is done in a way that the already filled observers names cells are compared to the names on observers reports. If the observer name on the report matches the observer in the table, then the correct row is located. Subsequently, in the located rows cells are filled with the marks scrapped from the reports with use of Google Sheets API [8].

8   Software Implementation

After the software have been developed it is needed to be set up on the employee workstation. These applications where created as double-click ones, what means that to execute the application the employee double-clicks on the icon representing an.exe file or its shortcut. To transform the set of.dll files which represent the application as compiled source code, the publish operation needs to be performed. The Microsoft Visual Studio IDE was used to achieve it [9, 10]. The imperative for setting a double-click application is Target runtime which needs to be set to win-x64 to create the applications in.exe file. On the employee's workstation and shortcuts icons were created. The Fig. 17 shows the shortcuts to the applications'.exe files already installed on the employee's workstation.

Now, the only work which is needed to be done by the employee is to download business specific files and execute developed applications.

Fig. 17. Shortcuts to the applications'.exe files.

9 Summary

Susceptible for automation processes leveraged in Referees' Committee of the Polish Foot-ball Association had been localized, examined in detail, and automated with use of the software developed in C# programming language. The automation's main aim was to maximally decrease the workload of an administrative employee who performed specified actions repeatedly, spending the least possible resources to complete the automation.

To sum up, the average total time value was taken from the measures of three complete manual and automated employee's sessions. The average time spent to perform one manual session was 2 h 36 min 30 s per week. Now the same processes with use of C# desktop applications developed in this chapter, require on average only 15 min 30 s per week to achieve the same results. It shows that the employee saves around 2 h 11 min per week, what in the year perspective reaches around 100 h. These results prove that the automation was needed, and its implementation profitable.

Nowadays, time is starting to be more valuable than money, the search for automation should be carried out in every possible way. The programmed solutions are crucial for almost any business operation. Therefore, the trend of automation can be clearly seen throughout the last decades and is rapidly accelerating, what is the only right direction on the way to improving business performances.

References

1. Polish Football Association. https://www.pzpn.pl/en. Accessed 19 Jan 2021
2. Maurer, M., Scherzinger, A.: Performance and Capacity Planning Method for BPEL4WS-based Business Process Automation. OmniScriptum GmbH & Co. KG, Saarbrücken (2007)
3. Introduction to Google Docs API. Developers Google. https://developers.google.com/docs/api
4. Introduction to Google Drive API. Developers Google. https://developers.google.com/drive/api/v3/about-sdk. Accessed 19 Jan 2021
5. Google API Developers Console. https://console.developers.google.com/. Accessed 19 Jan 2021
6. Html Agility Pack Documenation. https://html-agility-pack.net/documentation. Accessed 19 Jan 2021
7. Shvets, A.: Dive Into Design Patterns. Refactoring. Guru, Kamianets-Podilskyi (2018)

8. Introduction to Google Sheets API. Developers Google. https://developers.google.com/she ets/api/guides/concepts. Accessed 19 Jan 2021
9. Paszkiel, S.: Computer game in UNITY environment for BCI technology, analysis and classifi- cation of EEG signals for brain-computer interfaces. In: Studies in Computational Intelligence, vol. 852, pp. 101–110 (2020). https://doi.org/10.1007/978-3-030-30581-9_12
10. Slapek, M., Paszkiel, S.: Detection of gestures without begin and end markers by fitting into Bezier curves with least squares method. Pattern Recognit. Lett. **100**, 83–88 (2017). https:// doi.org/10.1016/j.patrec.2017.10.006

Modern Devices and Software Solutions as a Tool for Education on Local Biodiversity: A Case Study

Andrzej Olczak[1][(✉)] [ID] and Jan M. Kaczmarek[2] [ID]

[1] Department of Electrical, Control and Computer Engineering, Institute of Computer Science, Opole University of Technology, Prószkowska 76, 45-758 Opole, Poland
[2] Department of Zoology, Poznan University of Life Sciences, Wojska Polskiego 71c, 60-625 Poznań, Poland

Abstract. The global biodiversity crisis is becoming one of the key challenges for the global community. At the same time, hands-on knowledge of local biodiversity is decreasing among young people, replaced by information about global issues and/or artificial products provided by the media. While information and computer technologies (ICT) are blamed for the alienation of people from the natural world surrounding them, they may also be used for education about wildlife. The article describes a case study of the use of ICT for education about local biodiversity. A progressive web app (PWA) as a universal software solution for all types of devices was used for creating a virtual educational path within the biodiversity-rich urban park. The PWA complements and expands existing educational facilities, provides a multimedia experience when learning about local biodiversity, and introduces elements of gamification to increase the motivation of users.

Keywords: ICT in education · Ecological education · Biodiversity · Urban ecology · PWA (Progressive Web Apps) · Cross-platform software

1 Introduction

The global biodiversity crisis is one of the key challenges for the global community [1, 2]. Its impacts already affect or will soon affect a plethora of human activities bene-fiting from the so-called ecosystem services, provided by living organisms [3, 4]. Ecosys-tem services are a broad scope of processes, that include, for example, the global circula-tion of the elements, which is crucial for the Earth system to function as we know it [5], storage of carbon (e.g. in wetlands and forests), which mitigates human-induced global warming [6], natural crop pollination [7], and biological pest control in cultivated crops [8]. Sustaining such services on both a global and regional scale is crucial for the sur-vival of human civilization [9, 10]. Additionally, biodiversity provides cultural services as well, with living organisms fuelling all forms of human creativity, from ancient myths to contemporary popular culture [11]. A growing number of studies investigate the role of biodiversity in sustaining human health and well-being [12]. Examples include the

S. Paszkiel (Ed.): ICBCI 2021, AISC 1362, pp. 307–321, 2021.
https://doi.org/10.1007/978-3-030-72254-8_32

benefits of access to natural, diverse green spaces for both physical and mental health [13] and the notion that a majority of contemporary pharmaceuticals were originally derived from living organisms – with many potential medicines remaining unknown [14]. On the other hand, degradation of natural ecosystems leads to the emergence and spread of new diseases, with COVID-19 being the strongest example [15].

Currently, global development leads to loss of biodiversity at an unprecedented rate – for example, the global populations of wild animals have decreased on average by 68% since 1970 [16]. The current rate of species loss is dramatically higher than known from paleontological data; as a consequence, the current situation has been termed 'the sixth mass extinction', and placed in line with past global extinction events, like the one that erased dinosaurs [17]. Because of the enormous impact that biodiversity has on human civilization, halting the widespread extinctions of species and simplification of ecosystems will enable humans to retain the benefits of biodiversity that they do already enjoy – which they will likely notice only when gone.

Education is crucial for building positive attitudes for rich biodiversity and healthy ecosystems [18]. Importantly, humans apparently show innate interest in other living organisms, a term known as 'biophilia' [19]. However, the abundance of stimuli provided by information technology may alienate people from the outside, natural world: after finishing early school education, children are able to identify many more Pokémon characters than actual species of wildlife that they could spot in the immediate vicinity of their own homes [20]. Even in the context of actual wildlife, contemporary children recognize exotic species that they encounter through mass media better than local species [21–23]. As a consequence, the lack of knowledge about local wildlife may lead to erroneous conviction that biodiversity is 'somewhere else' – with potentially devastating consequences for conservation of local ecosystems. Therefore, successful education on plants, animals and other organisms living in the local environment is one of the factors crucial for halting current biodiversity crisis [24, 25] – especially considering that schools all too often fail to provide such knowledge [20].

It is hard to imagine the present world without modern electronic devices, such as computers, smartphones, or tablets. This worldwide expansion of information and communications technology (ICT) is often blamed for the observed disconnection with nature, especially in younger people [26, 27]. However, new technologies can also be used for education and raising awareness on biodiversity [28–32], especially in people already strongly oriented towards digital stimuli [33]. The variety of devices and system platforms makes the software development process for such solutions complicated. Here, we present a progressive web app (PWA) designed to complement the existing educational facilities in a biodiversity-rich urban park. The use of PWA allows the development of a cross-platform application, similar in capabilities to native solutions, which can be run on any device supporting modern internet browsers [34]. This type of solution has the advantage of speeding up the programming process, because the written code is universal for all modern devices.

2 Study Area

Bolko Island park is located in the town of Opole, in southwest Poland (N: 50° 39'05.4″ E: 17°55'26.9″). The area of the park is approximately 130 hectares. In the past, the area used to be an island on the river Odra, but the channel separating it from the mainland was partly filled during the middle of the 20th century, and later replaced by a new, wider channel built in the 1990's [35]. However, the original island area is still clearly separated from the surrounding urban and farmland landscape (see Fig. 1).

Fig. 1. Aerial view of the Bolko Island park (Source: [36]).

Originally, the area of the park was covered by old-growth deciduous forest dominated by oak, locally replaced by riparian forest or traditional farmland; in 1913, the area was turned into an English-style landscape park, which led to the creation of open meadows and ponds, as well as the planting of numerous exotic species of trees [35]. Importantly, the original deciduous forest was partially retained. This, as well as ecological succession during the second part of the 20th century, led to the current situation in which the majority of park area remains a biodiversity-rich, deciduous forest, dominated by native tree species. As a consequence, the diversity and abundance of birds in Bolko Island park is significantly higher than in other urban parks in the region [37], and it includes species strongly associated with natural deciduous woodlands like the middle spotted woodpecker *Dendrocoptes medius* or collared flycatcher *Ficedula albicollis* [38]. In general, Bolko Island Park is one of the most interesting urban parks in Poland in terms of avian diversity [38]. Additionally, the area hosts species of invertebrates typical for relatively undisturbed forest habitats, e.g. the large predatory beetle *Carabus scheidleri* [39] or flower chafer *Protaetia speciossima* [40] – both species are legally protected in Poland. The system of meadows within the park hosts a large diversity of plants and insects preferring open landscapes [41], including protected species like the large copper butterfly *Lycaena dispar* [39]. Rare and protected species of plants, like hyssop loosestrife *Lythrum hyssopifolia* and eggleaf twayblade *Listera ovata*, are also present in the park [39, 42].

Despite the presence of numerous legally protected species, Bolko Island park has never been subject to a thorough biodiversity assessment. Additionally, the existing facilities designed for ecological education are limited to a series of information boards, listing plant species characteristics for each habitat type within the park. However, the boards do not contain any photos or illustrations, which constrains their attractiveness to visitors having no prior knowledge in the subject. The aim of the PWA presented in this paper is to provide an attractive source of information on local biodiversity to visitors of the park that would include a broad scope of knowledge, various multimedia, as well as elements of gamification.

3 Modern Software Solutions

There are many different operating systems today. On computers, systems like Windows, MacOs and Linux are commonly installed. Android and iOS systems are most often installed on smartphones and tablets [43]. Such a variety of platforms has implications in the complicated process of software development.

3.1 Native Applications

Native applications are software created directly on a given system, which induces the necessity of implementing the same software for each platform. In this case, developing code in different programming languages is often required, which is a time consuming and costly process. The advantage of this solution is the possibility of using the full hardware potential available from the system level.

3.2 Cross-platform Solutions

Apart from native applications, cross-platform solutions can be run in each operating system using one software solution. That means that there is no need to create many source codes, so this type of software can be developed more quickly and cheaply than native applications. The main disadvantage is that, because of its universal nature, this type of solution cannot use the full hardware potential of each device on which it runs. Progressive Web Applications (PWA) are a modern example of this type of solution.

3.3 Hybrid Solutions

Hybrid solutions represent a compromise between developing one application, as in cross-platform solutions, and utilizing the possibilities of true native applications. In this case, the written source code can be applied on each platform by compiling solutions directly in native applications. The main disadvantages of this type of solution are compatibility and debugging issues, as well as the lack of some modules [44]. The most popular hybrid solutions are: Flutter, Ionic, React Native, Apache Cordova and Xamarin [45].

3.4 Solutions Comparison

The choice of the appropriate technology really depends on the needs that the application is to meet. Most solutions can be realized within cross-platform programming. If there is a need to develop a universal application that uses the hardware potential, or if the application is to be introduced to the store of a given manufacturer, e.g. Google Play Store or App Store, it is worth using the hybrid applications. If you want to use the full hardware potential, the highest performance, and to be able to place the application in the store of a given manufacturer, then native programming is the best choice.

Comparing the size of installation, PWA would have the smallest size, hybrid application size would be greater and a native solution would have the biggest size. A comparison of the speed of running applications, from tapping on an app-icon up to toolbar render, indicates superiority of native solutions, while the PWA has a slightly slower result, and the hybrid applications are the slowest [46].

4 Progressive Web Application

Progressive Web Applications (PWA) are cross-platform solutions that bring a native app-like user experience. They are an extension of web applications that meet certain requirement, such as:

- correctly filled web manifest file – it stores parameters and information about the website,
- application must be served from a secure domain (HTTPS),
- app icons for each device must be applied,
- a service worker must be registered [47].

After meeting the rules, our application gains access to many hardware capabilities and features, e.g. geolocation, camera, file access, bluetooth and more, specified in the HTML5 API standard and depending on the hardware platform on which the application is running. Features and embedded devices sensitive from a security point of view are blocked in this type of solution, e.g. NFC sensor, serial port, contacts and SMS/MMS on mobile devices, geofencing and more [48]. Other advantages of this kind of application are:

- installability – besides the ability to run in a web browser, the application can also add its own icon to the device home screen, and tapping this after running the app will cause it be released in a native container,
- linkability – the app can be shared via URL address,
- network independence – the PWA app can work offline by using a combination of different technologies, like Service Workers, Cache API, Web Storage and IndexedDB,
- re-engageability – content can be updated dynamically, users can be engaged by receiving push notifications from Web Push API and system notifications from the Notifications API,
- responsiveness – user interfaces can be fitted to each screen size,

- safety – applications work over HTTPS and their URLs are furthermore verified with a start_url parameter in the manifest file,
- discoverability – PWAs have better representation in search engines [49].

5   Software Project

5.1   Design Assumptions

The main design assumption is to create an educational software with which participants will learn about the biodiversity of Bolko Island. Application users should be able to access the application from computers, smartphones and tablets. Thus, there is a difference regarding the various operating systems that the application should run on. Moreover, due to the division between computers and mobile devices, it is possible to utilize the additional possibilities of the second group of devices, primarily geolocation and camera. Therefore, it is desirable to develop two versions of the application – outdoor and virtual.

In the first, the outdoor version created for mobile devices, the application scenario will be based on the movement of users through the park to the destination points displayed on the map. In the second, the computer version, participants will be able to click on these points on the screen using a mouse. After discovering a point, information about the place will be displayed on the screen, including descriptions and multimedia materials in the form of photos, videos and audio. In order to increase the involvement of participants, the application will contain embedded, interactive themed games.

Users will be able to gain score points for specific tasks, such as getting to a destination point, winning interactive game, or reading content. By using score, the gamification factor will increase. In addition, additional points may be scored by uploading players' own photos of the species encountered in the field, using the built-in interface with camera support. The photos are then evaluated by a moderator (Fig. 2).

Fig. 2. Main screen of "Zagraj w BIO z LO" application.

Based on the above assumptions, the progressive web application – "*Zagraj w BIO z LO*" ('*Play BIO with the high school*') was developed [50].

5.2   Content Management System

In order to manage the content in the application and the multimedia materials, it is desirable to develop a content management system (CMS). Such a system should be characterized by a two-tier scope of management. In the first level, content moderators should be able to choose a map on which to place points and add a new one. Using the second level, it should be possible to embed target points on the map area and assign content to them. In addition, the moderator should have access to photos sent by the participants of the application and should be able to describe the facilities in order to send information back and award the appropriate amount of points to users' scores.

5.3   User Interface

User interface (UI) is a set of visual elements that are a part of the user's communication with the application. The interface elements may include, for example, the application menu, buttons, multimedia elements, and different colors. Important features of a good UI are: readability, clarity and intuitiveness. Such an interface increases the positive user experience (UX) and makes people more willing to return to use the application.

When developing a cross-platform application, it is difficult to implement the UI, due to the different sizes of the screens on which it will be run. Therefore, the responsive web design (RWD) rules should be applied. It means that the page layout should automatically adjust to the size of the application window. In the case of web applications, this goal is achieved through an appropriately adapted HTML structure and specific rules in cascading style sheets (CSS). The figure below shows an example of a UI layout for a smartphone (Fig. 3).

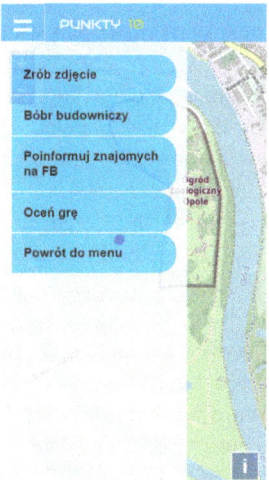

Fig. 3. Main menu of "*Zagraj w BIO z LO*" application.

5.4 Maps

There are several map solutions that can be used within applications. One of them is OpenStreetMap®. It is an open data project, which is licensed under the Open Data Commons Open Database License (ODbL) by the OpenStreetMap Foundation (OSMF). Developers are free to use this solution in their own projects; however, they are obliged to attribute OpenStreetMap on the map area. Another issue is the prohibition of using the OSM map server [51]. Therefore, it is important to run an own map server or to use an existing commercial one.

5.5 User Position Handling

Modern mobile devices, such as smartphones and tablets have built-in satellite navigation devices that support GPS, Galileo, Glonass and Beidou systems. These modules allow devices to recognize current position within a few meters' accuracy. Due to this, it is possible to track user location on the map.

Map support can be accomplished via OpenLayers library. It is a JavaScript open-source solution released under the 2-clause BSD License which supports display of maps, provides an API for building dynamic content, and makes the maps interactive [52].

By setting points on the map area in CMS using OpenLayers solution, we can easily get their geographical coordinates and save them to the database. Then, inside the application it is simple to get these coordinates and draw points dynamically on the users's map.

In order to handle user position tracking, Geolocation API must be used. It enables setting tracking options, such as high accuracy, timeout of return to a position, and maximum age of a cached position, to which it is acceptable to return. This method comes from the HTML5 API [53]. The next step is to display user position on the map by adding points and dynamically changing the user's position by using the method "setGeometry" with actual coordinates acquired from the method "getPosition" on position change from Geolocation API. In addition, the map can be centered based on the user position. In that case, the "setView" method with the center parameter on a current position must be applied.

Checking whether the user is within the range of a point marked on the map is also carried out in the position change method. User coordinates are verified with the "getClosestFeatureToCoordinate" method, which returns coordinates of the nearest point on the map. Then, the distance difference is calculated, and when the user is close, it returns a response and displays the educational content. Using the coordinates of a point, it is possible to check its contents in the database.

The map support for the desktop version is similar. The difference is in the method of obtaining coordinates. In that case, user coordinates are acquired by clicking on the map using the "pointerdown" method from map API (Fig. 4).

Fig. 4. User position tracking (blue point) as it moves to red point, displayed on OpenStreetMap® map.

5.6 Educational Content

Descriptions for points contain educational content. In order for application users to read the content with pleasure, it is important that the descriptions should be interesting, engaging, and entertaining. In order to increase readability, the text should be legible with important elements highlighted by the size and boldness of the font and a selection of appropriate colors. Additionally, descriptions should be enriched with multimedia materials, such as photos, videos and audio. The figure below shows an example of an educational content window with the above-described elements (Fig. 5).

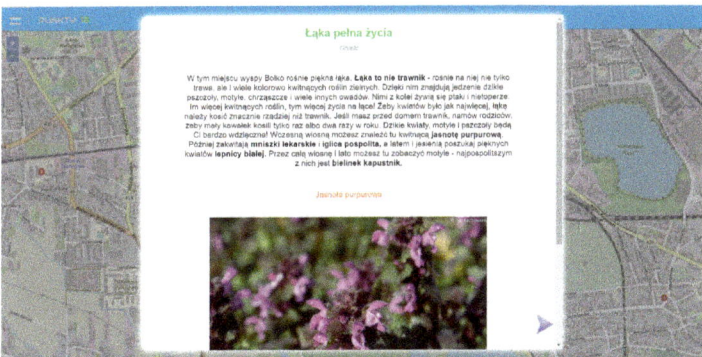

Fig. 5. Sample window containing educational content.

5.7 Interactive Built-In Games

The factor that will allow for greater user involvement is the implementation of interactive built-in games. These games should be short and concise, with a story that fits into the educational character of the application. At the end of each game, short educational content should appear. Users will be able to compete with each other by gaining points. Moreover, games should be available as a reward, e.g. after reaching a given user score or passing a given number of points on the map. In such cases, games can be added to the application menu and remain available permanently.

The following interactive games have been built-in as part of the application "*Zagraj w BIO z LO*", the first three appear in the spring-summer version of the application, while the rest are used in the autumn-winter version:

- Beaver the builder (pol. *Bóbr budowniczy*) – this game allows users to get acquainted with the life of the European beaver. The participants must collect tree branches and twigs in order to create a dam on the river; during movement, they also have to cross busy roads where car traffic poses a danger;
- Bee race (pol. *Pszczeli wyścig*) – users play the role of a bumblebee that collects nectar from flowers appearing in the meadow;
- Toad love vs. cars (pol. *Ropusza miłość kontra auta*) – the users takes on the role of a European common toad migrating towards a breeding pond, separated by a busy road – the goal is to avoid collision and reach water;
- Winter of a duck (pol. *Kacza* zima) – users play the role of a wild duck that forages in a park pond during winter; the goal is to consume edible objects (e.g. aquatic plants, seeds) and avoid inedible or unhealthy objects (e.g. moldy bread, plastic waste). Consumption of the latter objects leads to reduced health score. To mirror the actual situation of birds in winter, the energy score decreases with time and must be supplied with appropriate food;
- Plant the forest with the jay (pol. *Posadź las z sójką*) – this is a memo game. In the first stage (autumn), the users hide food (acorns) under various types of objects (twigs, rocks, leaves etc.). In the second stage (winter), acorns must be retrieved from the same places.
- It's an owl's life (pol. *Sowie życie*) – the users plays the role of a tawny owl that catches mice appearing on the board in a winter setting. The energy score decreases with time, thus it needs to be constantly supplemented with food. Inadequate energy levels at the end of a game result in failure – overcoming the acute stress inflicted by the New Year fireworks requires high energy levels (Figs. 6 and 7).

Fig. 6. Spring-summer versions of interactive built-in games

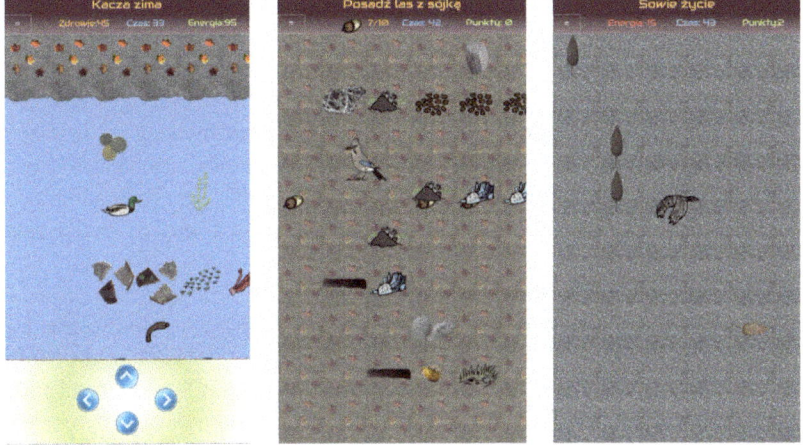

Fig. 7. Autumn-winter versions of interactive built-in games

5.8 Photos from Participants

Additionally, if the user would like to gain extra points, he or she may take a photograph of an animal, a plant or a fungus via the application camera tool, and upload it into the application. The photo is then assessed by a moderator inside the CMS, who will identify the species, add a number points to the user's score, and send this information back to the user.

6   Outdoor Software Tests

After completion of the programming work, outdoor software tests on Bolko Island were performed on the beta version of application. For the purposes of testing, four groups of young people were involved – three from the primary schools and one from the secondary school (164 people together). The whole process was divided into three days. Outdoor tests consisted of verifying the correct operation of the application and checking the educational issues with the help of a final survey. As a result of verification, the key elements of the application were assessed. The following Table 1 gives a summary of the questions rating among all users.

Table 1. User survey summary

User survey questions	Rating [1–5]
How do you rate the difficulty in finding a point from the map area?	3
How did you like the descriptions for the points?	3,7
How do you rate the multimedia materials (photos, videos and audio)?	3,7
How did you like the interactive games?	3,8
How do you rate the game overall?	3,8

The participants had moderate difficulty in finding their destination on the map. This is because the users need to identify the next point of destination according to their own location on the map, indicated by a centered light blue point. The application has been designed in such a way with the purpose of developing the ability to read a map and orient oneself in the area.

Another issue is the evaluation of the educational content of applications and interactive games and the overall evaluation of the game. The average rating of each of the above elements is similar and close to good. In the process of analyzing the sample of participants, it was evident that there existed a dependence of the low rating of individual components with errors in the application, which were indicated by users in the optional field with comments. The most common error that appeared during the tests was the inability to use the navigation module on some models of mobile devices. In some cases, there was a need to manually set the permission to use location mode inside the browser in the phone settings, which caused initial difficulties. In a few phone models, during the navigation initialization, the system returned a message about the lack of such a module; in this case, the users were not able to perform the tasks provided by the application.

7   Summary

The combination of modern technologies, knowledge provided in a medium different from usual classroom learning, and outdoor activity may be a promising form of complementing the education process concerning biodiversity. The use of built-in games as a reward for reading educational material may support the students to actively, tangibly participate in the learning process. In addition, the mobile version of the application is a field area game in which students can be engaged through outdoor activities, which will increase their interest in the subject and may improve their physical condition. Another important point is that participants need to orientate themselves in the area, thus they can improve their skills in reading maps.

An important aspect is the availability of devices which enable this type of education. Nowadays, computers and mobile devices are common. The advantage of mobile devices over computers is the potential to link direct, outdoor experience with learning: various built-in sensors and modules can be utilized, such as geolocation, accelerometer, gyroscope, and many more, which significantly expand the capabilities of smartphone-based educational applications.

The development of cross-platform applications such as PWAs, on the one hand, allows the use of various possibilities of system platforms. On the other hand, they are associated with certain challenges. The biggest challenges include the need to support different screen sizes as well as errors related to the geolocation module in some models of mobile devices.

As knowledge of local biodiversity decreases, linking the educational process in certain places where discussed species can actually be encountered may be a step towards improved environmental education. The *"Zagraj w BIO z LO"* application allows the acquisition (through educational content) and consolidation (through built-in games) of knowledge about the species of wildlife present in the area and potential threats to their survival.

References

1. IPBES: Global assessment report on biodiversity and ecosystem services of the Intergovernmental Science-Policy Platform on Biodiversity and Ecosystem Services. IPBES secretariat, Bonn, Germany (2019)
2. McElwee, P., Turnout, E., Chiroleu-Assouline, M., Clapp, J., Isenhour, C., Jackson, T., Kelemen, E., Miller, D.C., Rusch, G., Spangenberg, J.H., Waldron, A.: Ensuring a post-COVID economic agenda tackles global biodiversity loss. One Earth **3**(4), 448–461 (2020)
3. Chaplin-Kramer, R., Sharp, R.P., Weil, C., Bennett, E.M., Pascual, U., Arkema, K.K., Brauman, K.A., Bryant, B.P., Guerry, A.D., Haddad, N.M., Hamann, M.: Global modeling of nature's contributions to people. Science **366**(6462), 255–258 (2019)
4. Mooney, H., Larigauderie, A., Cesario, M., Elmquist, T., Hoegh-Guldberg, O., Lavorel, S., Mace, G.M., Palmer, M., Scholes, R., Yahara, T.: Biodiversity, climate change, and ecosystem services. Curr. Opin. Environ. Sustain. **1**(1), 46–54 (2009)
5. Norris, K., Terry, A., Hansford, J.P., Turvey, S.T.: Biodiversity conservation and the earth system: mind the gap. Trends Ecol. Evol. **35**(10), 919–926 (2020)
6. Soto-Navarro, C., Ravilious, C., Arnell, A., De Lamo, X., Harfoot, M., Hill, S.L.L., Wearn, O.R., Santoro, M., Bouvet, A., Mermoz, S., Le Toan, T.: Mapping co-benefits for carbon storage and biodiversity to inform conservation policy and action. Philos. Trans. R. Soc. B **375**(1794), 20190128 (2020)
7. Aizen, M.A., Garibaldi, L.A., Cunningham, S.A., Klein, A.M.: Long-term global trends in crop yield and production reveal no current pollination shortage but increasing pollinator dependency. Curr. Biol. **18**(20), 1572–1575 (2008)

8. Woodcock, B.A., Bullock, J.M., McCracken, M., Chapman, R.E., Ball, S.L., Edwards, M.E., Nowakowski, M., Pywell, R.F:. Spill-over of pest control and pollination services into arable crops. Agric. Ecosyst. Environ. **231**, 15–23 (2016)

9. Guo, Z., Zhang, L., Li, Y.: Increased dependence of humans on ecosystem services and biodiversity. PLoS One **5**(10), (2010)

10. Kubiszewski, I., Costanza, R., Anderson, S., Sutton, P.: The future value of ecosystem services: global scenarios and national implications. Ecosyst. Serv. **26**, 289–301 (2017)

11. Clark, N.E., Lovell, R., Wheeler, B.W., Higgins, S.L., Depledge, M.H., Norris, K.: Biodiversity, cultural pathways, and human health: a framework. Trends Ecol. Evol. **29**(4), 198–204 (2014)

12. Hough, R.L.: Biodiversity and human health: Evidence for causality? Biodivers. Conserv. **23**(2), 267–288 (2014)

13. Van den Berg, M., Wendel-Vos, W., Van Poppel, M., Kemper, H., van Mechelen, W., Maas, J.: Health benefits of green spaces in the living environment: a systematic review of epidemiological studies. Urban For. Urban Greening **14**(4), 806–816 (2015)

14. Herndon, C.N., Butler, R.A.: Significance of biodiversity to health. Biotropica **42**, 558–560 (2010)

15. Schmeller, D.S., Courchamp, F., Killeen, G.: Biodiversity loss, emerging pathogens and human health risks. Biodivers. Conserv. **29**, 3095–3102 (2020)

16. Almond, R.E.A., Grooten, M., Petersen, T.: Living planet report—bending the curve of biodiversity loss. WWF, Gland, Switzerland (2020)

17. Ceballos, G., Ehrlich, P.R., Barnosky, A.D., García, A., Pringle, R.M., Palmer, T.M.: Accelerated modern human–induced species losses: entering the sixth mass extinction. Sci. Adv. **1**(5), (2015)

18. Van Weelie, D., Wals, A.: Making biodiversity meaningful through environmental education. Int. J. Sci. Educ. **24**(11), 1143–1156 (2002)

19. Wilson, E.O.: Biophilia. Harvard University Press, Cambridge (1984)

20. Balmford, A., Clegg, L., Coulson, T., Taylor, J.: Why conservationists should heed Pokémon. Science **295**(5564), 2367 (2002)

21. Almeida, A., García Fernández, B., Strecht-Ribeiro, O.: Children's knowledge and contact with native fauna: a comparative study between Portugal and Spain. J. Biol. Educ. **54**(1), 17–32 (2020)

22. Genovart, M., Tavecchia, G., Enseñat, J.J., Laiolo, P.: Holding up a mirror to the society: children recognize exotic species much more than local ones. Biol. Cons. **159**, 484–489 (2013)

23. Schuttler, S.G., Stevenson, K., Kays, R., Dunn, R.R.: Children's attitudes towards animals are similar across suburban, exurban, and rural areas. PeerJ **7**, (2019)

24. Menzel, S., Bögeholz, S.: The loss of biodiversity as a challenge for sustainable development: how do pupils in Chile and Germany perceive resource dilemmas? Res. Sci. Educ. **39**, 429–447 (2009)

25. Palmberg, I., Berg, I., Jeronen, E., Kärkkäinen, S., Norrgård-Sillanpää, P., Persson, C., Vilkonis, R., Yli-Panula, E.: Nordic-Baltic student teachers' identification of and interest in plant and animal species: the Importance of species identification and biodiversity for sustainable development. J. Sci. Teacher Educ. **26**(6), 549–571 (2015)

26. Longbottom, S., Slaughter, V.: Direct experience with nature and the development of biological knowledge. Early Educ. Develop. **27**(8), 1145–1158 (2016)

27. Truong, M.A., Clayton, S.: Technologically transformed experiences of nature: a challenge for environmental conservation? Biol. Cons. **244**, (2020)

28. Buettel, J.C., Brook, B.W.: Egress! how technophilia can reinforce biophilia to improve ecological restoration. Restor. Ecol. **24**(6), 843–847 (2016)

29. Callahan, M.M., Echeverri, A., Ng, D., Zhao, J., Satterfield, T.: Using the phylo card game to advance biodiversity conservation in an era of Pokémon. Palgrave Commun. **5**, 79 (2019)
30. Coghlan, A., Carter, L.: Serious games as interpretive tools in complex natural tourist attractions. J. Hospitality Tour. Manage. **42**, 258–265 (2020)
31. Dorward, L.J., Mittermeier, J.C., Sandbrook, C., Spooner, F.: Pokémon go: benefits, costs, and lessons for the conservation movement. Conserv. Lett. **10**(1), 160–165 (2017)
32. Sandbrook, C., Adams, W.M., Monteferri, B.: Digital games and biodiversity conservation. Conserv. Lett. **8**, 118–124 (2015)
33. Edwards, R.C., Larson, B.M.H.: When screens replace backyards: strategies to connect digital-media-oriented young people to nature. Environ. Educ. Res. **26**(7), 950–968 (2020)
34. Can I use. https://caniuse.com/?search=pwa. Accessed 09 Jan 2021
35. Adamska, A.: Odrzańskie wyspy Opola – Bolko i Pasieka. Historia, rozwój przestrzenny i walory krajobrazu. Prace Komisji Krajobrazu Kulturowego **33**, 137–157 (2016). (in Polish)
36. Główny Urząd Geodezji i Kartografii. https://mapy.geoportal.gov.pl/imap/Imgp_2.html. Accessed 12 Jan 2021
37. Kopij, G.: Wstępne wyniki badań nad składem gatunkowym, rozmieszczeniem i liczebnością ptaków lęgowych na Wyspie Bolko w Opolu. Przyroda Śląska Opolskiego **24**, 15–24 (2018). (in Polish)
38. Kowalski, M., Zawadzki, M.: Bolko – wyspa ptasich skarbów. Ptaki Polski **57**, 38–41 (2020). (in Polish)
39. Mazur, M., Berlik, Ł.: Występowanie *Carabus (Morphocarabus) scheidleri scheidleri* Panzer, 1799 na wyspie Bolko w Opolu. Fragmenta Naturae **48**, 1–5 (2015). (in Polish)
40. Global Biodiversity Information Facility. https://www.gbif.org/occurrence/2311280456. Accessed 28 Jan 2021
41. Mazur, M., Ślusarska, E.: Ryjkowce (Coleoptera: Curculionoidea) wybranych zbiorowisk roślinnych Wyspy Bolko w Opolu. Fragmenta Naturae **42**, 87–97 (2009). (in Polish)
42. Nowak, A., Nowak, S.: Nowe stanowisko krwawnicy wąskolistnej *Lythrum hyssopifolium* L. na Śląsku Opolskim. Chrońmy Przyrodę Ojczystą **64**(2), 64–69 (2008). (in Polish)
43. Statcounter GlobalStats. https://gs.statcounter.com/os-market-share. Accessed 09 Jan 2021
44. Chrzanowska, N.: React Native Pros and Cons - Facebook's Framework in 2021 (Updated), Netguru. https://www.netguru.com/blog/react-native-pros-and-cons. Accessed 10 Jan 2021
45. Google Trends. https://trends.google.com/trends/explore?q=react%20native,flutter,xamarin,cordova,ionic. Accessed 10 Jan 2021
46. Biørn-Hansen, A., Majchrzak, T.A., Grønli, T.-M.: Progressive web apps: the possible web-native unifier for mobile development. In: Proceedings of the 13th International Conference on Web Information Systems and Technologies, WEBIST, Porto, vol. 1, pp. 344–351 (2017). https://doi.org/10.5220/0006353703440351. ISBN: 978-989-758-246-2
47. MDN Web Docs. https://developer.mozilla.org/en-US/docs/Web/Progressive_web_apps. Accessed 10 Jan 2021
48. What web can do today? https://whatwebcando.today. Accessed 10 Jan 2021
49. MDM Web Docs. https://developer.mozilla.org/en-US/docs/Web/Progressive_web_apps/Introduction#advantages_of_web_applications. Accessed 11 Jan 2021
50. Zagraj w BIO z LO. https://zagrajwbiozlo.pl. Accessed 11 Jan 2021
51. OSMF Operations Working Group. https://operations.osmfoundation.org/policies/tiles. Accessed 12 Jan 2021
52. OpenLayers. https://openlayers.org. Accessed 13 Jan 2021
53. MSD Web Docs, PositionOptions. https://developer.mozilla.org/en-US/docs/Web/API/PositionOptions. Accessed 12 Jan 2021

Influence of Program Architecture on Software Quality Attributes

Rafał Mzyk[✉] and Szczepan Paszkiel[iD]

Faculty of Electrical Engineering, Automatic Control and Informatics,
Opole University of Technology, Prószkowska 76, 45-758 Opole, Poland

Abstract. Day by day grow of technology allow programmers to solve more sophisticated problems and meet wide range of business requirements. Bigger expectations are also bigger responsibilities towards software engineers to make their implementations scalable, reusable and extendable. Consciously made trade-off's during development phase of software creation can emphasize some of its quality attributes but it is also very easy to make a piece of software which is nearly impossible to be, for instance, scalable. Even though there is still no perfect architecture, it is worth looking for solutions that will bring us more benefits than losses in the future.

Keywords: Software architecture · Software attributes · Decision making · Software development · Trade-offs in software architecture

1   Introduction

Software is everywhere, it doesn't matter if we run small grocery shop or we are a leader in big banking industry. We need a software to connect to our customers and to face with their needs. Depending on business we are running we would be interested in different software quality attributes. Quoting Jeff Offutt's "Quality Attributes of Web Software Applications" [2] - "[…] three of the most important quality criteria for success of web applications (and thus, the underlying software), were given as: 1. Reliability 2. Usability 3. Security An additional four important criteria are: 4. Availability 5. Scalability 6. Maintainability 7. Time-to-market". As Jeff surveyed there are various of quality attributes that are important. The order of it will be changed depending on business specific criteria. For banking industry, it will be security and reliability. We want to transfer our money 24/7 and it cannot have any errors because when it comes to money there is no place for mistakes. In other hand for e-commerce it will be availability and scalability. Online shop that doesn't work in Christmas time would be big money loss for business. It would be also big loss when software cannot deal with big number of customers taking order in the same time. But what does not depend on some business specific factor is that we can choose which of software attributes are crucial in scope of our project by making design decisions by building software architecture. According to "Design It" by Michael Keeling [3] - "A system's software architecture is the set of significant design decisions about how software is organized to promote desired quality

© The Author(s), under exclusive license to Springer Nature Switzerland AG 2021
S. Paszkiel (Ed.): ICBCI 2021, AISC 1362, pp. 322–329, 2021.
https://doi.org/10.1007/978-3-030-72254-8_33

attributes and other properties.".". Undoubtedly, wise design decisions will guarantee that some of quality attributes will be naturally exposed by our software.

2 Scalability

When talking about scalability we need to remember that there are two different ways of scaling our applications. Charles B. Weinstock and John B. Goodenough [4] defined application scalability like: "1. Scalability is the ability to handle increased workload (without adding resources to a system). 2. Scalability is the ability to handle increased workload by repeatedly applying a cost-effective strategy for extending a system's capacity.".". In modern software development we have many tools to prove those definitions. When we want our application to work with increased workload without adding more resources it is possible by using, for example, load balancers. Of course, we can always add more hardware or run another instance of our application. But it is not optimal way to deal with scalability and hardware scalability is not infinite. We may get to the level when another CPU or RAM won't improve our application performance. There come our design decisions that can make it much easier. It is very hard topic because we can literally scale different parts of our application. Everything depends on type of software we run. Some parts of our software are nearly impossible to scale – for instance presentation layer. We can work on optimization to make working with our GUI very smoothly, but it's always rendered on our client's side, so we don't really can do much with it that relates to scalability. What we can do is caching some of the data or remember state of our application in our client's side, so we don't need to call our API or database as many times. But the trade-off here is security – not every data should be stored in cookies or local storage. It also complicates our presentation layer a lot of which costs hours and slows time-to-market. It is very similar with desktop GUI's or mobile devices. When it comes to data layer there is more options. Sometimes architectural decision is not only how we would write our software but also which pieces we connect with each other. Depending on our needs we can use technology that resolving our problems. There are some popular databases like Elasticsearch which are supporting great number of events per seconds. It will scale with number of requests much better than Postgres. We can also implement some caching database which significantly will improve response time when there are many clients. When it comes to data layer, we can choose not only a technology. We have also pattern like CQRS. CQRS is sometimes called good practice but for me it is just architectural pattern. It is acronym of 'Command Query Responsibility Principle' and we can implement it in our application on different levels. The core clue behind CQRS is to separate commands from queries. After some development we often see that there are much more queries than commands. It is natural – in most programs we want to read data more than inserting it. It is also worth to mention that when analyzing our application business logic there are much more constraints on how to insert data than how to read it. It is obvious that if something starts to be complicated in software development world, we decide to aggregate it or fragmentize it. When we choose to implement this approach on data layer level, we would introduce different databases for commands and queries. Normally with very big projects it is NoSQL database for queries and classic relational database for commands. In most NoSQL databases data structure is looser and there is no transactions so fetching data is much faster. Although transactions are helpful while inserting data. The trade-offs are: two data bases to maintenance, latency, data synchronization, more technologies to learn, higher time to market, longer development phase. It seems that we pay a lot for scalability. But CQRS is not only used for scalability. It improves other attributes like code maintenance – it is easier to work with software that has organized structure with separated logic. Application layer is something we can scale the most by applying patterns and approaches that supports it. Lately popular are microservices architecture which one of the best if not the best advantage is scalability, but it has also a lot of disadvantages that should be taken into consideration. Microservices architecture pattern is a way of developing software that every big module should be independent or as little dependent as possible on other modules. It should be deployed as separated deployment unit. Paolo Di Francesco defines microservices in "Architecting Microservices" [5], as: "Microservice Architecture (MSA) has recently emerged as an architectural style particularly suitable to the cloud infrastructures. The MSA style is an approach to developing a single application as a suite of small services, each running in its own process and communicating with lightweight mechanisms. Although the set of MSA principles aim for high degree of flexibility, modularity and evolution, adopting, operating and maintaining microservice architectures in practice is challenging and time consuming". It is really a lot of infrastructure to build and hard work to achieve this. Our microservices must communicate via some message broker, there should be load balancer implemented, probably service discovery, API gateway, event sourcing, domain events. Software built on microservices architecture should be also deployed with some containerization and orchestration mechanism like Kubernetes. As we can see there is a lot of stuff we need to build and take care of but how does it improve scalability? Properly implemented microservices uncovers all disadvantages of maintenance attribute because of this whole infrastructure to take care of but when it comes to deal with massive number of users and events it is very easy to scale application as much as we need. Furthermore, all those applications are lightweight due to API gateway pattern and they are killed as soon as they are not necessary. It makes this architecture kind of perfect with cost-efficiency balance because no resources are wasted. Scalability problem can be easily resolved by proper software architect's design decisions. It is always trade-off. We get something, we lose something else. That's why it is very important to wait with decisions that may produce long term consequences if possible. In most not huge business cases using different databases for commands and query or microservices would be over engineering that consumes huge amount of money. That's why software architecture decisions should be always made by more than one person.

3 Maintainability

Maintainability is software quality attribute that I think is the most challenging in application lifecycle. The nature of software is to change. We, as a software engineers, invent more and more approaches and way to deal with it. The whole agile culture which the merits is adopting for changes is the answer for modification of software. The maintenance in context of computer programs are time when software is delivered to the customer according to the basic requirements but still needs to be polished. It is the fact that when somebody is using our software it always will be necessary to change something in it. Sometimes we need to migrate the library that we are using inside our project for security issues, sometimes business changes and that makes software deprecated. There is also possibility that competition gives pressure on we to add some missing features. Foremost architectural decision we can take in project to make it easier to maintenance is to use best coding practices and style guides. The architectural decisions include every detail of how our software is going to be developed and coding styles are no exception. The differences between code that applies SOLID principles (introduced in Robert C. Martin books [7, 16]) and those who don't are indescribable. But there is not only SOLID though people rarely mention other practices. There are plenty of rules that can be applied to make software more readable and working with less side-effects. We have rules like DRY(introduced in "Pragmatic Programmer" by Andrew Hunt [8]), GRASP(introduced in "Applying UML and Patterns – An Introduction to Object-Oriented Analysis and Design and Iterative Development" by Larman Craig [9]), Functional programming principles, KISS, YAGNI. But even if our code has great quality it still might be hard to maintain without style ruleset like PIP-8 in python or C# Coding Conventions. Besides those the impact is also made by set of architectural patterns we choose. For example, it is very easy to change our data layer when our application domain follows infrastructure ignorance principle. It might be near impossible task that could be only achieved by rewriting half of an application if we have tight coupling between our application modules. According to modules it is important to choose application structure that will be clean and logical. Not only by using meaningful names for modules but also choosing right place for it. It is worth to mention that even technology we use might impact on maintenance. If we would choose thirty years old technology to use these days probably there would be significant problem finding professionals that know this technology and want to work with it. Most of fresh products on the market offer LTS versioning some period. It ensures user that the technology will be supported in long term. The trade-offs in maintainability is often the balance between quality and time. Sometimes we have no time or money to implement all those principles. Sometimes it is unnecessary to use them in project which covers narrow scope of problem or is not enterprise wide. Use of new technology is often a high risk because we never know if it will be forgotten in few months. What is worth to remember is that this is recommended to always create software with will to reuse it and with consciousness that somebody might be changing it after long period of time.

4   Time to Market

It is very common in industry that we make presentation and mock-up, then we show it to the business and sell only the concept before we even start to create an application. This is very understandable, from marketing point of view the biggest profit we can get when there is no competition on the market, when our product is unique. Sometimes the trigger to make an application are some law adjustments. There is really short period like two or three months to make an application and sell it to the wide audience before somebody else do it. In those kinds of situation, we can literally forget about any good practices because all that matter is time. It is so common that we have got special terminology that we measure how many shortcuts we took by making software and how much does it cost in the future. It is called technical debt. According to "An exploration of technical debt [10]" there is poor understating of this term because of lack defined terminology and different definitions across academic literature. Static code analyzers often define technical debts as number of code smells which are places where written code is redundant, we can split abstractions etc. Nevertheless, it is always measured by hours that we need to put into the project to pay this debt. Systems with high technical debts are hard to extend and maintenance. This is trade-off that we pay for fast development. Our maintenance, scalability, security – we can make everything in weeks not in months but if project will succeed, we will surely need to rewrite a lot of code. Sometimes when we read somebody's code and we need to maintain it we wonder how somebody could write a piece of code with careless and perfunctory but when we look at wider context, we see that maybe time was key factor. Conscious software developers can take technical debt by choice not by mistake and treat it like another architectural decision. There are some nice patterns that we can follow up like architectural decisions records. It is a practice when we write down all architectural decisions with possible consequences and triggers to make it. It is widely used to show the developers team and the business why something may take a lot of time in the future and what can be treated like shortcut to make profit. We can also see how different time-to-market is coupled with software reliability attribute. As programmers we know a lot of techniques like test driven development (Kent Beck - "Test driven development" [11]). Writing and making all kind of tests like unit tests, end to end tests, integration tests etc. Make our code much more reliable. When module is well tested it surely works. But once again writing tests are time-consuming and money-consuming (hiring good QA), so it is another decision that is made during software development – writing test or we have no time for this? It depends on software architect if he decides to write test for every method, happy path or none. Another great tools that we can use in software development process to increase our time-to-market attribute are continuous integration and continuous delivery pipelines. There are still companies that are copying files with some file transfer protocol or other unsafe and easy to fail way. According to Martin Fowler's article "Continuous Delivery [15]" it makes deployment more frequent. CI/CD are mature feature at moment of writing this chapter, so It doesn't produce any disadvantages excluding some small money costs and time for configuration.

5   Security

Computer and web industry are making more and more data. Year by year we are asked for more information about ourselves and we create dozens of accounts on different platforms. As pointed by Bernard Marr in "How Much Data Do We Create Every Day? The Mind-Blowing Stats Everyone Should Read" [12] for Forbes - there are daily 2.5 quintillion bytes of newly created data. Every service that is important in our life is moved to the web and all of it stores sensitive data about us. Security is very important topic and people start to see it. Databases leaks or thefts organized with help of those leaked data are not surprising anybody. It is so important as we hire pen-testers to break our systems and OWASP top ten susceptibility ranking is something that every developer should be aware of. Most of enterprise application contain sort of payment possibility so program security attribute is very desirable these days. When making our software we need to have a lot of knowledge to make our application secure in every aspect. Every piece of our code can be the target of an attack. Like with other attributes by making good design decisions we can avoid most of them. Sometimes even very experienced programmers forget about encapsulation and rich domain models and use simple POCO or anemic models to execute some business logic. Basing on "Secure by design" we can model our domain models in a way that invalid objects are never created. Ninety percentage of security issues are going away by using domain driven design approach in our application. Security attribute should be also taken into consideration when choosing technology. Newly created frameworks are more probable to be insecure or have some backdoors. As a software architect we also choose way of authentication or authorization in our program. There are many conventions and identity providers including different hashing algorithms and sum checks. Every additional way of checking user's identity or his actions (logging systems) can be useful in context on making our application secure. The disadvantages of making security attribute exposed is higher time cost to implement those mechanisms. Another issue may be lower performance of our application. Another component that stays between our logic and user is always slowing down the whole business processes. Although disadvantages can be also lower if architecture is well implemented. It would be big difference if we check user identity every time, he hits our application or only once.

6   Usability

Checking out available options we can see that we have plenty design styles on the market. Google Material Design or Microsoft Design are only examples of many different styles that not only design from esthetical but also functional point of view. User experience is one of the most important factors that can help customer to choose between our software or competition. Introducing design thinking into the process of developing our software and teaching the team about importance of thinking about problem before solving it would make our software even better. From architectural side of the coin it is very important to make consistent systems. Choosing one design across all the applications is easy way to start building brand around our software. Users are learning our style – interfaces, way of working with problems etc. For long term project software should be built with same build blocks to gain another advantage which is reusability. It is very common with UI applications and thing called Web Components. Defined by MSDN Contributors [17]: "Web Components is a suite of different technologies allowing we to create reusable custom elements—with their functionality encapsulated away from the rest of our code—and utilize them in our web apps.". Using web components in scope of our application make it very reusable and independent from any java script framework. Although it is not so popular in small projects and may be overkill there.

7   Reliability

Who would use software if it would not always work? It is the greatest advantage of technology, in opposite to humans it is not failing over time. Reliability attribute lays between all other software attributes. It mostly depends on business kind that we build software for, but it all still is about architectural decisions. Different businesses have different requirements. All factors mentioned before having its impact on reliability. When we want our software to be used by thousands of people in the same time? We will build great infrastructure with microservices and virtualization. When we want our software to be always correct, we will focus on writing unit tests and build fast CI/CD pipelines. Independent from the business case and application case although is once again understating of the problem we are solving. Without it even the best technology and the best engineers won't create useful piece of code. Software is useful when is widely used by people whose needs it. Every of architectural decisions have final impact on software reliability. But the most important part is to understand and empathy with end user to fulfil his needs.

8 Conclusions

There is a lot of examples in literature about how software architecture is impactful on software quality attributes. It is independent. When we build software with consciousness, we will always build programs that resolve real problems and help real people. The above rules have been used in many implementations [18–23]. There is no perfect architecture that solves every problem that we may find during development. We always make trade-off. We need to know the prize - some of decisions may be uninvertible, thus it is recommended to make hard architectural decisions late. Every part of our application and every process we use from work methodology, through design is an architectural decision and might make our application better or worse. Software development these days is very tight coupled with business and we always should focus on understanding people that we try to solve issue for. In future projects awareness of cause and effect relationship between architectural decisions and software attributes will entirely change whole development process. Even though there is still no perfect architecture, it is worth looking for solutions that will bring us more benefits than losses in the future.

References

1. Moses, J.: Should We Try to Measure Software Quality Attributes Directly? Springer (2009). https://doi.org/10.1007/s11219-008-9071-6
2. Offutt, J.: Quality Attributes of Web Software Applications, Institute of Electrical and Electronics Engineers (2002)
3. Keeling, M.: Design It. The Pragmatic Bookshelf (2017)
4. Weinstock, C.B., Goodenough, J.B.: On System Scalability, CMU/SEI-2006-TN-012, Software Engineering Institute (2006)
5. Di Francesco, P.: Architecting Microservices. Gran Sasso Science Institute (2017). https://doi.org/10.1109/ICSAW.2017.65
6. Holvitie Sherlock, J., Licorish, A., Spínola, R.O., et al.: Technical debt and agile software development practices and processes: An industry practitioner survey. Elseiver (2018)
7. Martin, R.C.: Clean Code, Pearson Education (2009)
8. Hunt, A.: Pragmatic Programmer, Pragmatic Bookshelf (1999)

9. Larman, C.: Applying UML and Patterns – An Introduction to Object-Oriented Analysis and Design and Iterative Development, Pearson Education (2004)
10. Edith, T., Aybüke, A., Vidgen, R.: An exploration of technical debt. J. Syst. Softw. **86**, 1498–1516 (2013)
11. Beck, K.: Test Driven Development, Addison-Wesley Professional (2002)
12. Marr, B.: How Much Data Do We Create Every Day? The Mind-Blowing Stats Everyone Should Read, Forbes (2019)
13. Johnsson, D.B., Deogun, D.: Secure by Design. Manning (2019)
14. Bell, S.J., American Libraries; Design Thinking, Jan/Feb 2008; 39, 1/2; Social Science Premium Collection, University of Kentucky Libraries (2018)
15. Fowler, M.: Continuous Integration (2004). https://www.martinfowler.com
16. Martin, R.C.: Clean Architecture. Pearson (2017)
17. MSDN Web Components, 4.11.19 MSDN Contributors, https://developer.mozilla.org/. Accessed 06 Jan 2021
18. Paszkiel, S., Szpulak, P.: Methods of acquisition, archiving and biomedical data analysis of brain functioning, biomedical engineering and neuroscience. In: Hunek, W.P., Paszkiel, S. (eds.) Book Series: Advances in Intelligent Systems and Computing, vol. 720, pp. 158–171 (2018). https://doi.org/10.1007/978-3-319-75025-5_15
19. Paszkiel, S., Hunek, W.P., Shylenko, A.: Project and simulation of a portable device for measuring bioelectrical signals from the brain for states consciousness verification with visualization on LEDs, challenges in automation, robotics and measurement techniques. In: Szewczyk, R., Zielinski, C., Kaliczynska, M. (eds.) Book Series: Advances in Intelligent Systems and Computing, vol. 440, pp. 25–35 (2016). https://doi.org/10.1007/978-3-319-29357-8_3
20. Paszkiel, S., Sikora, M.: The use of brain-computer interface to control unmanned aerial vehicle, automation 2019: progress in automation, robotics and measurement techniques. In: Szewczyk, R., Zielinski, C., Kaliczynska, M. (eds.) Book Series: Advances in Intelligent Systems and Computing, vol. 920, pp. 583-598 (2020). https://doi.org/10.1007/978-3-030-13273-6_54
21. Paszkiel, S.: The use of facial expressions identified from the level of the EEG signal for controlling a mobile vehicle based on a state machine, automation 2020: towards industry of the future. In: Szewczyk, R., Zielinski, C., Kaliczynska, M. (eds.) Book Series: Advances in Intelligent Systems and Computing, vol. 1140, pp. 227–238 (2020). https://doi.org/10.1007/978-3-030-40971-5_21
22. Paszkiel, S.: Using neural networks for classification of the changes in the EEG signal based on facial expressions, analysis and classification of EEG signals for brain-computer interfaces. In: Book Series: Studies in Computational Intelligence, vol. 852, pp. 41–69 (2020). https://doi.org/10.1007/978-3-030-30581-9_7
23. Paszkiel, S.: Augmented reality of technological environment in correlation with brain computer interfaces for control processes, recent advances in automation, robotics and measuring techniques. In: Szewczyk, R., Zielinski, C., Kaliczynska, M. (eds.) Book Series: Advances in Intelligent Systems and Computing, vol. 267, pp. 197–203 (2014). https://doi.org/10.1007/978-3-319-05353-0_20

Development of Internet Communicator

Adam Baron[✉] ⓘ

Opole University of Technology, 45-758 Opole, Poland

Abstract. The chapter is devoted to the project and implementation of Internet application designed to allow communication between its users. The application is based on the REST API software architecture pattern, communication between the backend application and frontend applications is achieved with help of the JSON data format. The first part describes used technologies as well as programming environment preparation. The core part is the description of development of the application backend project which is created in ASP.NET Core framework accompanied by Entity Framework Core, MySQL, and other commonly used and readily available technologies. The subsequent sections are focused on the frontend applications. The mobile application project is realized in the Flutter SDK, while the web application project is developed in Angular framework. Implementation of the REST API allows the frontend applications to send and fetch data to the backend server. The fetched data is displayed in the application' interface to allow users' interaction.

Keywords: Internet communicator · Moblie application · Web application

1 Introduction

The Internet means of communication are inseparable part of modern society, many generations are actively using this form of interaction to connect with friends and relatives. Making connections between people should be main priority when creating application for communication over the internet [1]. The market of communication applications is mainly dominated by big corporations because of that privacy is a concern. Often every user activity in application is being logged and send to external servers to process. User behavior patterns are being sold to third party companies [2, 3]. As a result, taking into account the need of communication in human nature, and increased lack of privacy in the internet, it seems necessary to create an application which can help people to communicate, while their message history and actions in the internet remains private. Applications can be developed simultaneously for different operating systems, with use of multiplatform tools [4]. Utilization of RESTful API allows for flexibility, once developed API can be used by different services. REST application programming interface uses HTTP request with methods such as GET, POST, PATCH, to fetch, send and update data available on the server. APIs are great way to expose web services to the internet, with use of REST exposed data is easily controllable [5–7].

Access to exposed data must be authorized [8], in some cases to save server resources by preventing too many requests in other cases to prevent disallowed access to sensitive or

S. Paszkiel (Ed.): ICBCI 2021, AISC 1362, pp. 330–338, 2021.
https://doi.org/10.1007/978-3-030-72254-8_34

private data. The authorization of user' requests can be realized with use of authorization tokens. One of the most popular forms of authorization tokens is JSON Web Tokens (JWT) Bearer. The tokens use JSON notation to send data to client, this data include public key, type of token, and some optional information [9].

2   Description of Development Process

2.1   Description of Applied Technologies

ASP.NET Core is a cross-platform, open-source framework, for building cloud-enabled, internet-connected applications. ASP.NET Core supports development of web applications, Internet of Things applications and other types of applications. ASP.NET Core uses .NET as a base framework [10]. Entity Framework Core enables .NET developers to work with database with use of .NET objects, it also eliminates need for writing data-access code. The correct version of Entity Framework Core must be installed in backend project, every database engine has some differences, for that reason different versions are created [11].

TypeScript is an open-source language which builds on JavaScript. JavaScript is one of the most popular languages used for web development, the main syntax difference of TypeScript compared to JavaScript is strict data type definition. Type Script allows use of JavaScript code and regard it as valid alternative to native TypeScript [12]. Angular is a cross-platform framework which is highly flexible. Angular uses TypeScript language as a basis and HTML and CSS as means for displaying content in internet browsers [13].

Dart is a programming language optimized for creating user interfaces, Dart' focus is to enable user interface development iteratively. The process relies on providing developers with complier which supports hot reloading. Hot reloading compiles only the part of code recently changed by developer [14]. Flutter is a Software Development Kit (SDK) which uses Dart language as a base. Flutter provides libraries with premade user interface elements named widgets [15].

MySQL is an open source relational database management system; it uses Structured Query Language (SQL) as a communication mean for making changes in databases and server' settings.

REpresentative State Transfer (REST) is an architectural style for providing standards of communication between systems on the web. RESTful web services allow the requesting systems to access and manipulate web resources by using a uniform and pre-defined set of operations. REST provides means of data transfer by use of different data formats, some of them are: JSON, XML, HTML [5].

Docker is a virtualization software enabling installation of previously made software bundles named containers. The containers include operating systems and require services, for example: database servers, file servers, and many different types of software.

2.2   Configuration of the Development Environment

Process of configuration of the previously mentioned environments is described in this chapter. Additional packages were installed for every environment, a brief description of installed packages is further mentioned.

Backend Project Configuration. The .NET project was developed in the Rider IDE. The project files were setup with help of .NET built-in tools for creating projects, simple to use GUI overlay for these tools is available in Rider.

Database Configuration. For the development purposes the database server was set in Docker container. The container configuration was saved as docker compose file. Docker compose file is a YML formatted file with specific set of options, recognized by Docker.

.NET Project Configuration. The first element was an empty .NET solution as a base for .NET projects. The Api project was created as ASP.NET Core Web Application. All projects created while developing the application were using .NET 3.1.6. Infrastructure and Domain projects were created as Class Libraries.

Installation of NuGet Packages. NuGet is a package manager for .NET. The NuGet client tools provide means to produce and consume packages.

The first added package was Entity Framework Core. To use Entity Framework with MySQL the MySQL Data Entity Framework Core was added to the project. The packages required for Swagger were added in the next step. Swagger is an open-source tool for simplifying API development. The focus of Swagger is to help making requests from automatically generated graphical user interface based on API controller's functionality. Swagger also provides simple to use yet powerful means for documentation of the API controllers. The next installed package was AutoMapper. AutoMapper is an open-source modest sized library which helps with data mapping. The data mapping is realized between two different objects for example: data model and data transfer object. CloudinaryDotNet package was installed in the project. This package is required to provide access to an external hosting server for the pictures uploaded by users. The package enables simple to use means for accessing the Cloudinary API. DataEncryption package was installed in the project. The DataEncryption package is a small open-source project which provide a straightforward approach for data encryption in the database tables. The final installed packages were IdentityModel Tokens Jwt, AspNetCore Authentication JwtBearer and IdentityModel Tokens. These packages provide support for JWT Bearer tokens applied for authenticating users of the application.

Flutter Frontend Project Configuration. The Flutter project was developed in Android Studio, Dart language with Flutter SDK was used while creating the application.

Android Studio Configuration. The Flutter and Dart plugins were installed with help of the built-in plugin manager. The Dart plugin brings Dart language support for the Android Studio IDE. The Flutter plugin brings Flutter SDK specific code autocompletion to Android Studio IDE it also adds functionality to create project from template.

Installation of Additional Dart Packages. The newly created Flutter project required installation of additional Dart packages for the development process. The installed packages are http - provides HTTP functionality. It enables the project for sending HTTP requests and process the responses, jwt_decoder - provides means to decode received JWT tokens, the parts of encoded tokens can be easily extracted, date_format - enables straightforward decoding and conversion of the received date string to the Dart' built-in date data type, awesome_dialog - provides means to display beautiful dialogs for the user in simple manner. The dialogs are used to display information and errors, adaptive_dialog - provides means to display dialogs for user, the dialogs enable for quick user input into text field, modal_progress_hud - provides functionality for displaying progress indicators while preforming asynchronous operations, file_picker - provides file choosing functionality for the user, dio package provides extended HTTP functionality, for example file transfer in parts.

Angular Frontend Project Configuration. The Angular project was developed in WebStorm IDE.

Installation of Angular Environment. The Angular environment is installed with help of package manager built into Node.js. Node.js is an open-source, cross-platform JavaScript runtime. The Node Package Manager (NPM) provide quick and simple means to install desired packages. The angular Command Line Interface (CLI) is installed this allows for system wide access to Angular CLI. After the installation, the ng command is available which provides functionality such as creating new projects, starting the application server, and many more.

Installation of Additional Packages. The first added package was Bootstrap. Bootstrap is a free open source CSS framework enabling straightforward implementation of responsive design, and styling including typography, forms, buttons, navigation, and other user interface elements. The second package installed was AlertifyJS, a package which allows showing pop-ups in the down-right corner of the screen, pop-ups show information to user. The last package added to the project was Angular Material. Angular Material is an Angular package which brings Material design support to the application.

2.3   Development of the Back-End Project

This chapter is dedicated to description the development process of the backend project realized in the ASP.NET Core platform.

Users of the application can perform set of predefined operations. The use cases are shown on the diagram presented on Fig. 1.

Diagram of the database modelled for this application is presented on the Fig. 2. Relations between almost all the tables are of type one-to-many. The __EFMigratinsHistory table is not connected to any other table, its purpose is to store Entity Framework information about all previous modifications performed on the database structure. The Contacts table relates to two relations with User table to store information if both users added themselves to contacts.

Entity Framework Core Classes. The Entity Framework Core model is a specific type of class. In this type of class, the data fields are names of the tables of the database, the relations between tables is described in the models as well.

The following classes were created in development of the application: User model - created to store data created in process of registration of new users, Contact model - created to store information of user' contacts, MessageGroup model – created to store

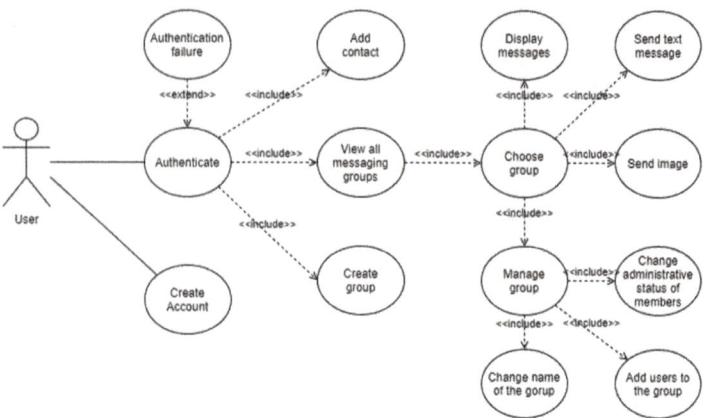

Fig. 1. Use case diagram.

data about groups, UserMessageGroup model – created as a linking table in the database, to realize many-to-many relation between User and MessageGroup tables, Message model – created to store information about text messages as well as images send by users.

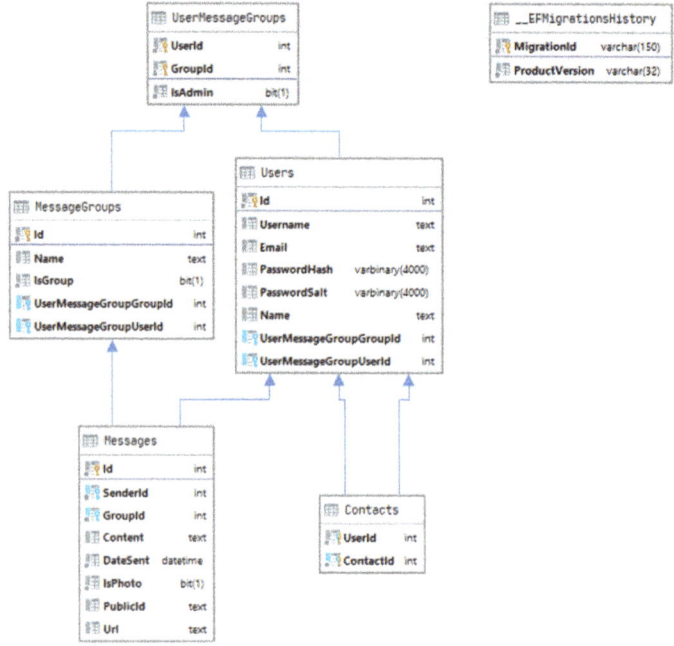

Fig. 2. Database diagram generated in the Rider IDE.

Database Context – AppDbContext. The database context is a unique Entity Framework class in which all the database models are defined, custom relations between tables are defined in the context. Entity Framework can recognise simple relations between tables, but in some more complex cases the relations are defined by the developer in the database context.

Data Transfer Objects. Data Transfer Objects (DTOs) are specially designed classes for one task – transferring data between API server and frontend applications.

In the API designed for the application there are more than ten DTOs. The objects are in the most part data models without fields which should not be exposed to the internet for example passwords.

API Repositories. Repositories are responsible for generating additional layer of abstraction between API controllers and database. With this approach any changes in the database will not affect controllers, and changes in controllers will not affect database. Every repository has an interface, the interface is referenced while any method of the repository is used in the application. Interfaces do not implement the methods, when interface' method is being called the implementation is being resolved with dependency injection.

All the repositories created for the application are: AuthRepository - responsible for creating new user accounts and authenticating users, UserRepository - handles database requests regarding information about users, MessageGroupRepository - responsible for connecting to the database to perform operations on the MessageGroup and User-MessageGroup tables, MessageRepository - designed to fetch data from the Message table.

API Controllers. Controllers are responsible for processing HTTP requests sent by the client application. After receiving request, a response is formulated and sent back to the client.

Controllers created for the API are: AuthController is responsible for authenticating already registered users and creating new unique user accounts, UsersController has three important functions: enables finding users by providing identification number, e-mail address or username, provides functionality to add another user to contacts, allows for displaying user' friends, MessageController' main functionality is to provide means of communication for application' users. Functions of the methods present in the MessageController are fetching a messages for a specified group, creating a new message either with text content or image content, fetching all the conversations for a specified user, GroupController is responsible for creating new groups, adding users to existing groups, renaming specified group, modifying administrative permissions for users, returning information about group' members and administrators.

2.4   Development of the Flutter Project

Application Screens. Screen in Flutter application is a part of the user interface. Screens are displayed on the device, and have buttons, input fields, and other elements allowing interaction with user. Application can have multiple screens, and each screen can be designed to provide single functionality.

Screens are Dart classes which are inheriting from StatefulWidget class or StatelessWidget class. The main difference between stateful and stateless widgets is that the state of the widget is expected to change in stateful widget while the content of stateless widget is expected to remain unchanged.

All the screens designed for this application are: WelcomeScreen was designed as screen which will be greeting users after starting the application, LoginScreen was created to authenticate already existing users of the application, RegistrationScreen allows new users of the application to create an account, ConversationsScreen displays all the groups currently logged in user is in. This screen allows users to choose the a group, ChatScreen is designed to display all the messages which were sent in a specified group, AddFriendScreen is an application' view which enable users to add other users to contacts by providing a username or e-mail address, AddFriendsToGroupScreen was designed to enable group administrators to add their contacts to a existing group, SetAdminScreen is for group administrators to change the administrative permissions of the user.

Application Dialogs. Dialogs are pop-ups shown in front of the application' screens designed to display information to user in easy to understand manner. Dialogs are also used to ask for user' input for example dialog with text input field. One of the dialogs is designed for creating new groups, it asks user for a name of group to create. Another dialog' function is to change name of the specified group.

Connection to the API. API connection classes are present in the Flutter application to send and receive information to and from the ASP.NET application. The classes are responsible for sending HTTP requests to the API controllers. The classes are divided in the same manner as in the backend application to maintain clarity of code.

2.5 Development of the Angular Project

Application Pages. Pages are Angular components designed to display information in a user-friendly manner. Users can interact with elements on pages. Design of the screens was created in the HTML and CSS files, method used to send the request was created in the Type Script (TS) files.

Pages designed while developing the application are: LoginComponent was designed to authenticate existing users, it is a part of AuthComponent, RegisterComponent provides functionality for registering new user accounts. The RegisterComponent is a part of AuthComponent, as is the case with LoginComponent, MessagesComponent is a screen designed to display all the user' message groups and messages sent in a specified group, ChatComponent is a child of the MessagesComponent it was designed as the window for displaying messages sent in a selected group, MessageComponent is designed to display single message from a selected group, ConversationsComponent enables users to select a group which will be displayed in the MessagesComponent, GroupComponent is designed to display group names and information about last message sent in the specified group.

Application Dialogs. Dialogs are components displayed with help of MatDialog module included in AgularMaterial library. Dialogs are used to display information to user and ask for user' input.

All the dialogs used in the application are: AddFriendDialog provides functionality to add new contact, AddGroupDialog is a dialog which enable user to create a new messaging group, AddFriendToGroupDialog provides functionality for the administrators of a group to add new members to it, ChangeGroupNameDialog enables administrators to change name of the selected group, SetAdminsInGroupDialog provide functionality to change administrative status of selected group' members, PictureDialog displays an image from a selected message in higher quality.

API Connection. Similarly, to the Flutter project, Angular application connects with the API with use of specially designed classes called services. Services provide access to all functionalities available in the API controllers [16, 17].

3 Summary

Some of the functionalities which are expected of more developed applications are: support for choosing emojis, applying emoji "reactions" to a selected message, replying to messages, applying usernames only for specific conversations, voice, and video calls. All these functionalities could be delivered one-by-one while the application remain in operational state with scheduled maintenance break. Deployment of these new functionalities should not impact in any manner existing application.

The desired functionality of the application was to provide its users with means of text-based communication. The backend project is completed for the desired functionality to be available for use. A new version of the API can be created to support all the features mentioned in the previous paragraph. As far as frontend part is concerned the user interface design of both Angular and Flutter projects is adequate to be named modern and functional. Design also allows access to all the backend API functionalities. All the technologies utilized in this project are in heavy development, a new version of the.Net framework was released while the project was in development. For the project to be safe and secure a migration to the latest version of the used frameworks is advisable before end of support from developers.

References

1. Bargh, J.A., McKenna, K.Y.A.: The internet and social life. In: Annual Review of Psychology, vol. 55, no. 1, 2004, pp. 573–590. Annual Reviews, Palo Alto (2004)
2. Gross, R., Acquisti, A.: Information revelation and privacy in online social networks. In: Proceedings of the 2005 ACM Workshop on Privacy in the Electronic Society, 2005, pp. 71–80. Association for Computing Machinery, New York (2005)
3. Acquisti, A., Gross, R.: Imagined communities: awareness, information sharing, and privacy on the Facebook. In: PET'06 Proceedings of the 6th International Conference on Privacy Enhancing Technologies, 2006, pp. 36–58. Springer, Heidelberg (2006)
4. Corral, L., Janes, A., Remencius, T.: Potential advantages and disadvantages of multiplatform development frameworks–a vision on mobile environments. In: Procedia Computer Science, vol. 10, pp. 1202–1207. Elsevier, Amsterdam (2012)
5. Troelsen, A., Japikse, P.: RESTful Services with ASP.NET Core. Apress, Berkeley (2020)

6. Fielding, R.T., Taylor R.N.: Principled design of the modern web architecture. In: ACM Transactions on Internet Technology, vol. 2, no. 2, pp. 115–150. Association for Computing Machinery, New York (2002)

7. Pautasso, C., Zimmermann, O., Leymann, F.: Restful web services vs. 'big" web services: making the right architectural decision. In: Proceedings of the 17th International Conference on World Wide Web, 2008, pp. 805–814. Association for Computing Machinery, New York (2008)

8. Franks, J., Hallam-Baker, P., Hostetler, J., Lawrence, S., Leach, P., Luotonen, A., Stewart, L.: HTTP authentication: basic and digest access authentication. In: RFC2617, vol. 2617, 1999, pp. 1–34. Internet Engineering Task Force, Fremont (1999)

9. Hardt, D., Jones, M.: The OAuth 2.0 authorization framework: bearer token usage. In: RFC, vol. 6750, p. 18. Internet Engineering Task Force, Fremont (2012)

10. ASP.NET Core Documentation. https://docs.microsoft.com/en-us/aspnet/core/introduction-to-aspnet-core?view=aspnetcore-5.0. Accessed 14 Jan 2021

11. Entity Framework Core Documentation. https://docs.microsoft.com/en-us/ef/core/. Accessed 14 Jan 2021

12. TypeScript Homepage. https://www.typescriptlang.org/. Accessed 14 Jan 2021

13. Angular Homepage. https://angular.io/. Accessed 14 Jan 2021

14. Dart Homepage. https://dart.dev/. Accessed 14 Jan 2021

15. Flutter Homepage. https://flutter.dev/. Accessed 14 Jan 2021

16. Paszkiel S.: Computer game in UNITY environment for BCI technology, analysis and classification of EEG signals for brain-computer interfaces. In: Book Series: Studies in Computational Intelligence, vol. 852, pp. 101–110 (2020). https://doi.org/10.1007/978-3-030-30581-9_12

17. Slapek, M., Paszkiel, S.: Detection of gestures without begin and end markers by fitting into Bezier curves with least squares metho. Pattern Recogn. Lett. **100**, 83–88 (2017). https://doi.org/10.1016/j.patrec.2017.10.006

Anonymity in Terms of Technologies

Joanna Zagozda(✉)

Faculty of Electrical Engineering, Automatic Control and Informatics,
Opole University of Technology, Prószkowska 76, 45-758 Opole, Poland

Abstract. The following paper investigates such topics as privacy in the Internet social media, mobile tracking and privacy-enhancing solutions. Its aim is to investigate the extent of the invigilation of private online activities, applied by the corporations and institutions, with addition of the context of the so-called "bulk collection". It shows the impact of the online users on the remained activity traces. Nevertheless, according to the included conclusions, in order to feel more secure when browsing the Internet, the reasonable attitude and paying more attention to the information shared should be a good habit.

Keywords: Surveillance · Privacy · Anonymity

1 Introduction

"On the Internet, nobody knows you're a dog". This is the caption visible under the Peter Steiner satiric comics, published in June 1993 for "The New Yorker" magazine. In reality, the sensation of anonymity is just an illusion. According to Kaźmierska and Brzeziński, the corporations do know not only that you are a dog, but also what type of food you like, what type of collar you wear and where are you walked [1]. The technology encountered a huge development since 1993. In fact, the Internet affects most part of people's lives. Modern trends, including Internet of Things, smart and wearable devices collect enormous amount of data concerning the everyday life and habits. In fact, it is possible to consider the technology as inseparable part of humans and a new generation of home-dwellers. The advanced solutions cause different reactions. There are of course technology enthusiasts, but on the other side it causes concerns and the skepticism. This problem is commonly reflected in the media and literature. Series like Mr Robot or Black Mirror are popular productions which at some point refer to the aspects investigated in this document. The dystopian future, described in the classic literature – Orwell's "1984", is the common association of the group members, which raises doubts concerning the speed of technology development and its access to personal information.

The described phenomenon was one of the main reasons for investigating the topic of anonymity on the Internet. Another factor which contributed in the creation of the following text is the uniqueness of the problem and lack of similar information bundle. The main core which was the basics for the idea development was the Edward Snowden case in 2013. In fact, as the topic investigation progressed, the more implicit idea of the thesis appeared. Apparently, the following thesis describes an interesting aspect of

S. Paszkiel (Ed.): ICBCI 2021, AISC 1362, pp. 339–346, 2021.
https://doi.org/10.1007/978-3-030-72254-8_35

the relation between technology and the society. Despite the legal protection from the law and EU regulations, the services are still able to discover a new way of receiving personal information. Therefore, the mechanisms of the IT corporations should be more clearly specified. Apart from that, the other intention of the research was to discover the possible alternatives to the common platforms, which still can compete in terms of the usability and the general user impression, in comparison with the original service.

2 Cambridge-Analytica

Modern technology and the Internet highly depend on data. It is said, that among all the resources, data is the most valuable one. Data mining, data privacy, data breaches – those are first Google predictions, when meeting phrase 'data'. Ordinary John Smith has no idea about how powerful might be the information he generates, even though it might seem trivial for him. He is also not aware of the business and money engaged based on his social media activity. In fact, technology no longer concern the IT field on its own. Other branches, including business, economy or politics also involve technological solutions. What is more, the vision of the future seems to be even more IT – oriented.

However, as the traditional means started to migrate in the IT cloud, the issue of the user privacy and safety started to appear more frequently as well. Another aspect which might cause the concern of the users is the issue of data abuse.

As a matter of fact, the issue of abusing the privacy policy appeared not so long ago, when in 2018 the world discovered the data migration to the Cambridge Analytica center. The CA is an organization, which specializes in data science and data mining. It also helps in creating the psychographics of the social media users, which helps in specifying the content targeting – both commercial and political [2]. The organization which owned the CA was the SCL Group (formerly known as Strategic Communication Laboratories). Founded in 1990, focused on data analytics and "conducting the behavioral change programs in 60 countries." [3]. However, it was mostly associated with its dependent infamous institution. According to its official website, the SCL group claimed to operate for a few decades, however due to the Cambridge-Analytica scandal, it became much less transparent. In fact, in the media it is usually referred to as "the parent of the Cambridge-Analytica".

The problem appeared in 2018, when the major data breach from Facebook was found. It was estimated, that information about 87 million Facebook users was breached out to the servers of Cambridge-Analytica institution. As already mentioned, its algorithms created the psychographic profiles based on the collected data.

Former employer of the Cambridge-Analytica, Christopher Wylie, 31-year old data scientist from Canada, exposed the information in March 2018. He shared his knowledge with The Guardian department – The Observer. As Wylie said, CA exploited Facebook to harvest millions of people's profiles. They built models to exploit what they knew about them and targeted their inner demons. That was the basis the entire company was built on [4]. In the interview he also mentioned, that the base approach of the company, was to separate and fragment the society, so that it would be much easier re-design the population [5]. According to his point of view, the users are trapped within the web content specially chosen, to fit the psychological picture generated by the CA algorithms.

Apparently, the origins of the scandal took place in 2014, when Cambridge student, Aleksadr Kogan developed an application – basically an online survey – named 'ThisIsYourDigitalLife' [6]. The application was created using one of the Facebook APIs. The API (Application Programming Interface) can be referred to as an 'application gateway'. It enables the developer to access the service internal data, without the need to learn its principles and technological mechanism. The survey was similar to other personality tests. It was estimated, that within two months, application was downloaded by 270,000 users. The tricky part was that the users had to login to Facebook first or use the application directly via Facebook, to submit the answers. The application of Kogan had access not only to the surveyed person, but also to her/his friends. In total, the data from around 87 million users was collected via the 'ThisIsYourDigitalLife'. What is more, the data was then forwarded to the Cambridge – Analytica. This event started so far, the biggest scandal in the Facebook history. Mark Zuckerberg claimed, that he was not aware of the way the application violated the trust and the right to privacy. On the other hand, the CEO of the CA, Alexander Nix states, that the company did not possess the user data. Apparently, the information gained by the application has not been removed. As the result, the application was removed from the Facebook list. Users, who were affected by the app were informed about the incident. Changes were also made in case of the APIs. Facebook made sure, that the developers cannot access the information belonging to users from the friends' list of the person, who is using the application. In case, when the user has not used the application for 90 days, she/he is automatically logged out.

Apparently, both Facebook and Cambridge-Analytica should share the consequences. According to the FTC (Federal Trade Commission), Facebook enabled the third-party websites access the user data already in 2012. What is more, the platform applied misleading privacy settings, which did not enable the complete protection. For example, even when the user applied the highest level of privacy settings, the third-party application used by one of his/her friends, could still access that protected information. Despite agreeing on the change of the indicated application flows, Facebook still silently allowed the third-party programmers to access the data of the already existing applications [7]. Therefore, as the information about Cambridge-Analytica was exposed, the platform was accused of violating the statements from 2012. As the result. Facebook was charged with the highest penalty fee in the social media history – 5 billion USD. It also had to implement a new model of privacy settings, which should be clear and straightforward for the users.

The Cambridge-Analytica went bankrupt and had to be shut down. Even its official website was closed soon after its operations were publicly exposed. However, new institution – Emerdata appeared in the place of old CA.

3   Mobile Tracking

It is estimated, that there are more mobile phones in the world, than the total human population. The phone invented in 1876 by Alexander Graham Bell, took a long journey of development in order to obtain the form commonly recognized today. One of the crucial moments happened in 1973, when Martin Cooper, the engineer at Motorola, introduced the first cell phone. The device enabled the wireless communication since the phone was supposed to be portable and did not need the constant wired connection. However, the phone was first used 10 years after its initial presentation. The reason for such long delay was the lack of the supporting infrastructure at that time. The design required the use of cellular towers.

The cellular network architecture is widely used today. Its basic principles are presented as follows. The geographical surface is divided into hexagonal areas, each with separate cellular tower. The tower is constantly listening, in order to receive a call from the mobile device. When phone is trying to establish a connection, reaches out to the nearest tower, which then redirects the signal to the home tower of the call destination. The default technology of the phones causes discussions in terms of the privacy concern. It is worth to note, that the phone is constantly sending information to the towers nearby, in order to find the closest one. The towers keep track on all the attempts of the communication constantly. As a matter of fact, multiple towers might ease the determination the position of the phone (and hence – the person) at any time. What is more, phone nowadays can be considered a mini – computer since it is not anymore used for making phone calls exclusively. Smartphones are crucial in social activities, like browsing the social media, watching videos, or checking the emails. In order to perform these activities, the device needs to have an access to the Internet. Smartphones are constantly searching for the reachable wireless access points, commonly called "Wi-Fis". As currently such access point can be found in almost every household, the device can be navigated with even greater proximity, as the "reference points" on the Wi-Fi map is dense.

4 NSA

June 2013 is a meaningful time in terms of privacy and online observation. It was also a very tense moment for Edward Snowden, the former NSA employee. He decided to share the confidential documents, concerning the secret operations of his workplace, with the journalists of The Guardian magazine.

4.1 Prism

One of the NSA projects which became known publicly after 2013 documents revealing was Prism. It was first launched in 2007. The program relied on the so-called data providers, big concerns – the IT companies. According to the official training slides shared by Snowden, companies which collaborated with the NSA included Google (with YouTube), Microsoft, Apple, AOL (America Online, one of the biggest Internet providers in the USA in the 1990s), Facebook and Yahoo. Apparently, the program had direct access to the companies' servers, from which it received all the possible data. Despite an extensive list of information providers, the companies claim that such collaboration did not take place, and additionally no one in the company have heard about such program. As it was discussed during the TedTalk event, Prism collected not necessarily the metadata, but it was focused on the content of the communication, including email content, live chats, or file transfers [15].

4.2 XKeyscore

Another NSA system revealed in June 2013 was XKeyscore. As it was claimed by Snowden, the NSA described XKeyscore as the tool of the widest range, used for searching the total Internet activity of the user [16]. It enabled the agents to search for such information as emails or even private chat messages. It was used to investigate both user online activity and metadata. In fact, it was even possible to monitor the social media activity of the target. The analyzer was actually advised to perform the investigation based on the specific metadata, as the system database was enormous and the displayed results could have been too difficult to view at once. It was estimated that approximately 1 – 2 billions of records have been added to the database daily. Apart from metadata, XKeyscore introduced to feature to search the database through the use of name, IP address, language used or phone number. When combining the system with other NSA solutions, it was possible to maintain almost the real-time monitoring of the target. It was estimated that by 2008, due to the XKeyscore functionality, about 300 terrorists have been captured.

5 How to Remain Anonymous?

5.1 Social Media Substitutes

Before Facebook, a website which allowed the user for the same performance was "Nasza Klasa" (www.nk.pl). It implemented such features as post sharing, commenting the posts, inserting the pictures, rating the pictures, etc. As Facebook was translated to Polish, nk.pl started to lose its users. The website is still active though. The substitute for the instant messaging application, could be found in another Polish service – Gadu Gadu. It was launched in 2001 and quickly became popular. It provides the encryption and security features. Users can communicate via text messages, send pictures and videos. Each user is identified by a special number. However, it is worth to mention, that Gadu-Gadu uses the Adware license. Even though the service itself is free of charge, the advertisements are displayed to the users. It also lacks one of the features present in case of Facebook Messenger or WhatsApp. The audio or video calls are possible only in case of using the webGG – the website application version. The case of alternative is slightly more difficult for the YouTube platform. Despite many options, YouTube still brings the greatest attention and defeats the competition. Yet, one of the promising websites is Odysee, a platform built via the LBRY protocol. As it is explained on its official website, the LBRY structure is similar to the BitTorrent architecture [17]. The content is stored not on the servers, but dispersed within different users.

5.2 Anonymity-Enhancing Search Engines

In order to remain private while using the Internet, the user can use web engines other than Google. Popular option is DuckDuckGo, which was founded in 2008. The search engine provides full anonymity and no tracking policy. It does not monitor its users and their online behavior. On the official website, it is possible to find an information saying, that all the income relies on the private advertisements, placed in the browser window

[18]. DuckDuckGo also implements the tracking prevention add-on, which provides the privacy when browsing the Internet. Another reason why DuckDuckGo is an interesting alternative to other searching engines is the impartiality. Since the company does not collect user data, the listed results are not reflected by the user profile or based on her/his preferences.

Another search engine, which enhances the anonymity is StartPage. It was first launched already in 1998 under the name Ixquick.com, rebranded later to Startpage.com. The search engine office is located in Netherlands. It is considered a suitable alternative in case when the user prefers the Google searching results. The search engine company pays Google for receiving its results [19]. Startpage provides the features which enable the user not only to search the Internet anonymously, but also enter the websites found with no fear of being identified. It is possible through the "Anonymous view" accessible next to every listed page found.

5.3 TOR Network

The Onion Router – provides anonymity, but not full security. It does not provide the protection from the malware or other malicious code. As the connection is transferred through the relay nodes, the connection is typically much slower than the direct one. The default number of the relay nodes is three. It is possible for the TOR user to configure the number of nodes manually. The nodes which are participating in the establishment of the connection are known as entry node, relay node and exit node. Entry node is the node used for entering the network. In case of relay node, it is used for the packet further transmission. The Exit node is the final node which leaves the network and redirects the user to the destination. Using the TOR might be considered problematic, not only because of the slower connection. It is worth to pay attention to the individual configuration. Perhaps the computer used might be allowed to perform as an exit node. It is also worth to remember, that not all TOR users are using this technology for the feeling of security and anonymity. There are still threat actors whose intentions are malicious and harmful. They might enter the Internet from their home machine and reach their destination using the relay of the guilty-free user.

5.4 Linux Tails

Tails is an uncommon Linux distribution. The acronym stands for The Amnesic Incognito Live System. It is Debian-based and a live OS. What it means is that the system might be implemented regardless of the host OS. It does not use the host hard drive, therefore it does not leave any traces of usability. It is usually provided via bootable USB stick or CD drive. The moment the external drive is removed, all data is erased. There is no evidence of the user activity. The memory is volatile and every time the drive is inserted into device, the system looks like a brand-new setup. The OS provides the pre-installed anonymity bundle. The package includes i.e. the TOR browser for the safe Internet access. In addition, the Tails implemented the NoScript plugin, so that the user can control the JS scripts, which also might have an impact on the anonymity on the Internet. Another installed application is Pidgin – the software allowing the user to introduce the encryption to the Instant Message conversations. What is more, the OS also includes the OnionShare – a platform used for private file sharing. It is directly connected to the TOR browser, which is needed for the file recipient do download the files.

6 Summary

Knowing what happens with the data shared by users might become a good habit. Apparently, the social media and online world keep violating the right to privacy, while on the other hand it is impossible to stop the technology development, stop living the normal life. In addition, the new gadgets are extremely tempting and seem to be a fancy life-experience extension, including mentioned earlier AI assistants or smart bands [20, 21]. Despite the privacy-friendly applications, the corporations will still find a new way of receiving personal data. Naturally, there are practices which eliminate the privacy threat almost completely, however they require a lot of patience and consistency. The average user would doubtfully consider those practices worth the effort.

References

1. Kaźmierska, A., Brzeziński, W.: Strefy Cyber wojny, Warszawa: Oficyna 4eM, s. 37 (2018)
2. Wayback Machine Archived website of the Cambridge-Analytica (2017): https://web.archive.org/web/20170710031857/https://cambridgeanalytica.org/
3. Wayback Machine Archived website of the SCL group (2019). https://web.archive.org/web/20190111195059/https://sclgroup.cc/home
4. The Guardian. https://www.theguardian.com/news/2018/mar/17/cambridge-analytica-facebook-influence-us-election. Accessed 11 Sept 2020
5. The Guardian. https://www.youtube.com/watch?v=FXdYSQ6nu-M. Accessed 11 Sept 2020
6. The Verge. https://www.youtube.com/watch?v=VDR8qGmyEQg. Accessed 11 Sept 2020
7. FTC. https://www.ftc.gov/news-events/blogs/business-blog/2019/07/ftcs-5-billion-facebook-settlement-record-breaking-history. Accessed 11 Sept 2020
8. Ted. https://www.ted.com/talks/malte_spitz_your_phone_company_is_watching#t-355385. Accessed 14 Dec 2020
9. Ted. https://www.ted.com/talks/mikko_hypponen_how_the_nsa_betrayed_the_world_s_trust_time_to_act. Accessed 10 Dec 2020
10. Snowden: Pamięć Nieulotna, Kraków: Insignis Media, p. 417 (2019)
11. IEEE Xplore. https://ieeexplore.ieee.org/document/7167361/authors#authors. Accessed 05 Jan 2020
12. CBR, https://www.cbronline.com/news/nsa-ciphers-iso. Accessed 05 Jan 2020
13. YouTube. https://www.youtube.com/watch?v=0hLjuVyIIrs. Accessed 10 Dec 2020
14. Snowden: Pamięć Nieulotna, Kraków: Insignis Media, p. 207 (2019)
15. Ted. https://www.ted.com/talks/edward_snowden_here_s_how_we_take_back_the_internet. Accessed 10 Dec 2020
16. Snowden: Pamięć Nieulotna, Kraków: Insignis Media, p. 423 (2019)
17. LBRY, https://lbry.com/faq/what-is-lbry. Accessed 05 Jan 2020
18. DuckDuckGo. https://help.duckduckgo.com/duckduckgo-help-pages/company/advertising-and-affiliates. Accessed 15 Dec 2020
19. Startpage. https://support.startpage.com/index.php?/en/Knowledgebase/Article/View/147/22/why-does-google-let-startpage-access-their-search-results. Accessed 15 Dec 2020

20. Paszkiel, S.: Using BCI and VR technology in neurogaming, analysis and classification of EEG signals for brain-computer interfaces. In: Book Series: Studies in Computational Intelligence, vol. 852, pp. 101–110 (2020). https://doi.org/10.1007/978-3-030-30581-9_12
21. Paszkiel, S.: Using BCI in IoT implementation, analysis and classification of EEG signals for brain-computer interfaces. In: Book Series: Studies in Computational Intelligence, vol. 852, pp. 101–110 (2020). https://doi.org/10.1007/978-3-030-30581-9_12

Author Index

S. Paszkiel (Ed.): ICBCI 2021, AISC 1362, pp. 347–348, 2021.
https://doi.org/10.1007/978-3-030-72254-8

Lightning Source UK Ltd.
Milton Keynes UK
UKHW020023150421
382005UK00002B/11

9 783030 722531